STATISTICAL METHODS FOR THE PROCESS INDUSTRIES

QUALITY AND RELIABILITY

A Series Edited by

EDWARD G. SCHILLING

Coordinating Editor

Center for Quality and Applied Statistics
Rochester Institute of Technology
Rochester, New York

W. GROVER BARNARD

Associate Editor for
Human Factors

Vita Mix Corporation
Cleveland, Ohio

RICHARD S. BINGHAM, JR.

Associate Editor for
Quality Management

Consultant
Brooksville, Florida

LARRY RABINOWITZ

Associate Editor for
Statistical Methods

College of William and Mary
Williamsburg, Virginia

THOMAS WITT

Associate Editor for
Statistical Quality Control

Rochester Institute of Technology
Rochester, New York

1. Designing for Minimal Maintenance Expense: The Practical Application of Reliability and Maintainability, *Marvin A. Moss*

2. Quality Control for Profit, Second Edition, Revised and Expanded, *Ronald H. Lester, Norbert L. Enrick, and Harry E. Mottley, Jr.*

3. QCPAC: Statistical Quality Control on the IBM PC, *Steven M. Zimmerman and Leo M. Conrad*

4. Quality by Experimental Design, *Thomas B. Barker*

5. Applications of Quality Control in the Service Industry, *A. C. Rosander*

6. Integrated Product Testing and Evaluating: A Systems Approach to Improve Reliability and Quality, Revised Edition, *Harold L. Gilmore and Herbert C. Schwartz*

STATISTICAL METHODS FOR THE PROCESS INDUSTRIES

WILLIAM H. McNEESE

McNeese Consulting
Clearwater Beach, Florida

ROBERT A. KLEIN

Continental Controls, Inc.
Houston, Texas

ASQC Quality Press   Milwaukee

Marcel Dekker, Inc.   New York • Basel • Hong Kong

ISBN 0-8247-8524-X

This book is printed on acid-free paper.

ASQC QUALITY PRESS
310 West Wisconsin Avenue, Milwaukee, Wisconsin 53203

MARCEL DEKKER, INC.
270 Madison Avenue, New York, New York 10016

Current printing (last digit):
10 9 8 7 6 5 4 3 2 1

PRINTED IN THE UNITED STATES OF AMERICA

ABOUT THE SERIES

The genesis of modern methods of quality and reliability will be found in a simple memo dated May 16, 1924, in which Walter A. Shewhart proposed the control chart for the analysis of inspection data. This led to the broadening of the concept of inspection from emphasis on detection and correction of defective material to control of quality through analysis and prevention of quality problems. Susequent concern for product performance in the hands of the user stimulated development of the systems and techniques of reliability. Emphasis on the consumer as the ultimate judge of quality serves as the catalyst to bring about the integration of the methodology of quality with that of reliability. Thus, the innovations that came out of the control chart spawned a philosophy of control of quality and reliability that has come to include not only the methodology of the statistical sciences and engineering, but also the use of appropriate management methods together with various motivational procedures in a concerted effort dedicated to quality improvement.

This series is intended to provide a vehicle to foster interaction of the elements of the modern approach to quality, including statistical applications, quality and reliability engineering, management, and motivational aspects. It is a forum in which the subject matter of these various areas can be brought together to allow for effective integration of appropriate techniques. This will promote the true benefit of each, which can be achieved only through their interaction. In this sense, the whole of quality and reliabilty is greater than the sum of its parts, as each element augments the others.

The contributors to this series have been encouraged to discuss fundamental concepts as well as methodology, technology, and procedures at the leading edge of the discipline. Thus, new concepts are placed in proper perspective in these evolving disciplines. The series is intended for those in manufacturing, engineering, and marketing and management, as well as the consuming public, all of whom have an interest and stake in the improvement and maintenance of quality and reliability in the products and services that are the lifeblood of the economic system.

The modern approach to quality and reliability concerns excellence: excellence when the product is designed, excellence when the product is made, excellence as the product is used, and excellence throughout its lifetime. But excellence does not result without effort, and products and services of superior quality and reliability require an appropriate combination of statistical, engineering, management, and motivational effort. This effort can be directed for maximum benefit only in light of timely knowledge of approaches and methods that have been developed and are available in these areas of expertise. Within the volumes of this series, the reader will find the means to create, control, correct, and improve quality and reliability in ways that are cost effective, that enhance productivity, and that create a motivational atmosphere that is harmonious and constructive. It is dedicated to that end and to the readers whose study of quality will lead to greater understanding of their products, their processes, their workplaces, and themselves.

Edward G. Schilling

PREFACE

Statistical Methods for the Process Industries can be used in three ways. First, it is a basic text to teach the fundamental quantitative tools of process improvement. Second, it can be used in seminar format to build a five- to ten-day short course in process improvement. Finally, it is a good reference. Each of the various quantitative process improvement techniques is broken down into distinct, freestanding chapters for learning or relearning a particular technique. Numerous examples show how to apply the techniques to the process industries.

Total Quality Management (TQM) improves the quality and productivity of one's business. TQM is based on these major philosophies:

- To continuously improve the process
- To discover and reduce or eliminate sources of variation
- To prevent, not detect, defects
- To use data in place of opinions for problem solving and process improvement
- To view the voice of the customer as the most important criterion for success
- To actively promote employee involvement

The systems of TQM include training, measurement and accounting, recognition and rewards, communications, and team processes. The focus of TQM is on satisfying the customer. The key to accomplishing this is through regular customer feedback. What does my customer need? How am I meeting those needs? What can I do to improve?

The major methods used in TQM include top management leadership, statistical methods, organized problem solving, and team involvement. This book emphasizes the statistical methods and organized problem solving.

Any process produces two things: (1) a product or service and (2) data. Data can be used to improve the process. This text focuses on how to use the data generated by the process to help improve the process. The improved process leads to improved products and services.

The book is organized around a seven-step problem solving model. The first three chapters provide an overview and discuss quality and variation. Chapter 4 introduces a strategy for process improvement for implementing TQM into our day-to-day jobs. This strategy focuses on how to improve what we do everyday. The problem solving model is then introduced (Chapter 5). The statistical techniques are introduced in the order they are encountered in the problem solving model. These techniques include the simple statistical tools (Pareto diagrams, process flow diagrams, control charts, etc.) as well as some more advanced techniques (comparing processes, nested experimental designs, etc.). Special topics include measurement systems and process capability. An appendix of useful tables and a short bibliography are found at the end.

This book was written to help the process industries implement the use of statistical techniques. The authors have over 30 years' combined experience in the process industries, with over 11 years' experience in TQM. It is our hope that this book, in helping these organizations overcome the barrier of "statistical process control applies only to widgets," will prove itself a very valuable resource.

William H. McNeese
Robert A. Klein

CONTENTS

Contents

Contents

Contents

Contents

Contents

Contents

1

INTRODUCTION TO STATISTICAL METHODS

Statistical Process Control (SPC), the use of statistical methods, is one of the major tools of a company's Total Quality Management (TQM) process. It represents a method of using data generated from processes to help us improve the quality of our products, services and information. SPC plays a major role in the concept of continuous improvement. SPC has two important parts. One part involves the use of simple statistical tools. The other part involves the business or management philosophy behind SPC. Both parts are equally important. This chapter presents an introduction to SPC as well as an overview of TQM.

OBJECTIVES

In this chapter you will learn:

1. Objectives of this book
2. Vision of TQM
3. Viewing processes as a system
4. Methods of TQM
5. History of TQM
6. Definition of SPC
7. Benefits of SPC

The purpose of this chapter is to present an introduction to the book. You will learn where the book is leading you, i.e., the objectives. In addition, you will learn the vision of TQM. This represents what we are trying to accomplish through the use of TQM. You will also learn how we view the overall system and what tools are available to help us improve that system. While the vision of TQM will tell you where we are headed, it is also important to know where we have been. To accomplish this, we will review the history of quality worldwide. In the last part of this chapter, we will define SPC and review the benefits associated with learning and using SPC.

STATING THE GOAL

Your objectives in using this book are:

1. To learn and develop an understanding of how SPC will help us attain the vision of TQM.

2. To learn and develop an understanding of how to use the SPC tools to help improve our processes.

3. To learn and develop an understanding of where the SPC tools fit into the problem solving model and an overall strategy for process improvement.

To help us achieve these objectives, this book is divided into chapters that cover the SPC tools and the philosophy behind them.

THE VISION OF TQM

Total Quality Management is designed to improve the quality and productivity of our business, i.e., we are trying to develop a new system or way of doing our business. Our goal is to become the highest-quality, lowest-cost producer of our products, services and information in our industry. Remember, quality refers not only to the product, but also to the service (on-time delivery, responsiveness, etc.) and the information we provide.

TQM is a collection of systems that works to direct a company toward two major philosophies:

- To continuously improve the process

- To view the customer as the most important criterion for success

Systems used include training, measurement and accounting, recognition and rewards, communications, and team processes. Policies implemented consist of quality, employment security, long-term compensation, and promotion. These systems and policies are combined through a quality planning process to achieve direction toward the new philosophies.

Three major elements combine to help us continuously improve processes to meet customer expectations. These three elements, as shown in Figure 1, are customer satisfaction, productivity, and employee involvement. They must all be present for TQM to be successful. Each element is discussed below.

Figure 1

ELEMENTS OF TQM

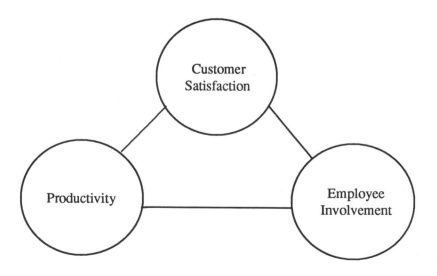

Customer Satisfaction

One key ingredient to quality improvement is realizing the importance of the customer. In TQM, the customer, whether external or internal, is number one. This means the focus of TQM is on the customer and on satisfying the customer. The key to satisfying the customer is obtaining customer feedback. What does my customer need? How am I meeting those needs? What can I do to improve? The goal is not just to satisfy the customer, but to exceed the customer's expectations.

Central to TQM is the concept of continuous improvement. By realizing the importance of the customer and getting feedback on what the customer needs, we can identify our processes that affect those customer needs. The next step is to begin working on continuously improving those processes to allow us to exceed customer expectations. Our drive to continuously improve our processes is based on the feedback we get from our customers, i.e., our system includes a method of regularly receiving customer feedback.

Another key ingredient is realizing that organizational processes are interdependent. No one works in a vacuum. Everyone has customers and suppliers. For many of us, our customers and suppliers are internal. One question we must ask ourselves is, "Who are my customers and my suppliers?" Your customers are those people or departments who receive your work or output. Your suppliers are those people or departments whose work or output you use.

Another key ingredient to our system is realizing the importance of suppliers. We, as customers, cannot be any better than our suppliers. Our suppliers, whether internal or external, must view us as we view our customers.

Important also is the network of internal supplier and customer relationships. Many of our processes are internal. We have to learn to look at internal customers and suppliers in the same way we look at external customers and suppliers. These internal supplier and customer relationships form a chain throughout the organization. The entire process will fail if there are any weak links in this chain.

Many companies today are developing methods to measure external customer satisfaction. This measure is often tied to a company's goals and objectives and/or compensation systems. Done properly, the measurement system can point out potential areas for improvement. Continuous improvement in this measure is the goal.

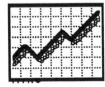

Productivity

While focusing on customer satisfaction is very important in TQM, one cannot look at customer satisfaction alone. We must be effective in meeting and exceeding customer expectations. However, we must do so efficiently. For example, suppose your customer's expectation is a typed memo with no mistakes. How can we satisfy the customer? We could type the memo without proofreading or checking the spelling. We could then return it to the customer. The odds are that the customer would return the memo with some corrections. We could continue this procedure several times until the memo contains no mistakes. We have satisfied the customer's expectations (i.e., no mistakes on the memo) but have we been very efficient in accomplishing this? The answer is no. We included a lot of rework in the process.

This concept also holds for manufacturing. Suppose we have a customer who has strict requirements for product color. Only half our production meets this requirement. Can we satisfy the customer? The answer is yes. However, to do so requires that we make twice the product we need. Again, we have satisfied the customer but we have not been very efficient in doing so.

The point here is that we need to improve the processes that produce our products, services or information for our customers. These processes must be improved so that the number of mistakes, amount of rework, wasted time, etc. can be decreased. This improves productivity. However, customer satisfaction must be maintained.

Various definitions of productivity exist. Productivity is normally defined as output divided by input.

The relationship between quality and productivity is often called the chain reaction (Deming, 1982). This is shown in Figure 2. As quality improves, costs decrease because of less rework, fewer mistakes and better use of time and materials. As a result, productivity improves. Our products capture a larger market share with better quality and a lower cost. As a result, we stay in business and provide jobs.

Figure 2

CHAIN REACTION

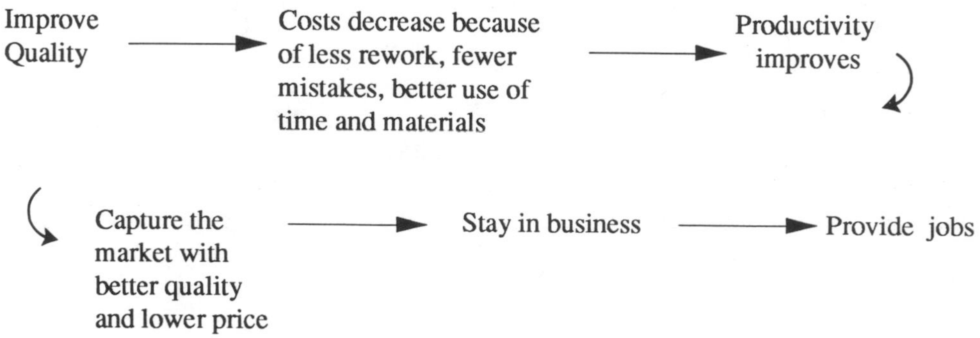

Improve Quality → Costs decrease because of less rework, fewer mistakes, better use of time and materials → Productivity improves

Capture the market with better quality and lower price → Stay in business → Provide jobs

Employee Involvement

The last key ingredient is realizing the importance of each individual's contribution to TQM. It is vital to the success of TQM that every employee participate. We realize that our organizational processes are interdependent and that we have many internal customer and supplier relationships. If one person or department does not participate, we can't realize the total benefits of applying TQM throughout our chain of processes.

There are several ways to increase the participation of employees in TQM. These methods include team management and performance management. Both of these methods are discussed in a later section. Both methods are aimed at assisting in creating an environment that supports employee involvement in TQM.

VIEWING PROCESSES AS A SYSTEM

The system we are trying to develop is essentially the same as advocated by Dr. W. Edwards Deming. Dr. Deming's method of viewing processes as a system is shown in Figure 3 (Deming, 1982). Output from the system is viewed as the result of a series of operations or transformations on inputs in various parts of the system.

Figure 3

VIEWING PROCESSES AS A SYSTEM

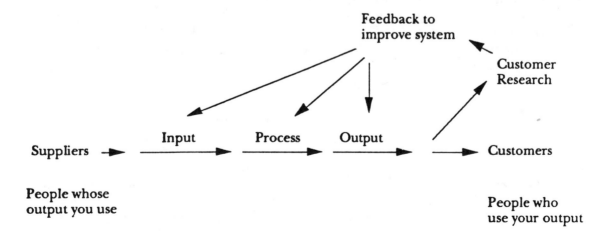

Input or raw materials for a process are received from suppliers. Examples of input include chemical products for use in our manufacturing plants, invoices to be processed, memos to be typed, requests for certain information, etc. After receiving the input, the next step in the process is to ensure that the input is adequate for its intended use. This could involve things like running tests on incoming chemical products, checking invoices for errors, checking memos to be sure the writing is clear, etc. The next step involves the process of converting the input to output. This could be producing a resin from chemical raw materials or producing a typed memo from a handwritten memo. The production process steps also include inspection, i.e., ensuring that our product meets quality standards.

The next step is distribution of the product or output to the customer. This may involve shipping a product in rail cars, carrying a typed memo to the original author, sending payment through the mail for an invoice, etc. The next two steps in viewing processes as a system are critical to the concept of continuous improvement. The first of these is customer research. This involves finding out if the customer is satisfied with our product. The next step is to use customer feedback to help us improve our processes. Remember, the goal is not only to satisfy the customer, but to exceed customer expectations.

An easy way to remember the method of viewing processes as a system is SIPOC. SIPOC stands for Suppliers, Input, Process, Output, and Customers. When using the term SIPOC, it is important to remember that the feedback from the customer is an important loop in the system we are trying to develop.

Note that this system contains the three major elements. Customer satisfaction is the reason we want to get feedback from the customers. Productivity is included by working on processes and realizing that processes are interdependent. Employee involvement is inherent since it is the employee who produces the product or service.

METHODS OF TQM

The methods we use in TQM are one way of achieving the vision discussed above. These methods are designed to give employees the help they need to accomplish the vision. There are four major methods in TQM. These are Top Management Leadership, SPC, Team Management and Performance Management. Each method is discussed briefly below.

Top Management Leadership: Quality is everyone's job but top management must provide the leadership for the process to be successful. Not only must management be supportive of and committed to TQM, but it must also be active participants in TQM.

How does top management participate in the process? It establishes the vision of where TQM will take the company and establishes the Corporate Quality Policy. It establishes constancy of purpose by allocating resources for long-term planning, research, education, maintenance, and improvement of product design and service. It also becomes involved in teaching, coaching and facilitating the use of TQM. Management also solves problems associated with its levels in the organization. One of the most important things management does is to provide reinforcement and recognition for those involved in TQM.

Statistical Methods: The use of statistical methods or SPC to monitor and improve our major processes is a vital part of TQM. The use of SPC provides a method of determining whether our processes are stable and capable and provides a mechanism for continuous improvement of our processes. A systematic method of solving problems also plays a major role in this element. The purpose of this book is to explain the techniques and philosophy of SPC. The simple statistical techniques you will learn to use are shown in Figure 4.

Team Management: Team Management is a performance- or data-oriented approach to running our business (Miller, 1987). It features the formation of teams throughout the company. Each team is given maximal involvement in and responsibility for performance in a given area. Team Management is one method of encouraging participation of all employees.

Team Management involves many things. It represents a blending of meeting skills, leadership, performance management and statistical techniques. Team Management focuses on natural teams. A natural team is composed of a supervisor and his/her direct reports. These teams are formed from the top of the organization down. Using this approach, everyone is on a natural team. If you are a supervisor, you are on two teams: your own natural team and your boss's natural team. The supervisor usually serves as team leader. A natural team determines its responsibilities, develops team measures that tell it how they it is doing, sets goals for its team measures, develops plans to reach their goals, is given the information it needs to run its business, is allowed to communicate with other teams as needed, meets regularly and identifies needs and problems.

During a team meeting, performance is reviewed, i.e., how are we doing versus our goals. Improvements or successes are recognized. The team format provides many opportunities to reinforce appropriate behavior and results. Any performance problems are addressed. Plans are set for the time before the next meeting. News and information are also shared.

There are many benefits to using Team Management. These include improved communication, increased employee involvement, a sense of belonging by team members, increased motivation, and creative problem solving.

Project teams also play an important role in Team Management. While natural teams focus on improving their major processes that are directly related to customers, project teams are formed to work on major problems that both affect customer satisfaction and require the combined skills of people from various departments to solve. A project team disbands once the problem has been solved. Many of the benefits achieved from the on-going natural teams are also achieved by the project teams.

Figure 4

STATISTICAL PROCESS CONTROL TECHNIQUES

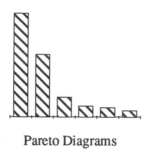

Pareto Diagrams

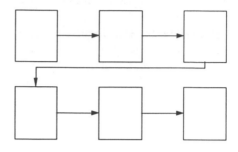

Process Flow Diagrams

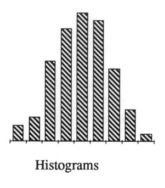

Histograms

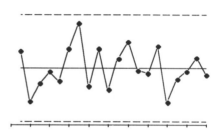

Control Charts

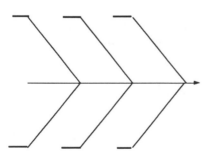

Cause and Effect Diagrams

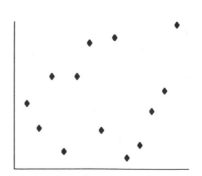

Scatter Diagrams

Performance Management: Performance Management (Daniels and Rosen, 1986) is a systematic, data-oriented approach to managing people at work. It emphasizes the use of feedback and positive reinforcement as the method to maximize performance. This approach leads to high and steady performance levels. The objective of Performance Management is to be able to identify areas of potential improvement and develop plans which allow these improvements to be obtained.

Before learning more about SPC, it is often helpful to look at the history of quality in the United States

HISTORY OF QUALITY

The concept of quality is not new. It has existed for as long as people have produced goods. For example, guilds in the Middle Ages insisted that apprentices go through long training periods. To become a master craftsman, these apprentices had to offer proof of their ability to produce quality products (American Supplier Institute, 1984).

Eventually, production moved to small owner-managed shops. Most of the products made by these shops were handmade, and there was no quality control program. In fact, there were no quality control problems. This does not mean that there weren't any quality problems. There were no control problems because the owner (management) knew what had to be done to produce a quality product. The owner was in daily, physical contact with every product his shop produced. These products could be shoes, gloves, or violins, for example.

Small owner-managed shops were the way of doing business until the advent of mass production. In 1798, the President of the United States, John Adams, realized that the military required 10,000 rifles in order to be adequately armed. No facility could produce 10,000 handmade rifles. In addition, since all were handmade, no parts were interchangeable among rifles. Eli Whitney, the inventor of the cotton gin, approached Congress with a proposal. He proposed making 10,000 rifles by exactly copying one rifle, part by part. In this manner, the parts between rifles would be interchangeable. Congress accepted the proposal and mass production was begun. No one thought to question what "exactly" meant.

The result of mass production was that these rifles did not work as well as the handmade rifles. In addition, the copied parts did not fit as expected. Of course, Whitney blamed this lack of fit on poor workmanship. He knew that his workers could produce exact copies if they tried hard enough. An understanding of variation was not yet developed. Whitney also established the precedent for running out of money, returning to Congress to ask for more and receiving it.

Following the Civil War, there was a period of rapid industrial development. For example, mass production was used to produce iceboxes and sewing machines. Mass production, at these levels, brought quality control problems that are still being resolved today.

9

This book focuses on the statistical aspects of quality. The statistical concepts, like quality, are not new. The "Father of SPC" was Walter Shewhart. During World War I, Shewhart, who worked for Bell Labs (at that time a subsidiary of Western Electric), was asked to design a radio headset for the military. One factor to consider when designing a headset is head breadth. Shewhart measured various people's head breadth. Shewhart discovered, as expected, that people have different head breadth. What he didn't expect to find was that the variation in people's head breadths followed a pattern. This pattern is shown in Figure 5. Most values were around a midpoint or average. As you moved away from the average, the frequency with which those head breadths occurred was less. Shewhart wondered if this type of variation was present in the production processes at Bell Labs. Shewhart found, after investigation, that this type of variation was present in many manufacturing processes.

Figure 5

HEAD BREADTH VARIATION

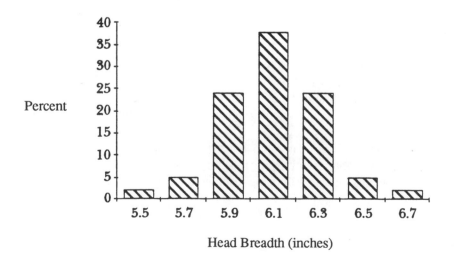

Head Breadth (inches)

By 1924, Shewhart had developed a control chart based on this variation. A control chart is a method of monitoring and analyzing variation over time. American industry made little use of control charts prior to World War II. During World War II, extensive use of control charts was made in the production of war time goods.

Two important events occurred at the end of World War II. First, Japan was destroyed economically. Second, the demand for American goods exploded. It is interesting to follow the development of quality in both these countries.

In the years following the war, some industrial production began in Japan. Products, however, had a terrible reputation for quality. In 1948, the Union of Japanese Scientists and Engineers (JUSE) was formed. JUSE wanted Japan to improve the quality of its products. In 1950, Dr. W. Edwards Deming was invited to Japan by General Douglas MacArthur. Dr. Deming worked with Japanese statisticians in studies of housing and nutrition. Meanwhile, JUSE was studying Shewhart's methods as a way of improving quality. Dr. Deming was asked to teach them more. Deming, and later Dr. J. M. Juran, initiated massive statistical training in Japan. This was the start of the quality revolution in Japan. The rest is history. Today, Japanese products, particularly in the automotive and electronics industry, are known for quality.

Why was Japan successful in improving quality? One reason was that upper management in Japan realized that building quality into products would increase their competitive position. Upper management also made sure that statistical methods were incorporated into their production processes. Upper management believed that its primary objective was to stay in business through quality and productivity improvements. Profits would then follow.

The statistical approach in the United States, however, dissipated in the years following the war. The major reasons for this included the lack of top management involvement and participation, the ability to sell American goods regardless of quality, the centering of statistical techniques in the quality control department, and short-term thinking of management.

Western industry is now declining as a result of this de-emphasis on quality. This decline started in the late 1960s when America began to lose markets to Japan. In the early 1970s with the onset of the oil crisis, the fuel efficient, high-quality, and dependable Japanese cars began to filter into the United States. The American consumer found these cars far superior to their American counterparts. American automotive companies began to lose market share. This caused plant shutdowns and the loss of many jobs. Overall, there has been a tremendous increase in the number of foreign goods being imported to America. This has made producers aware that their customers have a choice and are becoming more demanding in their quality expectations.

The emphasis on quality has recently increased in America. The roots of this resurgence can be traced to a documentary (produced by NBC) that reported what Japan had accomplished by using the techniques advocated by Dr. Deming.

In today's markets, it is no longer sufficient to do our best. We must know what to do and what tools we need to use to improve our quality, productivity, and competitive position.

WHY BECOME INVOLVED WITH TQM?

There are four major reasons why companies have become involved with TQM: customer demand, competition, resource utilization and the quality/cost relationship. (There are other reasons also.)

Customer demand is one reason to become involved with TQM. This demand started with the automotive companies. American automotive companies began using SPC in the early to mid 1980s to combat the competition from abroad. The automotive companies began requiring that their suppliers also use SPC. This is one feature of SPC. Customers also require their suppliers to use it. In this manner, SPC moves through the chain of customers and suppliers.

Another reason for becoming involved in quality is competition and the concept of sole supplier. Dr. Deming advocates reducing the number of suppliers for any one part or product. In fact, he favors using only one supplier. While we may never adopt this sole supplier concept, the demand for quality products from suppliers naturally leads to fewer suppliers for one product. As a supplier, this means we could lose some market share unless we become involved in quality. As competition grows, we need to find ways to differentiate our products and services from our competitors.

Still another reason for being interested in TQM is underutilized resources. The use of SPC allows you to make significant gains without major investments in capital or more people.

The fourth reason for becoming involved in quality is the relationship that exists between quality and cost. This relationship is called the chain reaction by Dr. Deming and was discussed earlier (see Figure 2). It states that as quality improves, costs decrease because there are fewer mistakes, less rework and better use of time and machines. Productivity improves and we increase our market share by offering better quality and lower prices. We stay in business and provide jobs.

DEFINITION OF SPC

Any process produces two things. The process produces either a product or a service. In addition, a process produces data. These data can be used to improve the process. SPC involves using data, generated by the process, to improve the process. This process improvement leads to improved products and services. One can look at the definition of SPC using the following:

Statistical	The use of numbers and data (to study)
Process	The things we do (to make them)
Control	Behave the way we want.

It is important to remember that the above definition focuses on the data part of SPC. There is also a business or management philosophy associated with SPC.

BENEFITS OF SPC

There are many benefits associated with using the SPC techniques and philosophy. Many benefits will become apparent through this book. Some obvious benefits include the following:

1. Improved product quality
2. Increased quality consciousness
3. Increased customer focus
4. Data based decisions
5. Predictable processes
6. Known process capabilities
7. Cost reduction
8. Increased involvement
9. Specific quality definitions
10. Accurate, precise test methods
11. Residual knowledge base

SUMMARY

In this chapter, the objectives of the book were introduced. These objectives are basically to learn how the SPC techniques and philosophy can be used to continuously improve the quality of our products, services, and information. The vision of TQM was also introduced. This vision is definitely based on customer satisfaction. The tools of TQM (top management leadership, statistical process control, team management and performance management) were introduced. A history of quality in the United States was discussed. The definition of SPC was briefly explained, and benefits associated with SPC were listed.

EXPECTATIONS

Based on this introductory material, what are your expectations for this book? Record your comments below.

2

QUALITY AND VARIATION

To be able to understand and successfully apply SPC techniques and philosophy, one must develop an understanding of variation. Why don't we always get to work at the same time each day? Why don't our shipments all arrive at the customer's plant at the same time, each and every time? Why do some invoices have errors and others don't? Why do some typed memos have more mistakes than others? This chapter introduces the concept of variation.

OBJECTIVES

In this chapter you will learn:

1. Definition of quality
2. Shewhart's concept of a process
3. Prevention vs detection mode of operation
4. Definition of variation
5. Assignable cause variation
6. Common cause variation
7. Statistical control

This chapter is designed to develop an understanding of variation. Quality is defined first. We will then define what we mean by a process and why it is important that we focus on improving our processes. The definition of variation is then introduced. This includes examining how we have viewed variation in the past and how we will view variation in the framework of SPC. SPC views variation as coming from two distinct sources: assignable cause and common cause variation. Both of these sources of variation will be examined in detail. This will permit you to identify which source of variation is present in your processes. The concept of statistical control is also introduced in this chapter. It is important to know if your process is in statistical control because the action you take to improve your process depends on whether or not it is in statistical control.

DEFINITION OF QUALITY

How do you define quality? Think about the last time you received a product or service that you considered to be of high quality. What made you think this was a quality product or service, i.e., what characteristics did the product or service have that made you think of it as quality? Think about the last time you received a product or a service that you considered to be lacking in quality. What made you think this was not a quality product or service? What is the difference in the way you feel when you have received a quality product versus when you haven't received a quality product? What effect does this have on what you do in the future?

People will answer the questions above in different ways, i.e., there are many ways one may define quality. So, what is quality? We can look at quality from different angles. For

example, we may view quality in terms of technology and look at
specifications; we may view quality in terms of time and look at reliability
and delivery; we may look at quality in terms of ethics and search for
courtesy and honesty; or we may look at quality in terms of profit. What do
these terms tell us about quality? The difficulty is that people view quality
from their own perspective. One person's view of reliability will probably
be different from another. The point is that quality is in the eye or perspective
of the customer.

We define quality as customer satisfaction or "fitness for use" (Juran, 1979). To provide quality
products, services and information, we must meet or exceed customer expectations.

SHEWHART'S CONCEPT OF A PROCESS

The definition of quality was introduced above. We have talked about improving the quality of
our products, services and information with the overall objective being to exceed customer
expectations. The key to making these improvements, however, does not come from focusing
on those products, services or information. The key to improvements comes from focusing on
the processes that produce those products, services, or information.

What is a process? A process refers to how we do things, such as making chemical or oil products,
typing a memo, processing an invoice, etc. There may be a few steps or many steps in a process.
In general, a process may be divided into six major elements. These elements are people,
materials, methods, machines, measurement and environment.

Walter Shewhart thought of processes in these terms. The output of a process was due to the
manner in which these six elements worked together. Shewhart's concept of a process is shown
in Figure 1.

Figure 1

SHEWHART'S CONCEPT OF A PROCESS

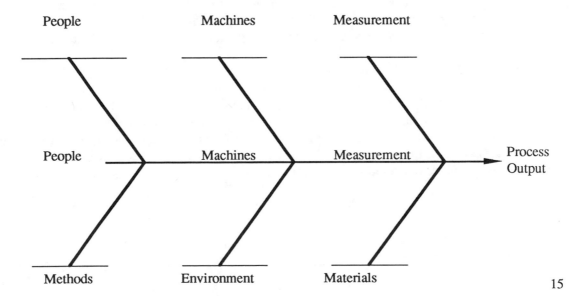

15

People refer to those individuals involved in making the product. These could be the front-line personnel in the process of producing gasoline or could be a secretary in the process of producing a typed memo. Materials refer to all material items used in the process of making the product. For example, water is a material used in the production of many chemicals, while a handwritten memo is material used in the production of a typed memo. Methods refer to the procedures we use to produce our product. In the process industry, this may include the manner in which a reactor is operated. In producing a typed memo, this would include the method of checking for spelling errors.

Machines refer to all devices we use in the process. For example, a machine in the production of some solid chemicals may be a spray dryer. For a typed memo, the machine may be a word processor. Measurement refers to the systems we have in place to measure what the process is doing or how well the process is doing it. In chemicals production, temperature gauges and lab tests are examples of measurement systems. The time it takes to produce a typed memo or recording the number of mistakes in a memo are examples of measurement systems for producing a typed memo. The final element, environment, refers to the conditions surrounding the process. In chemicals production, the environment may refer to outside weather conditions. For typing a memo, the environment may include the amount of lighting and room temperature.

The output from a process is determined by how these elements interact or work together. The key to improving the quality of our products, services, and information is to improve the quality of the processes producing them.

PREVENTION VERSUS DETECTION

Why should we focus on improving the process versus focusing on improving the product? To examine this, consider the diagram in Figure 2. The top part of Figure 2 demonstrates the approach of focusing on the product. In this process, the product is made using some process. Once the product has be made, it is inspected to see if it meets some set standard or specification. If the product meets this standard, it is ready for distribution to the customer. If the product does not meet this standard, it must be reworked, thrown away, or sold at a discounted price. If enough product does not meet the standard, the process may be adjusted to compensate for this. This mode of operation is called detection. Substandard product is being detected after it is made.

This detection mode of operation is present in many companies. The problem with this approach is that time, materials, etc. are being wasted on product that has already been produced. Once the product is made, there may be little you can do to correct the mistakes. At best, it would involve rework. You will have to spend additional time and energy on the product. Detection is not efficient and is costly. It makes more sense to make the product right to begin with. The slogan "Make it right the first time" is common today. How do you do that? The answer is to move from a detection mode of operation to a prevention mode of operation.

Figure 2

CONTROL SCHEMES

DETECTION CONTROL SCHEME

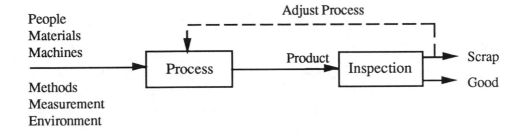

PREVENTION CONTROL SCHEME

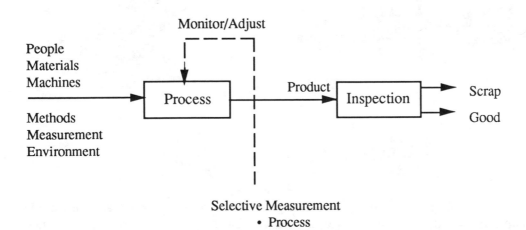

The bottom part of Figure 2 shows the prevention mode of operation. In this mode, selective measurements on the process and in-process material are made. This permits us to monitor how the process is behaving or performing. Problems, which may develop, are handled before the final product is made. Thus, we are preventing substandard product from being made. While this approach is obviously superior to the detection mode, it is not always easy to accomplish. Often, we may not know what in our process affects final product quality. SPC techniques help us discover these effects as well as provide a method for monitoring our processes over time. These techniques tell us when to leave the process alone and when to look for sources of problems.

VARIATION

We've talked about the quality of products, services, information, and processes. Why do we have problems with quality? There are three main reasons. One reason is that our customers often do not know their needs. Our customers may not know what characteristics in our products are important to them. They just want something that works. Another problem involves the quality of design. We sometimes try to sell our products into markets that the process and product were not designed to meet.

The major reason for quality problems, however, is variation. What is variation and why does it cause quality problems?

There is an old saying that no two things are exactly alike. This statement is true. The differences may be large, such as the differences between two types of cars, or the differences may be extremely small, such as the difference between a memo and a photocopy of the memo.

One objective of a manufacturing process is to produce material that is essentially the same. For example, the production of some chemicals involves batch reactors. The objective is to make each batch the same. In reality, however, there are differences between the batches. One objective of an invoice process may be to process each invoice in the same amount of time. In reality, the time is different for each invoice that is processed. These differences are called variation.

Variation exists in everything we do. What causes variation? Think back to Shewhart's concept of a process. Any variation in the six elements (people, machines, methods, measurement, material and environment) will cause variation in the output or product of a process. For example, variation in invoice processing time may be due to different people doing the work as well as the methods they use. Variation in chemicals production may be due to different reactors (machines), to different raw materials, to changes in environmental conditions, or to differences in the measurement systems.

We have defined variation and looked at some potential causes of variation. Variation is the net result of many factors that are constantly affecting a process. It is possible that the effect of some factors may not be measurable, but they are present. So, variability is always present and is inevitable in any process whether it be a manufacturing or non-manufacturing process. To control variation, we must begin to look at what causes it to occur, i.e., what are the sources of variation in the process?

Think about the length of your thumb. Is it the same length as everyone's? Of course not. Why doesn't everyone have the same thumb length? There are many reasons.

How have we attempted to handle variation in the past? With the advent of mass production, the problem was how to make parts interchangeable. Of course, the parts were not identical despite the mass production. The question was how similar do the parts have to be so that they are interchangeable. Specifications, which told how similar parts had to be, were developed. Thus, variation was divided into two classes:

Acceptable product: Within specifications

Unacceptable product: Outside specifications

This method of looking at variation is still common today. Most products, for example, have specifications (as they should). For example, one specification may be how pure an alcohol must be (for example, an alcohol must have 98% purity). Until recently, this was the method of looking at variation. If the product was in specification, we could ship it. If the product was out of specification, it had to be reworked or scrapped. Specifications are normally based on perceived customer needs.

The concept of specifications also applies to non-manufacturing areas. For example, the specification on a typed memo might be no misspelled words. If the memo meets that specification, it is acceptable and mailed. If it does not meet that specification, it must be reworked.

This approach to variation led to the detection mode of operation. The focus was on the final product. It also led to many problems. Manufacturing people were always trying to loosen the specifications to make it easier to meet the requirements. Customers were always trying to tighten the specifications. In addition, the problem of how good does the product have to be entered the

Chapter 2

MEASUREMENT OF THUMBS

Group Number Avg	Results	

Sources of Variation:

picture. Suppose the product was just barely out of the specifications limit. Could it still be used? None of these discussions focused on the process and what the process was capable of producing. In addition, the original idea of producing product that was essentially the same was lost.

In general then, it is preferable to produce a product with a small amount of variation than a large amount of variation. This increases the chance that we produce product that meets specification. So, production involves controlling the amount of variation present in the product. To do this, we have to control the sources of variation in the process. This is where Shewhart made his contribution to quality management. His contribution moves a process from operating in a detection mode to operating in a prevention mode.

To understand this approach, one must understand common cause variation and assignable cause variation. Neither of these types of variation has anything to do with specifications.

Common cause variation is the variation that exists because of the way the process is managed. It refers to the many sources of variation that are normally present within a process. For example, consider how long it takes you to drive to work (assuming your process for doing so is the same from day to day). Suppose, on average, it takes 20 minutes to get to work. It doesn't take, however, 20 minutes each day. Suppose you know that, in general, it will take between 15 minutes and 25 minutes to get to work. Some days it may take 22 minutes; other days 18 minutes; others 24 minutes, etc. The difference between each day's time and the average, most of the time, will not be alarming to you. You realize that the differences are due to the traffic, speed you drive, etc. All these causes are called common cause variation. They represent the normal variation in your process of getting to work. In general, there are a large number of common causes present at any time. Each cause has an effect that can't be measured. For example, suppose one day you get to work in 17 minutes. You can't relate that to driving at a certain speed, to traffic patterns, to the way your car was running, etc. As long as you keep the process the same, common cause variation is the same from day to day. Thus, common cause variation is consistent over time. We can predict the effect of common cause variation.

Everything varies. The amount of time it takes you to get to work each day is not the same day after day. In addition, you can't predict how long it will take you to get to work tomorrow. However, if only common cause variation is present, you can make predictions about how long the drive will take on average and how much variation there is about that average. Insurance companies make use of this knowledge. Figure 3 demonstrates this concept. People live to be different ages (i.e., everything varies). No one knows how long he or she will live (i.e., individual things are unpredictable). However, insurance companies can predict with great accuracy what percentage of people will live to be 60, 65, etc. (i.e., groups of things from a constant system of causes tend to be predictable).

Figure 3

CONSTANT SYSTEM OF CAUSES TENDS TO BE PREDICTABLE

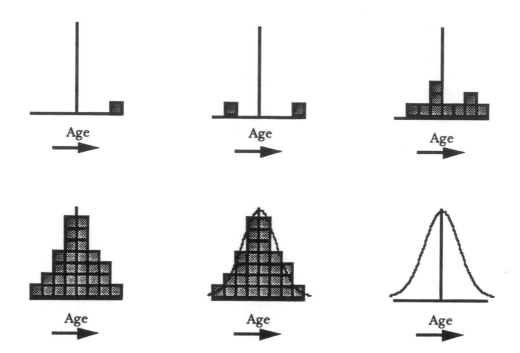

Assignable cause (also called special cause) variations are not part of the way the process is managed and are considered abnormalities. This means that something that normally doesn't occur has happened. Consider the driving-to-work example. If you get a flat tire on the way to work, you will not arrive at work in the normal span of 15 to 25 minutes. This is an assignable cause. Its effect on the process is measurable. For example, if you have a flat tire, it might take 60 minutes to get to work. This is definitely out of the normal range of 15 to 25 minutes and is directly attributable to having a flat tire. Other possible assignable causes for driving to work would include adverse weather conditions and traffic congestion due to an accident. Assignable causes are sporadic in nature and cannot be predicted. However, they are usually easily discovered.

The approach to variation in this case is to look at what variation is due to common causes and what variation is due to assignable causes. It is important to distinguish between the two because the responsibility for process improvement depends on what type of variation is present. In general, assignable causes are not part of the process and can be discovered. It is usually the responsibility of the front-line personnel to find and eliminate assignable causes. Assignable causes are often related to one machine, one person, one type of raw material, sudden change in environmental conditions, etc.

Common cause variation, on the other hand, is part of the system or process. To decrease common cause variation, you must change the system. This is management's responsibility. Only management, not the front-line personnel, has the authority to change the system or process.

Dr. Deming has estimated the percentage of time problems are due to common or assignable causes. Dr. Deming says that 85% of the problems are due to common causes and are management's responsibility. Only 15% of the problems are due to assignable causes and are the responsibility of the front-line personnel (Deming, 1982).

Thus, to determine how to improve a process, we must separate common causes from assignable causes of variation. How do we do this? There is only one way to effectively distinguish between common and assignable causes of variation, and that way is to use control charts. Control charts present a picture of a process over time and tell you when only common causes are present or when there are also assignable causes present.

STATISTICAL CONTROL

A process is in statistical control if only common cause variation is present. When a process is in statistical control, we can predict what the process will make in the near future. Once a process is in statistical control, the only economical way to improve it is to change the system.

Being in statistical control is not a natural state. Work must be done to permanently eliminate assignable causes over time.

Consider the driving-to-work example again. If the process is in statistical control, it means only common cause variation is present. Suppose (by using a control chart) you know that it will take from 15 to 25 minutes to arrive at work. This variation is due to common causes. When you leave for work, you don't know exactly how long it will take to get there. For example, you don't know that it will take 22 minutes today. However, you do know that you will arrive at work in 15 to 25 minutes, assuming there are no changes in the process or assignable causes that occur. You also know that, on average, it will take 20 minutes to get to work. Thus, by getting the process into statistical control, you know what your long-term average (20 minutes) will be and how much variation (15 to 25) to expect day to day.

How do common cause, assignable cause, and statistical control fit with specifications? Once a process is in statistical control, it has a natural tolerance associated with it. This natural tolerance is the variation due to common causes. In the driving-to-work example, the natural tolerance is 10 minutes (25 minus 15). Note that the process itself determines what the natural tolerance is. The natural tolerance is due to common cause variation. This is due to the way the process is managed. Specifications, on the other hand, are set by a person. Engineering tolerance is usually defined as the upper specification limit minus the lower specification limit. A process is capable of meeting specifications if the natural tolerance (common cause variation) is less than the engineering tolerance.

A process can be in statistical control and not meet specifications. A process can be out of statistical control and meet specifications. One goal of TQM is to have processes that are in statistical control and meet specifications.

To achieve this, in many cases, the common cause variation will have to be reduced. This is the primary reason for focusing on our processes and forms the basis of continuous improvement.

SUMMARY

Quality is defined as "customer satisfaction" or "fitness for use." A process refers to the way or manner in which we do things. It is important to focus on the process to prevent problems from occurring in the first place.

To effectively improve a process, one must understand variation. There are two types of variation. Common cause variation is the variation inherent or normally present in the process. Problems due to common cause variation are the responsibility of management. Assignable cause variation is an abnormal event; one that does not commonly occur. Problems due to assignable causes of variation are the responsibility of those closest to the process.

A process is in statistical control if only common cause variation is present. A process which is in statistical control is consistent over time, i.e., we can predict what the process will produce in the future.

APPLICATIONS

What do you feel are the three most important points in this chapter? How do these points affect what you do at work? Record your comments below.

3
INTRODUCTION TO CONTROL CHARTS

Processes, whether manufacturing or service in nature, are variable. You won't always get the same result each time. The reason for this is that there are sources of variation in all processes. There are two major sources of variation. One is common cause variation. It is the variation which is inherent in the process due to the way it is managed. Common cause variation can be reduced. To reduce common cause variation, the system must be changed. This is normally management's responsibility. There will always be common cause variation present in a process. The second type of variation is assignable cause variation. This variation is caused by things that don't normally happen in the process. Front-line personnel have the responsibility for finding and removing assignable causes of variation. A process is in statistical control if only common cause variation is present. How do we know if only common cause variation is present or if there are also assignable causes of variation present? The only way to determine this is through the use of a control chart. Control charts are introduced during this chapter.

OBJECTIVES

In this chapter you will learn:

1. Why control charts work
2. An example of a process in control

How does a control chart tell if only common cause variation is present? This is determined by the data itself. Using the data, we compute a range of values we would expect if only common cause variation is present. The largest number we would expect is called the upper control limit. The smallest number we would expect is called the lower control limit. In general, if all the results fall between the smallest and the largest number and there is no evidence of nonrandom patterns, the process is in statistical control, i.e., only common cause variation is present.

WHY CONTROL CHARTS WORK

A control chart is a picture of your process over time. Control charts help identity the magnitude and type of variation present. To determine why control charts work, we will examine two examples.

Suppose you are in charge of an operation which produces a kerosene stream. The kerosene is used as the raw material in another process in the plant. You typically test the kerosene for various properties, including API gravity. Lately, there has been some concern expressed about the quality of the kerosene. There is too much variation in the gravity. You are interested in determining how consistently the kerosene is meeting gravity specifications. The specifications for the API gravity on the kerosene are from 24 to 50. You also want to determine how to decrease the amount of variation.

Chapter 3

To determine if the kerosene is meeting specifications and is about the same all the time, you decided to take five samples each hour. The data collected are shown in Table 1. Is the kerosene meeting specifications and what can we do to decrease the amount of variation in the API gravity?

It is difficult to answer these questions just by looking at a table of numbers. Figure 1 is a plot of the individual measurements. As can be seen from the figure, essentially all the individual results are within specifications. In fact, there is only one sample out of spec. The remaining 99 are within specifications. The figure also indicates that most of the kerosene API tends to be toward the lower end of the specification. Thus, by plotting the individual results, we see that 99% of the samples are within specifications and the average result is closer to the lower spec limit than the upper spec limit. How do you start to improve this process, i.e., decrease the amount of variation present?

To determine this, we need to look at the same data in a different way. Suppose instead of plotting the individual values, we take the average of the five samples pulled each hour. The average, called a subgroup average, has been calculated for each subgroup and is shown in Table 1 also. For example, the five samples taken the first hour had API gravities of 36, 35, 34, 33, and 32. The average of these five results is 34. These subgroup averages could then be plotted, as shown in Figure 2. This figure details how the average hourly kerosene gravity is varying. Figure 2 is called a X control chart. The "bar" over the $\overline{X}$ means average. Control limits for these subgroup averages could also be calculated and plotted. This has been done and the limits are shown in Figure 2 also (don't be concerned with how the limits were calculated; we will cover this later). Remember, the upper control limit is the largest average hourly gravity we would expect to see if only common cause variation is present. The lower control limit is the smallest we would expect.

As can be seen in Figure 2, there are some points above the upper control limit and below the lower control limit. When this occurs, it means that there is an assignable cause present; the process is out of statistical control at that time. The front-line personnel should be able to find out what caused those points to be beyond the control limits and to correct the problem.

Figure 2 looked at the variation in subgroup averages from hour to hour. We can also examine the variation in individual samples within an hour. This is determine by finding the "range" of each subgroup. The range is defined as the largest value minus the smallest value in the subgroup. This has been done for each subgroup. For example, the largest result for the five samples pulled the first hour was 36. The smallest result was 32. The range is then 36 - 32 = 4. The range values are also shown in Table 1.

We can plot the ranges values (just as we plotted the subgroup averages). Figure 3 is a plot of the range values. This chart is called an "R" or range chart. We can also calculate control limits. This has been done and the control limits are shown in the figure. Again there is a point above the upper control limit. This indicates that there was an assignable cause present. The front-line personnel should be able to find out what caused this assignable cause and to correct the problem.

Table 1
KEROSENE API GRAVITY RESULTS

Subgroup Number	Five Samples Each Hour					$\overline{X}$	Range
	1	2	3	4	5		
1	36	35	34	33	32	34.0	4
2	31	31	34	32	30	31.6	4
3	30	30	32	30	32	30.8	2
4	32	33	33	32	35	33.0	3
5	32	34	37	37	35	35.0	5
6	32	32	31	33	33	32.2	2
7	33	33	36	32	31	33.0	5
8	23	33	36	35	36	32.6	13
9	43	36	35	24	31	33.8	19
10	36	35	36	41	41	37.8	6
11	34	38	35	34	38	35.8	4
12	36	38	39	39	40	38.4	4
13	36	40	35	26	33	34.0	14
14	36	35	37	34	33	35.0	4
15	30	37	33	34	35	33.8	7
16	28	31	33	33	33	31.6	5
17	33	30	34	33	35	33.0	5
18	27	28	29	27	30	28.2	3
19	32	27	29	36	35	31.8	9
20	33	35	35	39	36	35.6	6

(Data adapted from Grant and Leavenworth, 1980)

Figure 1

INDIVIDUAL SAMPLE API GRAVITY RESULTS

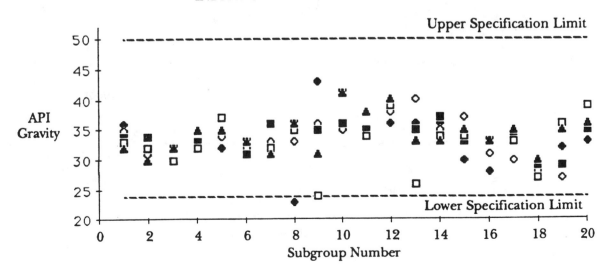

Figure 2

$\overline{X}$ CHART FOR API GRAVITY
(Subgroup size = 5)

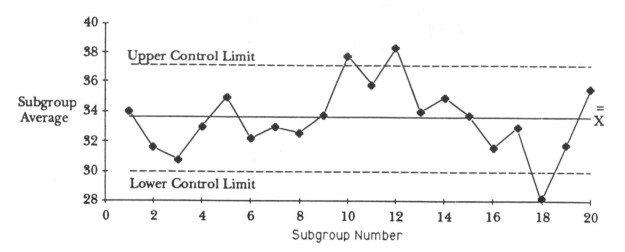

Figure 3

RANGE CHART FOR API GRAVITY
(Subgroup size = 5)

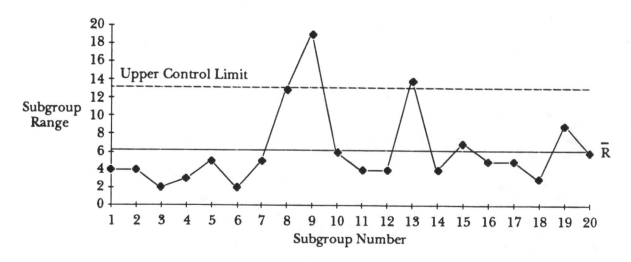

What conclusions can we reach based on this analysis? The process produces most of the material within specifications. The charts show lack of statistical control because there are points beyond the control limits. This indicates that there are assignable causes present. It is the responsibility of the front-line personnel to find and correct these problems. A process can be meeting specifications but not be in statistical control. The techniques required to develop and maintain these charts are learnable and are not highly mathematical.

The second example involves the production of a solvent. The solvent is produced in another part of the plant. The initial boiling point (IBP) of the solvent is of interest. Specification on the IBP is 139 - 142. We know we have been receiving material out of spec. To determine exactly what is happening in the process, we decide to take five samples per hour and measure the IBP. The results of taking this data are shown in Table 2.

Figure 4 is a plot of the individual results. As can be seen, most of the production is out of specifications. Whose responsibility is this and what must be done to decrease the amount of off spec material? To answer these questions, we must look at the data from a different perspective. Again, we can determine the subgroup average (average of the five individual samples each hour) and the subgroup range (the largest sample result minus the smallest sample result each hour). The subgroup averages and ranges are shown in Table 2.

The subgroup averages are plotted in Figure 5. The figure shows the variation in the average IBP from hour to hour. Control limits have been calculated and added to the figure. There are no points beyond the control limits. The process is in statistical control with respect to the subgroup averages. (Note: there are other tests for control besides points beyond the control limits; these will be introduced later.) This means that there is only common cause variation present. To improve this process (i.e., decrease the amount of variation present), the system must be changed. This is management's responsibility.

The subgroup ranges are plotted in Figure 6. This figure shows the variation in the individual sample results within an hour. Control limits have been calculated and added to the figure. As can be seen, no points are beyond the control limits. The process is in statistical control. Only common cause variation is present. To improve this process, the system must be changed.

What conclusions can we reach from this example? The process is in statistical control, but is not meeting specifications. Thus, in control does not mean within specifications or that the results obtained from the process are acceptable. To decrease the amount of off spec material, the system must be changed. This is management's responsibility.

A control chart is a picture of the process over time. It tells us when to take action, i.e., when to look for an assignable cause. It also tells us when to leave the process alone, i.e., when the process is in control since only common cause variation is present. The responsibility for improving a process (decreasing the amount of variation present) depends on the type of variation present. If

Table 2
RESULTS OF INITIAL BOILING POINT OF SOLVENTS

Subgroup Number	Five Samples Each Hour					$\overline{X}$	Range
	1	2	3	4	5		
1	140	143	137	134	135	137.8	9
2	138	143	143	145	146	143.0	8
3	139	133	147	148	139	141.2	15
4	143	141	137	138	140	139.8	6
5	142	142	145	135	136	140.0	10
6	136	144	143	136	137	139.2	8
7	142	147	137	142	138	141.2	10
8	143	137	145	137	138	140.0	8
9	141	142	147	140	140	142.0	7
10	142	137	145	140	132	139.2	13
11	137	147	142	137	135	139.6	12
12	137	146	142	142	140	141.4	9
13	142	142	139	141	142	141.2	3
14	137	145	144	137	140	140.6	8
15	144	142	143	135	144	141.6	9
16	140	132	144	145	141	140.4	13
17	137	137	142	143	141	140.0	6
18	137	142	142	145	143	141.8	8
19	142	142	143	140	135	140.4	8
20	136	142	140	139	137	138.8	6
21	142	144	140	138	143	141.4	6
22	139	146	143	140	139	141.4	7
23	140	145	142	139	137	140.6	8
24	134	147	143	141	142	141.4	13
25	138	145	141	137	141	140.4	8
26	140	145	143	144	138	142.0	7
27	145	145	137	138	140	141.0	8

(Data adapted from Grant and Leavenworth, 1980)

Figure 4

INDIVIDUAL SAMPLE IBP RESULTS

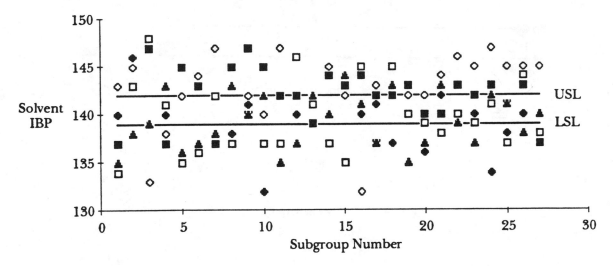

Figure 5

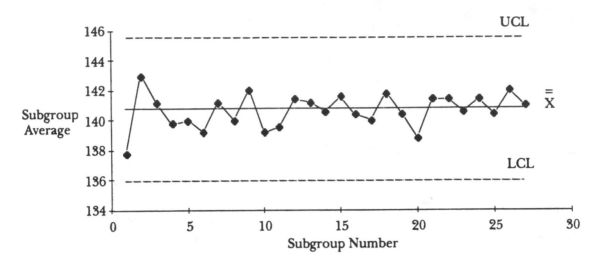

$\overline{X}$ CHART FOR IBP OF SOLVENTS
(Subgroup size = 5)

Figure 6

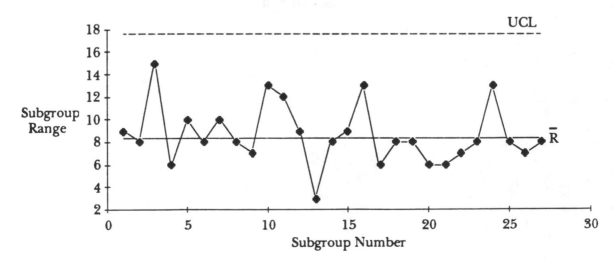

RANGE CHART FOR IBP OF SOLVENTS
(Subgroup size = 5)

assignable causes are present, it is the front-line personnel's responsibility. If only common cause variation is present, it is management's responsibility.

This concept seems relatively simple. However, in practice it is difficult. Many people don't understand the concept of variation. The tendency is to respond to whatever the last data point was. For example, suppose your process involves drying a product in a fluid bed type dryer. Moisture measurements are taken on a routine basis at the exit of the dryer. The tendency for many people is to increase the drying temperature if one result is higher than the previous result. This tendency actually increases the variation in the process. The correct approach is to use a control chart to monitor the moisture content. As long as the process remains in control, no action is taken on the up and downs in the moisture measurements. If a point goes beyond the control limit, action is then required. The action required is to find out what caused the point to be beyond the control limit. It may be necessary, depending on the investigative results, to increase the drying temperature.

The example in the next section highlights this tendency to respond to a single data point.

BEAD BOWL EXPERIMENT

You work in a factory which produces white beads for a customer. Five other people on your shift also make beads. Also on shift are two inspectors who check the beads and a chief inspector who checks the inspectors. There is also a recorder who records the number of defects (yellow beads) produced by each person on shift. You recently signed a card pledging to make "zero defects." In addition, the company has developed an award called "employee of the day." Whoever makes the least defects in a day gets to park next to the front door the next day. All pay and promotional opportunities will be based on how well you perform.

You make 50 beads per day. This is the quota. You can't make less or more. Once your quota has been produced, both inspectors count the number of defects and report their results to the chief inspector. The chief inspector then reports the results to the recorder.

Record the results on the chart in Figure 7 and plot them on the graph on the graph in Figure 8.

Figure 7

BEAD BOWL EXPERIMENT

Number of Yellow (Defective) Beads in the Sample

Person	Day				Total
	1	2	3	4	
Total					
Cum. Average					

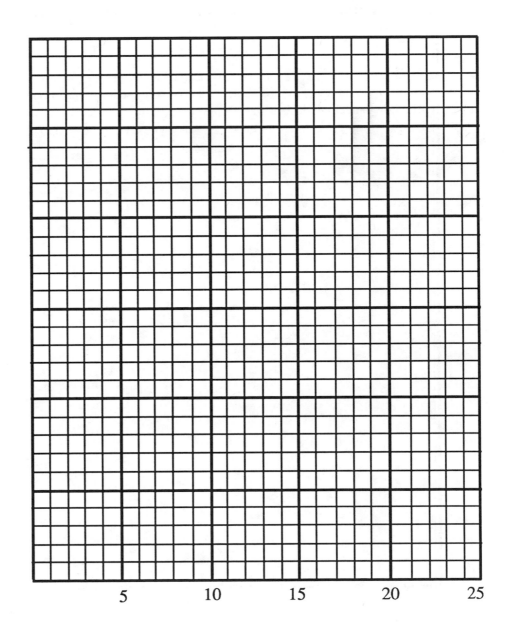

Figure 8

BEAD BOWL EXPERIMENT: GRAPH

Number of

Yellow

(Defective)

Beads

5   10   15   20   25

Sample Number

SUMMARY

In this chapter, the concept of why control charts work was introduced. It was seen that a control chart can separate assignable causes from common cause variation. Control charts serve as a basis for action. Specifications are not control limits. A process can be in control but the output from the process not be acceptable. An example of a stable or in-control process was introduced through the bead bowl experiment.

APPLICATIONS

Think about what you have been introduced to in this chapter. What do you think are the 2 or 3 most important items in the chapter and how do these items impact on what you do at work? Record your comments below.

1.

2.

3.

4

STRATEGY FOR PROCESS IMPROVEMENT

How can you use information about the Total Quality Management process, the SPC tools, and the problem solving model back on the job? How can you determine what impact you have made by working on process improvement? These are typical questions. Your objective in using this book is to learn various tools to help improve the quality of our products, services, and information through process improvement. A large amount of information is provided. To effectively use this information, it is helpful to have a road map. A road map tells you where you are and where you want to get to eventually. The biggest advantage of a road map is that it tells how to get there. This chapter provides a road map for process improvement. The problem solving model is included as part of this road map.

OBJECTIVES

In this chapter you will learn:

1. Strategy for process improvement
2. How to measure progress and results
3. How to work on the process

The objective of this chapter is to describe an overall strategy for process improvement. This is a multi-step strategy, which focuses primarily on customer satisfaction. This strategy provides the road map for process improvement. Two methods of measuring the results from process improvement are also included. One method allows you to monitor your progress in process improvement even though results are not yet available. The other method involves determining the impact to the company in dollars. Finally, suggestions for how to work on the process are given.

PROCESS IMPROVEMENT STRATEGY

This section describes the overall strategy for process improvement in detail. This strategy is a multi-step process and is shown in Figure 1. Each step is described in detail below. The strategy holds for individuals or teams working on improving processes.

Step 1: Mission Statement

The first step in process improvement is to define the mission of your department, team, or job. The mission consists of the most important results of a department, team, or job. It is important that the mission be stated as specific results and not behaviors. Table 1 lists some points for distinguishing between results and behaviors. In addition, the mission must not be a measure. For example, no team's sole mission is to make a profit. The mission statement includes the results that lead to the profit.

Figure 1

STRATEGY FOR PROCESS IMPROVEMENT

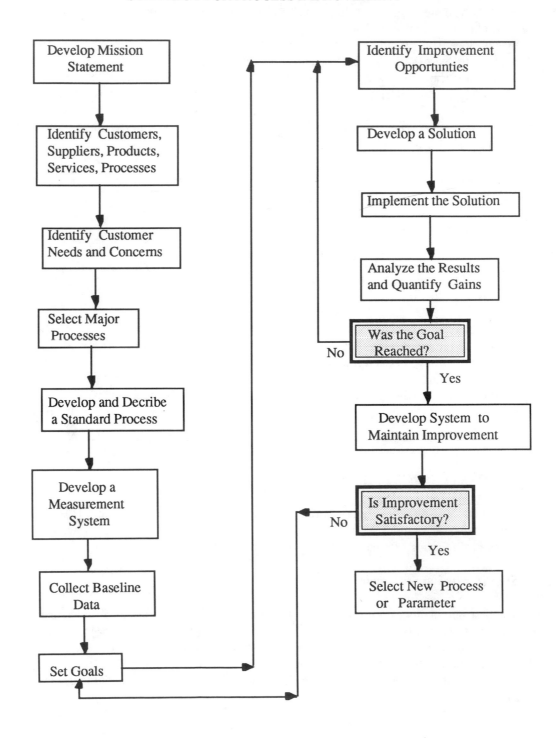

Table 1

POINTS FOR DISTINGUISHING BEHAVIORS FROM RESULTS

	Behavior		Results
1.	What people are doing	1.	What people have produced
2.	What you see people do while they are working	2.	What you see after they stop working
3.	Must see people working	3.	Not necessary to see people working
4.	Tends to be expressed by present tense verbs, ending in "ing"	4.	Tend to be expressed by noun-adjective pairings, such as a "painted chair"
5.	Cue words: by, through	5.	Cue words and phrases: to, in order to, so that, to achieve, to effect, to be able to
6.	Commonly used terms: input process activity means	6.	Commonly used terms: output product outcome ends
7.	Examples: inspecting designing attending taking breaks conducting meetings	7.	Examples: rejects specs completed yield run time submitted suggestions

(Daniels and Rosen, 1986)

Various methods exist to develop mission statements. One method involves three steps. The first step is to list all the results you can think of. The next step is to develop a statement describing the overall purpose of the department, team or job. Each listed result is then compared to the mission statement. Results that support the mission statement are important results. Mission statements, in general, should be only one or two sentences in length.

There are five general criteria for evaluating a mission statement. These criteria are referred to as the ACORN test (Gilbert, 1978). ACORN stands for Accomplishment, Control, Overall Objective, Reconcilable, and Numbers. Each of these is described below in terms of a natural team's (supervisor and direct reports) mission statement. The same test holds for department or individual job mission statements.

Accomplishment: A mission must be stated as results, not as behaviors or as measures.

Control: The mission must be under the "control" of the team. The question to ask yourself is "If the team did everything in its power and authority and used all the resources available to it, would the team significantly impact the desired results?" The answer to this question must be yes. Note that this does not mean there must be direct control over the mission. The key is that the team must have more influence over the mission than any other team. The question of "influence" will not always be easy to determine. For example, a corporate manufacturing vice-president's natural team will be interested in production. However, this team should not have more influence over production than a plant manager's natural team.

Overall Objective: The basic question to ask here is "If the results were completely achieved, would anything else be expected?" If the answer is no, then the mission statement represents the team's overall objective. If the answer is yes, the mission statement does not meet this criterion.

Reconcilable: The basic question to ask here is "If the mission were accomplished perfectly, would the mission of other teams be hampered?" The answer to this question must be no. This ensures that various teams' mission statements are compatible. We will not have teams working against one another.

Numbers: You must be able to measure the mission. This is the only way you can determine whether or not you are fulfilling the mission of the team. If the mission can't be measured, you must find a new mission statement. If the mission statement satisfies the first four criteria, it is possible to develop a method of measurement. In some cases, measurement systems will already exist. In other cases, a measurement system will have to be developed.

The end product of this step is a written mission statement.

This step has involved developing a mission statement for a department, team or individual job. The next step begins to look at the processes involved in achieving the mission.

Step 2: Identification

This step involves identifying customers, suppliers, products, services, and processes. Based on your mission statement, who are your customers? That is, who receives and uses your products and services? Who are your suppliers, i.e., whose products and services do you use as input to your processes? What are your processes that directly affect your mission statement? What products and services do those processes produce?

The end product of this step is a list of customers, suppliers, products, services and processes. Often steps 1 and 2 in this process are done at the same time. It is difficult to decide what your mission is without also deciding who your customers are. The next step in our strategy is to begin to identify customer needs and concerns.

Step 3: Identifying Customer Needs and Concerns

One goal of TQM is to exceed customer expectations, not merely meet them. To exceed these expectations, we must have regular feedback from the customer on how we are doing. A first step in receiving this feedback is to identify customer needs and concerns.

First, we must select the customers on which to focus. This will normally be major customers that represent a significant portion of our business. It is helpful to look at customers from a long-term focus instead of short-term focus. For example, suppose we have a customer who buys $2 million/year of products. If we consistently meet and exceed this customer's expectations, we will develop a long-term relationship with this customer. Defining long-term as the next ten years, this customer then represents $20 million in sales.

After deciding which customers are "major," we should speculate about the results. What do we think the customers will say about the quality of our products, services and information? What do we think the customers' needs are? What areas do we think the customers will mention as improvement opportunities? What do we think the customers will want in the future? We can then compare our speculation to actual results. This will provide us with a measure of how well we know our customers.

The next step is to determine how to gather the information. It may not be necessary to obtain data from all the major customers. A random sample may be appropriate. In addition, you may want to obtain data from potential or past customers. A major question is "What kind of information do I want to obtain?"

There are several methods of obtaining information from the customer. The most preferable is through direct conversation. Surveys and questionnaires are commonly used, but are seldom as effective in gathering information. Observing the customer using your product or service is another method of gathering information. Other decisions to be made include who will collect and analyze the data and how it will be analyzed.

The next step is to gather the information. It is usually best to begin with a few customers and then analyze the results. It may be that changes are needed to improve the quality of the information gathering.

Once the information has been collected, the results are analyzed. These results can then be compared with your initial speculation of results. It is important to check the validity of your conclusions, i.e., do the customers and others in the company agree with your conclusions? The final step is to take action on any obvious changes that need to be made.

In addition, plans should be made for regular contact with the customers. It is important that you follow up on any needs or concerns identified.

The end product of this step is a list of customer needs and concerns in order of priority. It may also include a "wish list" from the customer.

Step 4: Selecting Major Process(es)

The purpose of this step is to select the process or processes that are related to customer needs and concerns. These will be the processes that we focus on improving. It is important to remember that, in a prevention mode of operation, we want to work on improving the process that produces the product or service. This allows us to ensure that the product is produced correctly the first time. Our problem is not the quality of our products or services but the quality of our processes.

The end result of this step is list of major processes that affect customer needs and concerns.

Step 5: Developing and Describing a Standard Process

The purpose of this step is to increase the uniformity of a product or service by developing a standard process and a process flow diagram. This should eventually be done for each of the major processes selected above. Although you will eventually work on several processes at the same time, it is preferable to begin by working on one process only.

The first step is to develop the process flow diagram. It should include suppliers, input or raw materials, the production process, distribution of the output or product, customer feedback, and use of that customer feedback. This diagram is best constructed by those closest to the process. Consensus on what the process is must be reached. It may be beneficial to develop a diagram of the physical work flow. If there are obvious changes that could be made to improve the process, they should be made immediately.

The next step is to develop a standard operating procedure, i.e., based on our current knowledge, what is the best way to operate this process? Once this standard operating procedure has been developed, all people involved in the process use it. Once everyone is following the same procedure, it will be easier to make changes to the process and determine if improvements have been made.

The end product of this step is a completed, detailed process flow diagram and a written standard operating procedure.

Step 6: Developing a Measurement System

The purpose of this step is to develop a method of measuring the effectiveness of the processes identified above. It is important to be able to measure the progress made in improving a process.

Decisions must be made about what type of data should be collected. These data should be directly related to customer needs. In addition, decisions must be made about how the data will be collected, how often the data will be collected, who will collect the data, how the data will be recorded, and how the data will be displayed.

Operational definitions must also be developed. For example, if one measure is on-time, we must pinpoint exactly what on-time means.

The measurement system must be stable or consistent over time. This must be continuously checked throughout the process.

The end product of this step is a measurement system to monitor what the process is doing. It could be more than one measurement system. It may be an existing system or a new system.

Step 7: Collect Baseline Data

The purpose of this step is to determine how the process is currently behaving. Baseline data must be collected on all processes identified above. This permits one to determine what effect future changes have on the process.

A decision must be made about how much baseline data are needed. In general, if data are collected as frequently as daily, one should collect baseline data for about one month. This data can then be displayed in a control chart. At times, it may be more beneficial to display the data as a Pareto diagram. If data are collected weekly or monthly, the baseline data can have as few as three or four points. Data collected less frequently than once a month will be of limited use. Historical data can be used as baseline data if they are available.

The end product of this step is a graph or chart with the baseline data.

Step 8: Setting Goals

The purpose of this step is to set a goal for process improvement. Normally, this goal should be statistically based if a control chart is being used. A rule of thumb is to set goals at 1/2 to 1 standard deviation. These goals are normally reasonable and should be attainable. This helps increase the opportunities for reinforcement.

The end product of this step is a goal, often added to the graph of the baseline data.

Step 9: Identifying Improvement Opportunities

The purpose of this step is to begin to work on improving the process with a plan of reaching or exceeding the goal. The procedure for identifying improvement opportunities depends on the process. Is the process in control? Is the process out of control? Are there recurring problems that occur? What are the causes of these recurring problems?

If the process is out of control, the first step is to begin to remove assignable causes permanently. It is usually the responsibility of front-line personnel to identify and remove these causes. If the process is in control, the system must be changed to improve it. This is usually the responsibility of management, but front-line personnel often have excellent ideas on how to improve a process. Cause and effect diagrams, scatter plots, experimental design techniques, brainstorming, and selecting-in are useful techniques to help determine why a process behaves as it does.

If recurring problems need to be addressed, the problem must first be defined. How often does it occur? When does it occur? Why does it occur? How severe is it? What causes it to occur?

The end product of this step is that you have identified an improvement opportunity. This will normally be a list of causes of why the process is not operating at the level suggested by the goal set above. The list of causes should be prioritized.

Step 10: Developing a Solution

The purpose of this step is to develop a solution to permanently remove the causes identified above. Brainstorming, selecting-in and consensus decision-making are techniques that can be used to generate a solution.

The end result of this step is a solution to help reach the goal.

Step 11: Implementing the Solution

The purpose of this step is to implement the solution(s) identified above. It involves developing an implementation plan. An implementation plan has two components: one-time action steps and required on-going behaviors.

Action steps are one-time items that must be done prior to implementing the solution. The action steps should include who will do what and by when.

The on-going behavior requirements address what we want people to do differently in the future.

The end product of this step is an implemented solution.

Step 12: Analyzing the Results and Quantifying Gains

The purpose of this step is to determine how effective the solution is. Data must be collected to determine if the implemented solution improved the process. If the process was not improved at all, two possibilities exist. One is that the wrong cause(s) was selected. The other is that the wrong solution was selected. For either case, you must return to the steps above.

Another possibility is that the process improved, but the goal still has not been reached. In this case, there are other causes still affecting the process. You must return to step 9 in this case.

The last possibility is that the solution worked, and the goal was reached. If this is the case, you proceed to the next step.

Step 13: Developing a System To Maintain Improved Performance

The purpose of this step is to develop a system to ensure that the gains are maintained over the long term. Too often a goal is reached and performance subsequently falls back to original levels because no system was installed to maintain performance. This system could be continued tracking of the process over time, perhaps on a less frequent basis.

Step 14: Selecting New Process or Parameter

The first 13 steps have led to an improved process. A decision must be made concerning the improved process. Is it sufficient or are there more worthwhile gains to be made? If the improvement is sufficient, a new process or parameter should be selected and the procedure started over. If more improvement is worthwhile, a new goal should be selected and the process starts anew at Step 8.

MEASURING RESULTS

There are several methods for measuring progress in process improvement. Two methods will be introduced here. One method involves a matrix system for the steps involved in improving processes. The other method involves the benefit to the company, in dollars.

Matrix Method

The matrix measurement method focuses primarily on completing the various steps in the strategy for process improvement. The matrix is shown in Table 2. The purpose of the matrix is to permit you to improve your measure for improving processes even though the entire procedure is not yet complete.

Four preliminary steps must be completed before the matrix can be used. These steps involve an individual's, team's, or department's overall responsibility. These steps don't involve one process. The four steps that must be completed are:

1. Developing a mission statement
2. Identifying customers, suppliers, services, products, and processes
3. Identifying customer needs and concerns
4. Selecting major processes

Once these four steps are completed, the matrix can be used to measure the progress made in

Table 2

REVIEWING AND IMPROVING PROCESSES MEASUREMENT MATRIX

Points	Step
10	Optimum Performance Maintained
9	Process at Optimum
8	Performance Maintained
7	Goal Reached
6	Solution Implemented
5	Solution Selected
4	Goal Set
3	Baseline Data Collected
2	Measurement System Developed
1	Process Flow Diagram, Standard Operating Procedure Developed

Notes: This matrix applies to each process being reviewed and improved. To obtain the points for each step, the steps prior to that step must have been completed.

improving processes. The matrix represents a total of ten points for each process being improved.

For a given process, one point is awarded for completing a process flow diagram and developing a standard operating procedure. Two points are awarded for developing the measurement system. Three points are awarded for collecting baseline data. Four points for setting a goal and five points for selecting a solution.

Six points are awarded for implementing the solution. Seven points are awarded for reaching the goal. Reaching the goal is not the end of the improvement process. Performance should be maintained at that level or improved further. The matrix takes this into account. Eight points are awarded if the performance has been maintained. Nine points are awarded if this process represents the optimum process, i.e., the improvements made in the process now make it satisfactory and no further work is required. Ten points are awarded if the performance at this optimum is continued over time.

The ten point matrix applies to each process being reviewed and improved. There is no limit to the number of points that can be obtained. For example, if ten processes are being examined, a total of 100 points are possible.

In completing the matrix, each step must be done to be credited with the points. For example, suppose you are working on a process and have completed the process flow diagram, the standard operating procedure and developed a measurement system. This is worth two points. Suppose you have also set a goal. You do not get four points for this since you haven't collected baseline data. You still only get two points.

Dollar Impact

The purpose of this measurement is to begin to quantify the benefits of process improvement to the company. The objective is to determine the annual permanent improvement. While it may be difficult in some cases to determine a dollar amount, the effort should be made in all cases. The three main categories of dollar impact are increasing revenues, decreasing costs and cost avoidance.

1.   **Increasing Revenues**

Examples of increasing revenues are given below.

A. Increasing sales due to better quality in your products or services increases our revenues. It may be difficult to determine the importance of quality in sold-out markets. However, when improved product or services add incremental sales, this should be counted. One approach to estimating increased revenues would be annual incremental volume times netback.

B. Regaining lost business or maintaining business in jeopardy due to process improvement and better quality should also be counted in the dollar impact. Again, the formula would be annual regained or stabilized volume times netback for that product.

C. Price increases in products and services that are accepted in the marketplace over competitive products due to higher quality also add to our revenue. The increased revenue due to the new higher price would be the annual volume times the price differential.

D. Increasing plant capacity when you are sold out is another way of increasing revenues. This could be done by debottlenecking or by increasing the stream factor. This could be calculated by the netback times the annual incremental volume brought on line.

E. Often, a combination of the above will increase revenues.

2.   **Decreasing Costs**

There are essentially two types of decreased costs. One type is cost reduction that is reflected in the bottom line. The other type is alternate use cost reduction. For example, this involves savings in people's time. It is assumed that the time saved is put to productive use. This type of cost reduction will not necessarily impact the bottom line. The type of decreased cost should be identified. Examples of both types are given below.

A. Improving processes often leads to a reduction in amount of time needed to complete a process. Reduced manpower on a job is a significant way to reduce costs. We will assume that any time saved will be put to a constructive alternative use. To calculate the savings due to time savings, take the annual time savings (in hours) divided by 1880 hours per year times the midpoint annual salary for the job level effected. You may want to increase the salary effect if the incremental time saved is usually overtime. Care should be taken in determining this cost savings to ensure that the salary group levels of individuals arc kept confidential. If only one person is involved in an improvement, select the salary group above or below his/her current group (whichever is closest to the actual salary). Use the midpoint of this group to determine cost savings.

B. Costs of raw material per pound of product or per job can be reduced. This may be achieved by reducing scrap and waste or by using a cheaper but equal substitute.

C. Costs can be reduced by decreasing inventories.

D. Costs can be reduced by gaining more favorable prices or terms from vendors and suppliers.

E. Costs can be reduced by decreasing process variation. This could be the costs associated with off-spec material or costs due to sampling.

The examples above do not cover all of the possible items that could be included in terms of costs and revenues. In some cases, information from the business or marketing personnel may be needed to determine the actual impact. They can help put a value on certain improvements by considering factors like whether you are sold out and do integrated margins need to be considered.

An example of determining dollar impact is given below. The numbers are for demonstration purposes only.

> A maintenance team at a plant realizes that downtime due to failures of reactor valves is 40 hours per year per reactor. Assume there are four reactors. The team feels it can improve valve performance. It works using TQM tools and determine that using different seal materials and a different parts inventory will reduce downtime to 30 hours per year per reactor. Additionally, it reduces the time required to overhaul the valves from an average of 20 hours down to 15 hours. So what is the real dollar impact? First, incremental sales are generated because the reactors can run more of the time. Also, the time/cost to repair the valves has been reduced. A dollar value will have to be placed on the incremental product produced. The maintenance team can then calculate the savings due to the reduced repair time.

A second example is given below. Again, the numbers are for demonstration purposes only.

> A team has been working on the process of processing invoices. Suppose the company processes about 150,000 invoices per year. The team has determined that the cost of processing an invoice is $60 per invoice. This costs includes people's time, materials, etc. The yearly cost for processing invoices is $9 million. Through process improvements, the team decreases the cost per invoice to $54 (a ten percent reduction). Most of this improvement is due to increased efficiency (fewer errors, less rework, etc.) The annual dollar impact, assuming that the time savings was put to constructive alternative uses, is then $9 million times 10 percent or $900,000.

A third example involving a decrease in process variation is given below.

> Several teams have been working to improve the process capability associated with one product. The original process has a process capability of 0.8 (indicating 0.82% of the product is out of spec). Annual production for this product is 150 million lbs. Off-spec production is 1.23 million lbs per year. Cost associated with off-spec (e.g., resampling, reworking, blending, production penalties, etc.) has been determined to be $0.10 per pound. Total cost associated with off-spec is then $123M per year. Work on the process has resulted in decreasing the process variation. As a result, the process capability has increased to 1.17. This indicates that the process now produces 0.046% off-spec material or 69 thousand pounds per year. Total cost associated with off-spec is then $6.9 thousand per year. Cost savings associated with this improvement is $123 thousand - $6.9 thousand = $116.1 thousand per year.

3. Cost Avoidance

Cost avoidance represents the third major category. It will also be the most difficult to measure. Cost avoidance represents the costs the company avoids due to decreases in the probability that a rare event will occur. This rare event includes items such as environmental violations or legal lawsuits.

An example of cost avoidance is given below. The numbers are for demonstration purposes only.

> Suppose you are a member of the legal department and working on the process of avoiding a certain type of lawsuit. To determine cost avoidance, you must first determine the maximum amount of liability the company would have under this type of lawsuit. Assume you determine this amount to be $10 million. The next step is to determine the probability of this occurring. There are three types of probability: remote (5% chance), possible (50% chance), and probable (75% chance). Assume you assess the probability to be remote (5%). To calculate the cost savings due to avoidance, you simply multiply the maximum amount times the probability or $10 million x 0.05 = $500 thousand. In this case, assuming the avoidance does not occur, the savings is $500 thousand.

A second example of cost avoidance is given below. This example deals with safety.

> Suppose manufacturing is working on improving safety. At the start of the improvement process, manufacturing averages 15 recordable injuries per year. The average medical cost for a recordable injury is determined to be $10 thousand. Hidden costs (such as accident investigations) are determined, on average, to be twice the medical costs. After improving the safety process, manufacturing has decreased the average number of injuries per year to 5. The cost savings due to avoidance is then the improvement in recordable injuries (15 - 5 = 10) times the cost per recordable injury ($30 thousand) or $300 thousand.

There are many costs to consider when considering the dollar impact. A summary of some potential costs to consider is given in Table 3. These costs are divided into four categories: prevention, appraisal, internal failure and external failure costs. These categories are the normal categories used when developing an overall cost of quality.

HOW TO WORK ON THE PROCESS

There are numerous methods that can be used to work on processes. An individual may work on a process directly related to his or her job. A natural team (composed of supervisor and direct reports) may work on processes related to the team's performance. In addition, department or departments may form project teams to work on processes that cut across team or departmental boundaries.

With the growing emphasis on team management, natural teams provide a logical mechanism for working on team processes.

Table 3

COST DESCRIPTION SUMMARY

PREVENTION COSTS

Marketing/Customer/User
Marketing Research
Customer/User Perception Surveys
  /Clinics
Contract Document Reviews
Product/Service/Design Develop-
  ment
Design Quality Progress Reviews
Design Support Activities
Product Design Qualification Test
Service Design Qualification
Field Trials
Purchasing
Supplier Reviews
Supplier Rating
Purchase Order Tech Data Reviews
Supplier Quality Planning
Operations
Operations Process Validation
Operations Quality Planning
Design and Development of Quality
  Measurement and Control
  Equipment
Operations Support Quality Plan-
  ning
Operator Quality Education
Operator SPC/Process Control
Quality Administration
Administration Salaries
Administration Expenses
Quality Program Planning
Quality Performance Reporting
Quality Education
Quality Improvement
Quality Audits

APPRAISAL COSTS

Purchasing Appraisal Costs
Receiving or Incoming Inspections
  and Tests

Measurement Equipment
Qualification of Supplier Product
Source Inspection and Control Programs
Operations (Manufacturing or Service)
  Appraisal Costs
Planned Operations Inspections,
  Tests, Audits
Checking Labor
Product or Service Quality Audits
Inspection and Test Materials
Set-up Inspections and Tests
Special Tests
Process Control Measurements
Laboratory Support
Depreciation
Measurement Equipment Expenses
Maintenance and Calibration Labor
Outside Endorsements and Certifi-
  cation
External Appraisal Costs
Field Performance Evaluations
Special Product Evaluations
Review of Test and Inspection Data
Miscellaneous Quality Evaluations

INTERNAL FAILURE COSTS

Product/Service Design Failure
  Costs (Internal)
Design Corrective Action
Rework Due to Design Changes
Scrap Due to Design Changes
Production Liaison Costs
Purchasing Failure Costs
Purchased Material Reject Disposi-
  tion Costs
Purchased Material Replacement
  Costs
Supplier Corrective Action
Rework of Supplier Rejects
Uncontrolled Material Losses
Operations (Product or Service)
  Failure Costs

51

Table 3 (con't)

COST DESCRIPTION SUMMARY

Material Review and Corrective
    Action Costs
Disposition Costs
Troubleshooting or Failure Analysis
Costs
Investigation Support Costs
Operations Corrective Action
Operations Rework and Repair Costs
Rework
Repair
Reinspection/Retest Costs
Extra Operations
Scrap Costs

Downgraded End Product or Service
Internal Failure Labor Losses

EXTERNAL FAILURE COSTS

Complaint Investigations/Customer or
    User Service
Returned goods
Retrofit Costs
Recall Costs
Warranty Claims
Liability Costs
Penalties
Customer/User Goodwill
Lost Sales

From *Principle of Quality Costs*, ASQC Quality Cost Committee, J. T. Hagan, Editor

SUMMARY

This chapter presented a strategy for process improvement. This represents the road map to help us apply the techniques presented in this book. It is a multi-step process that focuses on improving customer satisfaction by improving our processes and our productivity. In addition, methods of measuring results were presented. One method involved a matrix to allow you to track the various stages of process improvement. The second method involved determining the dollar impact. Finally, a short description of how to work on the process was given. Natural teams provide a good method of working on team-related processes.

APPLICATIONS

Think about your major processes at work. In the space below, sketch out a plan of how you would apply the Strategy for Process Improvement to one process.

5

PROBLEM SOLVING MODEL

Have you ever sat down to work on a problem and didn't know where to start or how to approach the problem? Have you ever worked on a problem with a team where everyone was trying to solve the problem their own way? Have you ever solved a problem and found yourself confronted with the same problem at a later date? In this chapter, the problem solving model is introduced. This model is a method of solving problems which can be used by both teams and individuals. The model will guide us toward solving problems so that they don't come back again.

OBJECTIVES

In this chapter you will learn:

1. Philosophy of problem solving
2. Steps of the problem solving model

One of the most important activities that teams, as well as individuals, engage in is problem solving. All people need an effective set of skills that they can use with precision when confronted with a problem. Likewise, when a group of people get together and attempt to solve a problem, it is helpful to agree on a step-by-step process for solving the problem. The problem solving model incorporates an effective set of skills into a step-by- step process. The model combines the use of statistical tools, such as control charts and process flow diagrams, with group problem solving skills, such as brainstorming and consensus decision making. The statistical tools help us make data based decisions at various points throughout the model. The group problem solving skills help us draw on the benefits of working as a team. In this seminar, we will focus on the statistical tools that will assist us in making data based decisions. Figure 1 is a diagram of the problem solving model.

PHILOSOPHY OF PROBLEM SOLVING

Before we begin a discussion about the steps of the problem solving model, we should talk a little about the philosophy that good problem solvers have about problems. Here are a number of ideas that are part of the philosophy (Miller, 1987):

Problem solving should occur at all levels of the organization. At every level, from top to bottom, problems occur. Everyone is an expert in the problems that occur in his or her own area and should address these problems. Problem solving is a part of everyone's job.

All problems should not be addressed with the same approach. There are some problems which are easily and suitably tackled alone (using "Command" decisions). Not all decisions need to be made by teams nor do all problems need to be solved by groups. However, groups of people help to break mental sets (i.e., figuring out new ways of doing things). In addition, people are more committed to figuring out and implementing a solution to a problem if they are involved in the problem solving.

Figure 1

PROBLEM SOLVING MODEL

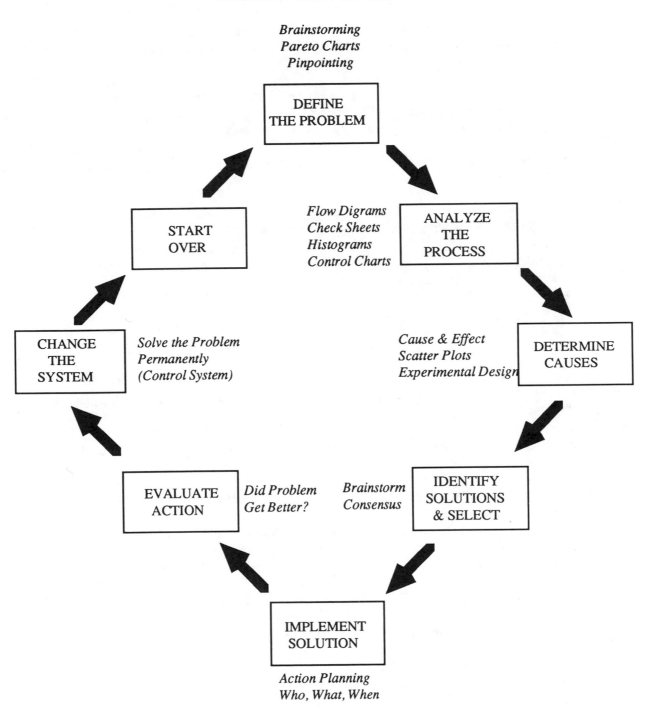

Problems are normal. Problems occur in every organization. In excellent companies people constantly work on solving problems as they occur. Problems are opportunities to make things better and should be viewed as such.

Be hard on the problem and soft on the people involved. When working on a problem, we should focus on solving the problem, not on whose fault the problem is. We should avoid personalizing the problem and blaming others.

People should address the problems in their own areas. Everyone has problems associated with their work area, and they should take ownership for trying to solve these problems instead of waiting for their supervisors or another team to tell them what to do.

Problems can be solved. This is true even if the problem has existed for a long time. Everything can be improved.

DESCRIPTION OF THE PROBLEM SOLVING MODEL

The problem solving model contains seven steps which should help you improve problems you will confront. The steps in the problem solving model are described below.

Step 1: Define the Problem. The objective of this step is to determine which problem to work on. The Pareto chart provides a method of prioritizing the major problems and selecting the most important one.

Step 2: Analyze the Process. The objective of this step is to determine how the process is behaving. That is, "What is the process doing right now?" This is done by collecting and analyzing data. Process flow diagrams give a picture of the process you are working with and help you identify possible areas for measurement of the process. Check sheets give you a format by which to collect data. Control charts and histograms help put the data collected into a form which gives you information about your process.

Step 3: Determine Causes. The objective of this step is to determine why the process is behaving as it does. Developing a Cause and Effect diagram helps you think about causes of a problem in terms of people, measurements, methods, materials, machines, and environment. Scatter plots give you some indication about the effect of one possible cause on the problem, and experimental design looks at the effect of several causes and their interactions on the problem.

Step 4: Identify Solutions and Select. The objective of this step is to identify potential solutions and select the best solution to implement. If you are working in a team, the brainstorming and consensus decision making techniques are very helpful in this step.

Step 5: Implement the Solution. The objective of this step is to implement the selected solution. This requires action planning. You have to ask yourself or the team "What are the steps I/we have to do to put this solution in place?"

Step 6: Evaluate Action. The objective of this step is to determine if the implemented solution has had an effect on the problem. Continuing use of control charts is helpful here.

At this point, if the implemented solution did not improve the problem, you have to go back to step 4 (possibly even step 3) and try again. If the implemented solution did work, you move on to step 7.

Step 7: Change the System. The objective of this step is to change the system if the solution you have implemented has solved or improved the problem. This involves the use of a control system to ensure that the problem remains permanently solved.

When problem solving, the problem solving model can be used as a "cheat sheet" to determine what techniques you should be using along the way. You may not use all the techniques in each step and you may not even use all the steps in the problem solving model. In addition, you may jump back and forth between two steps before moving on. What you want to avoid is skipping too many steps such as jumping from Defining the Problem to Identifying Solutions. You also want to make sure that you don't stop midway through the problem solving model. Always close the loop and start again. The techniques or steps you use are dependent upon the data you need in order to make data based decisions and the scope of the problem to be solved.

SUMMARY

In order to be successful at problem solving, we need to adopt a new philosophy about the way we view and approach problems. We need to become aware of the problems in our work area and tackle those problems at every level. In order to feel comfortable doing this, there has to be an environment where people feel like it is normal to have problems and that they won't be personally blamed for the problems they bring to light. In addition, we have to be open to addressing problems with different approaches. Most importantly, we have to keep in mind that there is a solution to all problems.

Once we adopt this philosophy about problems, we need a step-by-step approach to problems and skills necessary for problem solving. The problem solving model gives us that step by step approach which, if followed, ensures that our problem is truly solved versus just being remedied for the short term. The problem solving model gives us the statistical tools that allow us to make data based decisions. The problem solving model has us choose a problem within our process to work on, analyze what our process is doing now, determine why our problem is occurring, select a possible solution to our problem and implement this solution. After the solution has been in place for a while, we need to determine whether the potential solution has actually alleviated the problem. If it did, we need to establish a system that will keep that solution in place permanently.

APPLICATIONS OF THE PROBLEM SOLVING MODEL

Think about a problem that you have recently been involved in trying to solve. Did you follow the steps in the problem solving model? If so, how did these steps help you solve the problem? If not, did this make it more difficult to solve? What advantages do you see in using the problem solving model? What disadvantages? Record your comments below.

6

PARETO DIAGRAMS

There are always problems on which to work. There is not enough time in the day to work on all problems. How do we determine which problem to work on first? How do we get everyone to agree on what the major problem really is? Where will we get the most return for our investment? The Pareto diagram, which is introduced in this chapter, provides a data based methodology of doing this. The Pareto diagram is useful (1) for gaining consensus on what the major problem is and (2) as a communication tool.

OBJECTIVES

In this chapter you will learn:

1. What a Pareto diagram is
2. When to use a Pareto diagram
3. How to construct a Pareto diagram

The Pareto diagram is a special type of bar chart used to determine which problem to work on first to improve a process. The Pareto chart was developed in the late 1800s by an Italian economist named Wilfredo Pareto. It is based on what is now called the Pareto principle. Pareto found that 80% of Italy's wealth was held by only 20% of the people. This 80/20 rule is generally true for many things. For example, 80% of our problems are probably due to only 20% of the possible causes. The Pareto diagram allows us to separate the "vital few" from the "trivial many." This permits us to focus our time and resources where they will be most beneficial.

INTRODUCTION TO PARETO DIAGRAMS

A Pareto diagram is a special type of bar chart which is used to determine what problem to work on first. It can also be used to determine how often causes of problems occur. The problem or cause is listed on the x (horizontal axis). The frequency of occurrence or cost associated with each problem or cause is plotted on the y (vertical) axis. The problems or causes on the x axis are listed in decreasing order. The problem or cause that happens most frequently or costs the most is listed first. The problem or cause that happens least frequently or costs the least is listed last.

Figure 1 is an example of a Pareto diagram. A company wanted to determine the major reasons for off-spec product. Data over the past six months were collected. The data included the amount of off-spec material and the reason the material was off-spec. There were six major reasons for off-spec material: contamination, color, bulk density, purity, composition, and other. The "other" category contained a number of reasons for off-spec which did not occur very often. A Pareto chart was then developed from the collected data. The reasons for off-spec material were listed on the x axis. The height of each bar on the diagram corresponds to how many pounds of product was off-spec due to that reason. The major reason for off-spec material was contamination, which accounted for 2 million pounds of off-spec product. The second major reason was color, which accounted for 1.2 million pounds of off-spec product. The other categories accounted for only 120 thousand pounds of off-spec product. The Pareto diagram shows quite clearly that the major reason for off-spec product was contamination. This is the problem that should be worked on first.

Figure 1

PARETO DIAGRAM: REASONS FOR OFF-SPEC PRODUCT

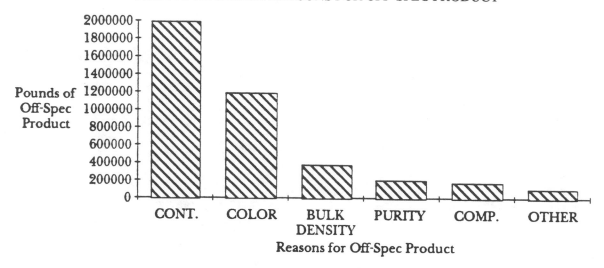

Reasons for Off-Spec Product

WHEN TO USE A PARETO DIAGRAM

Pareto diagrams can be used in many situations. They can be used to determine the major causes of why marriages fail or survive, what kind of gardening techniques are most successful, or why people are watching less commercial television than they used to. At work, Pareto diagrams can be used in all areas including production, maintenance, shipping, accounting, and safety. They can be used to determine the major reasons for problems in any of these areas. These problems may deal with waivers, injuries, delayed shipments, invoice errors, etc.

Pareto diagrams are very useful problem solving tools. They are used in the first step (defining the problem) of the problem solving model. The objective of this step is to determine which problem to work on first. Pareto diagrams permit a group to reach consensus on what the major problem really is. Since the diagram is based on data, it is very difficult to argue over which problems are most important.

Pareto diagrams can also be used to evaluate process improvements over time. Figure 1 showed a completed Pareto diagram for reasons for off-spec product. If the company worked on the major reason (contamination), a Pareto diagram done six months later should show the improvement. The height of the bar relating to contamination would be smaller.

Pareto diagrams are very good communication tools. They allow a person to quickly grasp the magnitude of the problem. They are very good tools to use in presentations to management. Since management is often interested in dollars, it is best if the y axis represents the cost or dollars.

HOW TO CONSTRUCT A PARETO DIAGRAM

The steps in constructing a Pareto diagram are listed below. A process flow diagram of these steps is shown in Figure 2.

1.   Determine the problems or causes to be used on the diagram.

2.   Select the time period to be covered on the diagram.

3.   Total the frequency of occurrence (or cost) for each problem during the time period.

4.   Draw the x and y axes, putting the proper units on the y axis.

5.   Under the x axis, write in the most important problem (largest frequency) first, then the next most important, etc.

6.   Draw in the bars. The height of the bar will correspond to the frequency of occurrence for each problem on the x axis.

7.   Title the graph and include any other important information.

Figure 2

CONSTRUCTION OF PARETO DIAGRAMS

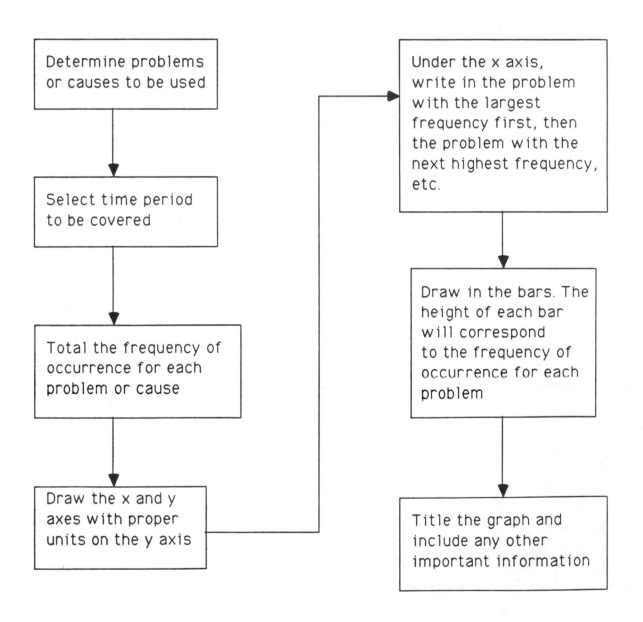

PARETO DIAGRAM EXAMPLE

Invoice Errors

A team in the purchasing department was working on reducing errors on invoices. Considerable efficiency has been lost lately due to the time it takes to correct errors on invoices received for payment. The team was given the assignment of determining what could be done to reduce the number of errors on invoices. The team decided to collect data on what the reasons for errors on invoices were. The team decided to collect data for four weeks. The data collected are shown below. Using this data and the chart on page 64, construct a Pareto chart. Which reason for invoice errors should be worked on first?

	Frequency of Occurrence by Week			
Cause	1	2	3	4
Quantity	1	0	1	1
Price	0	3	2	3
Wrong P. O. Number	4	2	4	3
Wrong Address	0	0	0	1
Wrong Description	1	1	1	0
Improper Signature	2	1	1	0

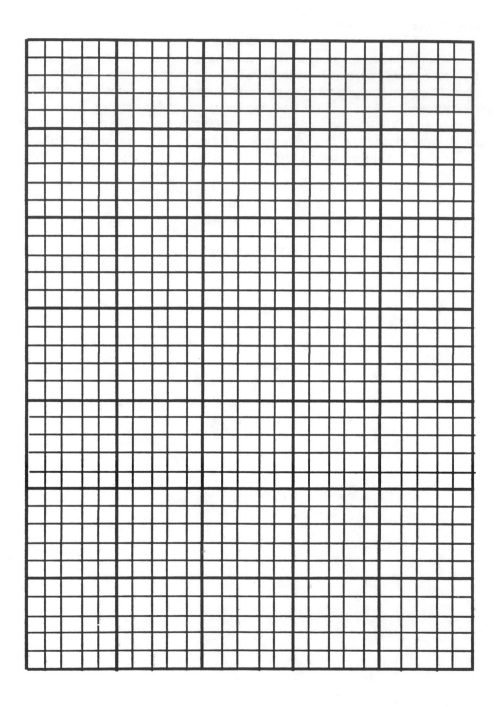

PARETO DIAGRAM EXAMPLE

Plant Injuries

Safety is always a number-one priority in a manufacturing plant. A company had been working for several years to decrease the number of injuries in its plants. The data for the past three years are given below by type of injury. Construct a Pareto chart (p. 66) based on these data and answer the following questions.

Type of Injury	Year 1	Year 2	Year 3
Hands/Arms	23	21	22
Feet/Legs	7	4	6
Head/Neck	3	1	1
Eyes	2	2	2
Chest	1	2	1
Back	0	3	3

Which type of injury occurs most frequently?

Has there been any improvement over the past three years in decreasing the number of injuries?

Have any type of injuries increased or decreased significantly in the past three years?

Which type of injury should be worked on first?

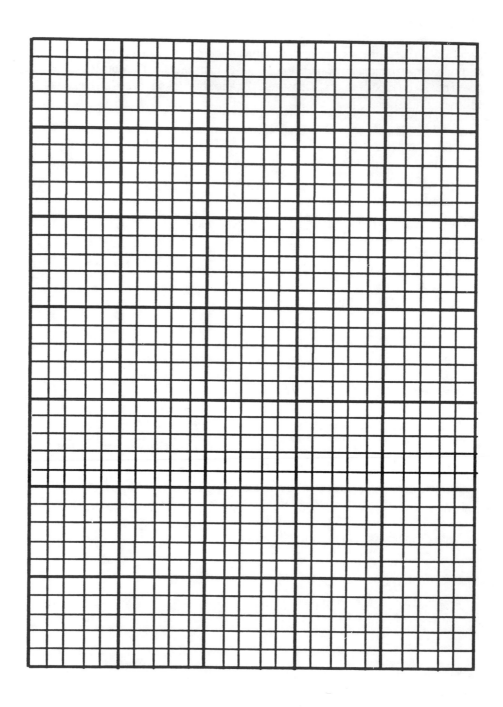

PARETO DIAGRAM EXAMPLE

Plant Downtime

A maintenance team in a plant was interested in working on the major causes of downtime in one operating unit. The team decided to look at data for the past six months to determine the various reasons for downtime. It decided to measure downtime in hours. The data for the past six months are given below. Construct a Pareto diagram (p. 68) based on these data and answer the following questions.

	Hours of Downtime					
Reason	Jan	Feb	Mar	Apr	May	Jun
Line Plug	18	0	0	32	0	0
Computer	9	10	5	6	10	5
Compressor	6	0	34	0	0	0
Pumps	30	24	45	115	56	60
Raw Material	5	30	0	0	0	0
Weather	23	7	0	0	0	0
Valve Leak	8	24	13	45	7	13

What is the major reason for downtime?

What percentage of downtime is caused by this major reason?

Which reason should be worked on first?

Are there other factors to consider besides hours of downtime when determining which cause to work on first?

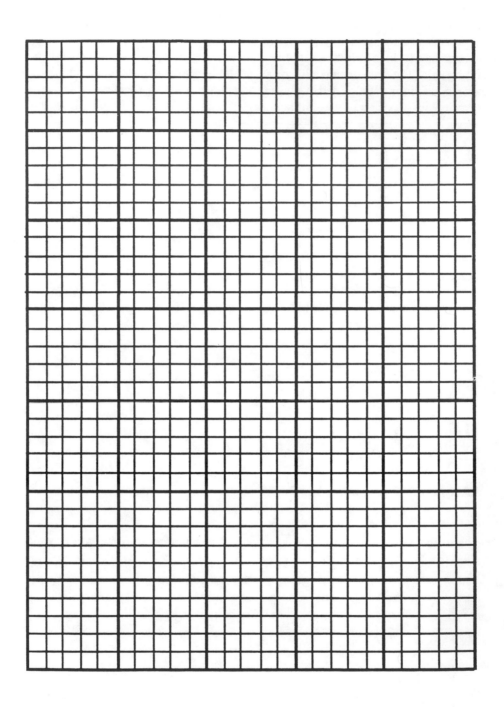

SUMMARY

The Pareto diagram is a useful problem solving tool. It provides a data based method of determining which problem to work on first. It helps a team reach consensus on where valuable time and resources should be spent. It is also a very effective communication tool. A Pareto diagram makes it easy to see the magnitude of various problems. Pareto diagrams can also be compared to show improvements from one time period to another.

APPLICATIONS OF PARETO DIAGRAMS

Think about your own job or work area. Write down three possible uses for Pareto diagrams and what you feel you could learn from these uses of Pareto diagrams.

1.

2.

3.

7

PROCESS FLOW DIAGRAMS

Have you ever wondered what the steps are in invoice payment, constructing an engineering proposal, estimating the cost of a design or approving check requests? Have you ever thought that your "way" of doing something was the best, only to find that someone else does it more efficiently? One way of understanding what the steps are in a process is to draw a process flow diagram. Process flow diagrams are introduced in this chapter.

OBJECTIVES

In this chapter you will learn:

1. What a process flow diagram is.
2. Why process flow diagrams are useful.
3. When to use process flow diagrams
4. How to construct a process flow diagram.

A process flow diagram should be drawn for every process being worked on. Construction of a process flow diagram should include the people actually doing the work in the process. They are a "living document" to be updated when the process changes.

INTRODUCTION TO PROCESS FLOW DIAGRAMS

A process flow diagram is a detailed picture of a process which represents the logical flow of activities from the beginning of the process to the end of the process. For example, receipt of a handwritten memo may be the beginning of the typing process, and an error-free typed memo may be the end of the process. When drawing a process flow diagram, it is important to consider what the starting point is and what the ending point is.

Figure 1 is the process flow diagram for a credit collection process -- how we might collect funds due us from our clients. In essence, it shows what to do when a customer does not pay an invoice on time. The process involves sending the customer a letter requesting payment of the invoice; it requires a telephone call to the customer if no response to the letter is received by us; it may (finally) require that the customer pay in advance for a catalyst shipment. By examining the process, we may determine ways to remove unnecessary steps.

Figure 2 is an example of the process we might use to price our products. This process uses Price Exception Requests or PER's. A PER is a legal document which indicates exceptions to standard price or standard terms (freight, rebate, etc.). Figure 2 shows the steps of preparing a PER by sales personnel and the documents generated as a result of the PER. Although Figure 2 may look different from Figure 1, the overall format is the same -- both figures have a beginning and an ending to the process.

Figure 1

CREDIT COLLECTION PROCESS

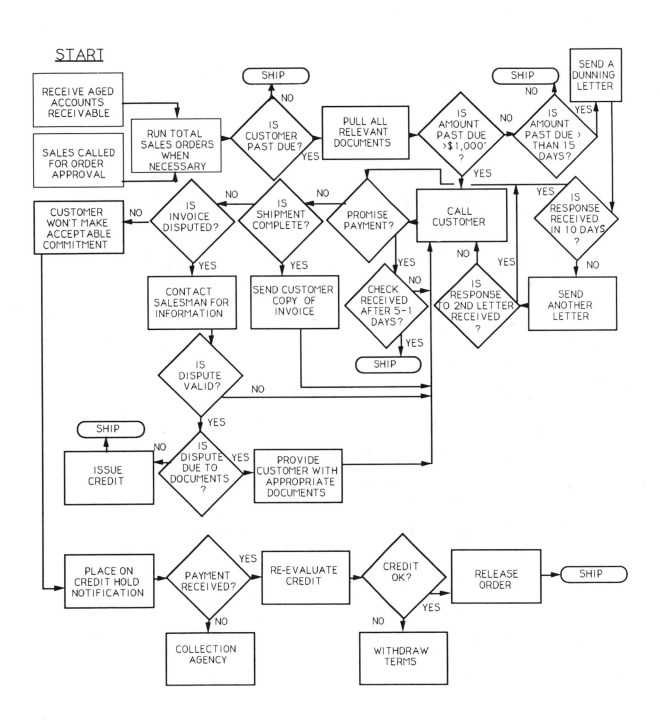

Figure 2

PROCESS FLOW DIAGRAM FOR PRICE EXCEPTION REQUESTS

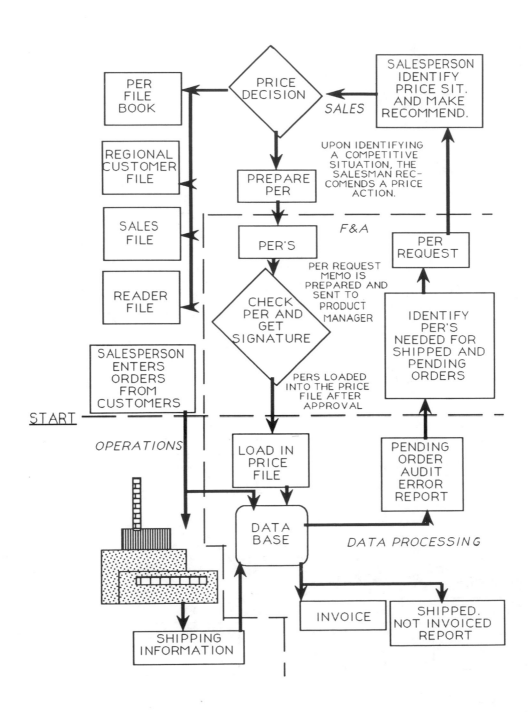

Process flow diagrams should follow the SIPOC format as shown in Figure 3. It should include suppliers and customer feedback -- two key components often excluded from the process.

Figure 3

VIEWING PROCESSES AS A SYSTEM

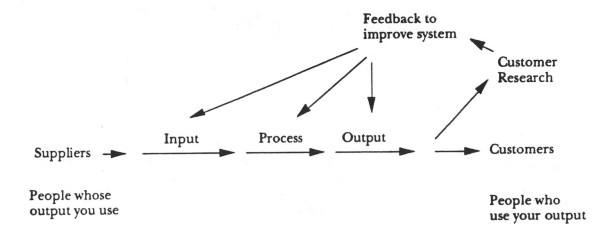

Process flow diagrams help a team or individual gain an understanding of how the process is behaving. Process flow diagrams are useful as training tools and as a method of gaining consensus on what the process actually is. If there are obvious changes that could be made to improve the process, they should be made immediately.

Figure 4 is an example of what a system with no quality problems looks like. It is very simple -- only three steps in the process. This is how the process should function.

Figure 4

PROCESS WITH NO QUALITY PROBLEMS

Now compare the diagram in Figure 4 with the process flow diagram in Figure 5. Notice how complex the system below looks as a result of quality problems being introduced--i.e. missing designs, no signatures, etc.

Figure 5

PROCESS WITH QUALITY PROBLEMS

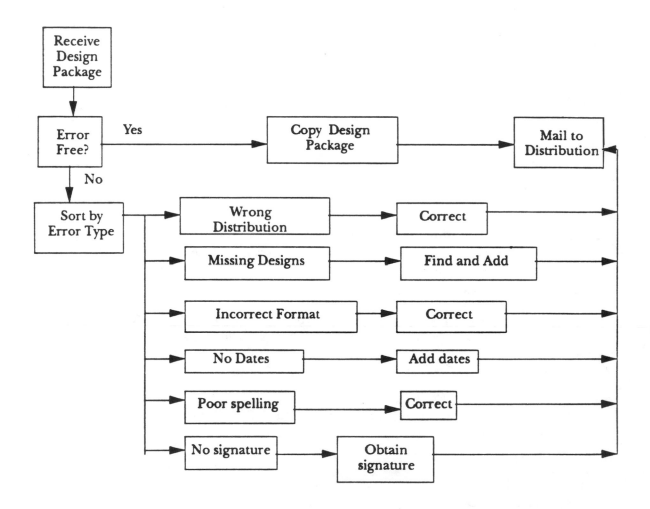

This demonstrates how process flow diagrams help us identify factors or unnecessary steps which affect a process. Process flow diagrams also help us to identify sampling points at which to measure the process.

WHEN TO USE PROCESS FLOW DIAGRAMS

Process flow diagrams can be used in accounting, financial, purchasing, marketing, manufacturing, engineering, transportation, research, etc.

Process flow diagrams are found in the "Analyzing the Process" step in the problem solving model. In this step, we are trying to determine how the process is behaving.

STEPS IN CONSTRUCTION OF A PROCESS FLOW DIAGRAM

The steps in constructing process flow diagrams are given below.

1. Decide on the process to study. Ask which process makes the product or service with which you are concerned.

2. Sketch out the step-by-step process -- be as detailed as possible. You may need several small process flow diagrams to make one large one. This can be done as a group activity. Usually the output of the process is placed on the right and the inputs on the left.

3. Review the diagram with others involved in the process -- get their input to make it complete.

4. Identify the sampling points and quality parameters on the process flow diagram. Be sure that suppliers, customers and customer feedback are included.

5. Brainstorm with those involved in the process as to where problems arise. This will help you focus your group's activities to improve the process.

PROCESS FLOW DIAGRAM GROUP EXERCISE

1. Think about a process that is familiar to all members in your group (i.e., planting a garden or changing a flat tire). Draw out the steps in the process, keeping in mind the steps of construction for process flow diagrams.

2. Determine ways to improve the process.

3. Each group will present its process to the rest of the class.

Take about thirty minutes to prepare your process flow diagram.

SUMMARY

A process flow diagram is a detailed picture of a process. One should be drawn for every problem on which the team works. Process flow diagrams are used to gain consensus on what is the true process; they are also used to determine problem areas to improve.

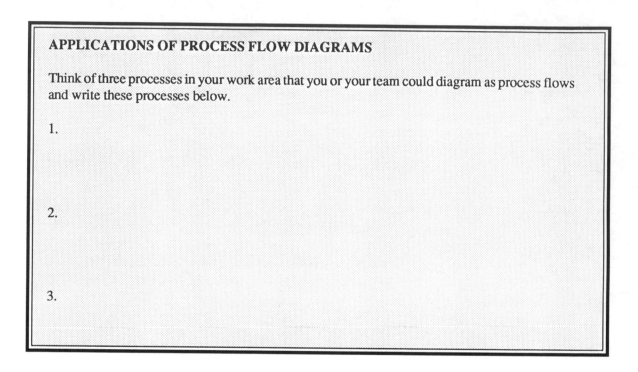

APPLICATIONS OF PROCESS FLOW DIAGRAMS

Think of three processes in your work area that you or your team could diagram as process flows and write these processes below.

1.

2.

3.

8

HISTOGRAMS

What result occurs most frequently from this process? How much variation is there about this most frequently occurring result? Is the variation symmetrical? Is the process producing any material that is out of the specification range? These types of questions are common when beginning to look at how a process is behaving or operating. This chapter introduces a tool which can help you answer these types of questions.

In addition to a product or service, a process also produces data. These data can be used to improve the quality of the process, and thus the quality of the product or service. Everything varies, including the data generated by a process. However, if the data are generated from a process in statistical control (only common cause variation), the data as a group tend to form a pattern called a distribution. Distributions are characterized by three parameters: location (average or typical value), spread (amount of variation) or shape (the pattern of variation - bell-shaped, symmetrical, etc.) These parameters of a distribution can be estimated by using a histogram. Histograms are introduced in this chapter.

OBJECTIVES

In this chapter you will learn:

   1. What a histogram is
   2. When to use a histogram
   3. How to construct a histogram

Control charts represent a picture of how a process varies over time. Histograms, on the other hand, present a picture of how the process "stacks up" over time. They illustrate how many times a certain data value or range of data values occurred in a given time frame. Histograms provide estimates of the location, the spread and the shape of a distribution.

INTRODUCTION TO HISTOGRAMS

Suppose a process is in statistical control and is being monitored by taking samples on a regular basis. The samples are tested for a certain product characteristic. The results from the testing will not be the same every time. This is because there are many sources of variation in a process. These sources of variation include variation in the way people do their work, in raw materials, in test methods, etc. Although each sample result is different, the results taken as a group will form a pattern or distribution. Figure 1 illustrates this concept. The individual sample results vary. However, if the results that are similar are "stacked up" over time, the pattern or shape of the distribution begins to form. Histograms provide a method of determining this pattern or shape.

Figure 1

HISTOGRAMS AND VARIATION

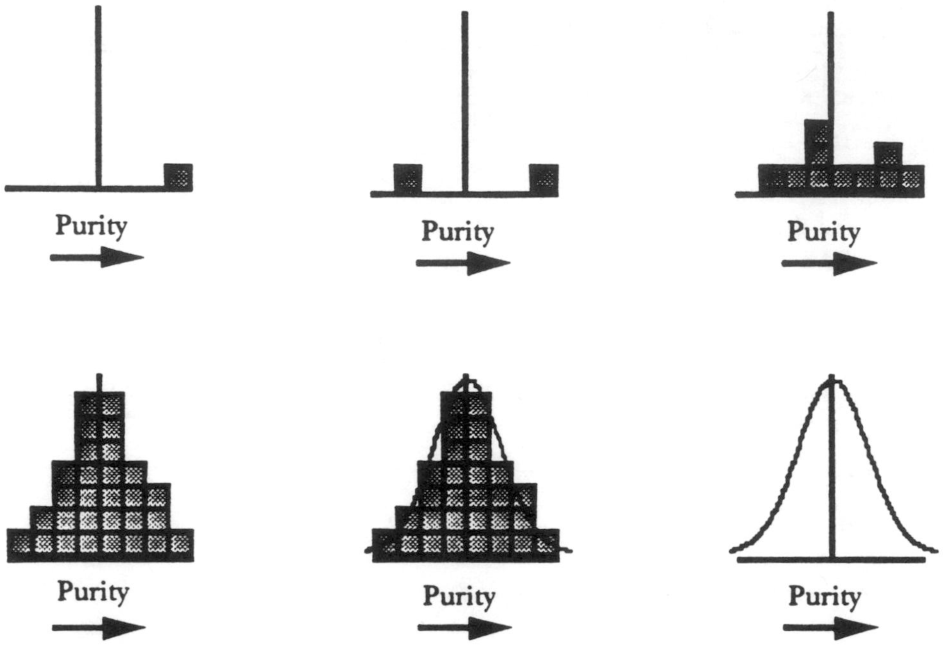

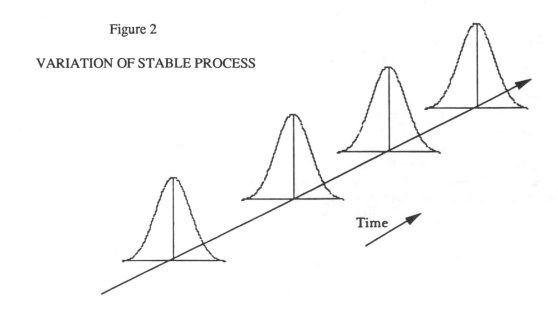

A process in statistical control has only common cause variation present. It is predictable in the near future. The average and the amount of variation present will not change as long as the process remains in control. In addition, the shape of the distribution will remain the same over time, as shown in Figure 2.

Figure 2

VARIATION OF STABLE PROCESS

Time

Histograms are used in the second step of the problem model. This steps involves "Analyzing the Process," i.e., How is the process currently behaving?

Figure 3 shows a completed histogram. This histogram displays the number of days it takes railcars to reach their final destination. The histogram includes data collected over a one-month period (the time frame or "history"). The number of days required for a given railcar to reach its destination is given on the x (horizontal) axis. The number of times this value occurred (number or frequency of occurrences) during the month is given on the y (vertical) axis. For example, railcars took 13 days to arrive at their destination 10 times during the month. They took 15 days to arrive at their destination 16 times during the month.

Figure 3

HISTOGRAM FOR DAYS TO DESTINATION

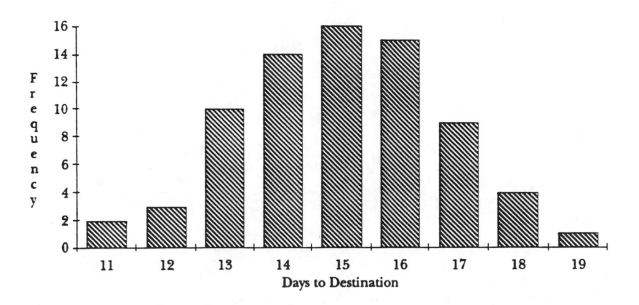

The histogram can be viewed as a summary of information over a certain time frame; in the railcar case, the time frame is one month. The histogram provides useful information. We can easily determine the value that occurred most frequently during the time frame. This value is the highest point on the histogram and is called the "mode." For this example, the mode is 15 days.

The histogram also gives an indication of how much variation is present. An indirect measure of this variation is found by comparing the smallest value and the largest value on the histogram. For the railcar example, it takes from 11 (smallest value) to 19 (largest value) days for the railcars to reach their final destination.

The histogram also provides an estimate of the shape of the distribution. For example, does the histogram represent a normal distribution (a bell-shaped curve)? Figure 4 shows a bell-shaped curve superimposed on the histogram. This histogram appears to fit this distribution. Many histograms will have this type of shape. Most values will occur around the highest bar (mode). As the values move away from the mode, the values occur less frequently. Normally there will be a gradual decrease as the values move away from the mode.

Figure 4

NORMAL DISTRIBUTION AND

HISTOGRAM FOR DAYS TO DESTINATION

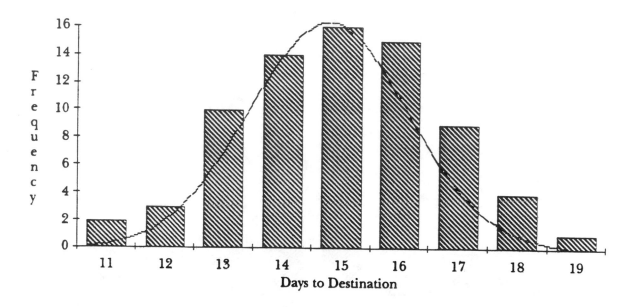

A histogram also permits us to compare the results with specifications. For example, suppose our specifications for delivery are 15 days +/- 3 days. This means we require the railcars to reach their final destination between 12 and 18 days from the time they leave the plant. These specifications can be added to the histogram. This has been done in Figure 5. It is easily seen that some of the deliveries are outside the specifications of 12 to 18 days.

A histogram then tells us four things:

1. What the most common value (the mode) is
2. The amount of variation present
3. The shape of the distribution
4. The relationship of the data to the specifications

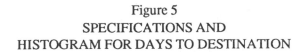

Figure 5
SPECIFICATIONS AND
HISTOGRAM FOR DAYS TO DESTINATION

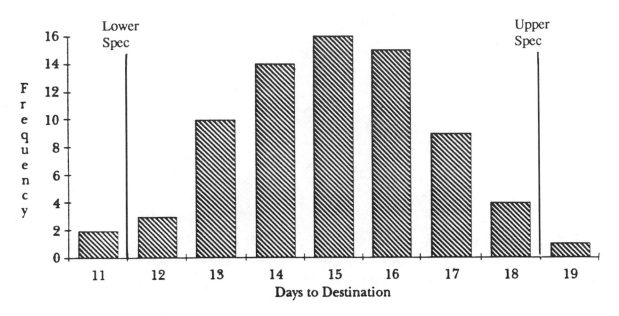

There is some additional nomenclature associated with histograms. A class is a data value or range of data values that are used to construct the histogram. For the railcar example, there are 9 different classes (data values). Each class is represented by a bar on the graph, as shown in Figure 6. The class width is the width of the bar, and each class width represents one day (see Figure 6). The overall range is the difference between the maximum data value (19 days) and the minimum data value (11 days). Thus, the overall range is 8 days.

Histograms should be used in conjunction with control charts. They provide additional information to us. It is important to remember that histograms do not tell us if a process is in statistical control. Only control charts can do this. The reason for this is that control charts look at variation over time (one data point followed by another). Histograms look at variation as a snapshot in time (for example, over one month). The order the data were generated in is lost. For example, when looking at the railcar histogram, you can't tell when, in time, the delivery took 15 days. We just know that it occurred sometime during the month.

The relationship between the normal distribution and the histogram was discussed above. There are other types of distributions. Examples of two of these are shown in Figure 7. These are examples of skewed distributions. The mode (highest point) occurs on one side of the data. Few values occur on one side of the mode; most values occur on the other side. Figure 7a is a positively skewed distribution. The average value occurs to the left of the center of the range. This shape often occurs in a process where a parameter is being minimized. It can also occur if the test

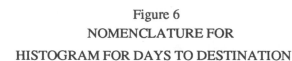

Figure 6
NOMENCLATURE FOR
HISTOGRAM FOR DAYS TO DESTINATION

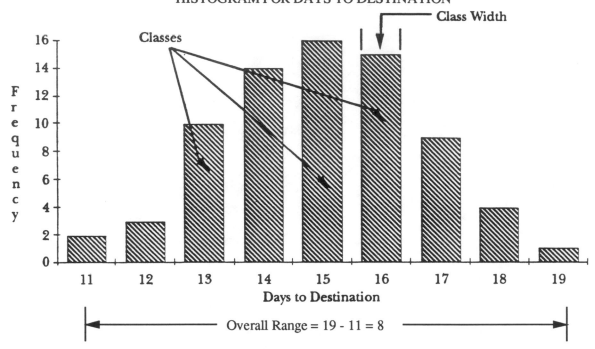

Figure 7

SKEWED DISTRIBUTIONS

7a) Positively Skewed

7b) Negatively Skewed

method is not capable of detecting a substance below a certain minimum. Figure 7b is a negatively skewed distribution. The average value occurs to the right of the center of the range. This type of histogram often occurs when a parameter is being maximized. These types of distributions are real and occur naturally; one should not normally be concerned if a histogram has these shapes.

Histograms can reveal problems. Figure 8 shows several histograms that are not naturally occurring. If your histogram looks like any of these, it is normally an indication that a problem exists. Suppose you are interested in determining how well a supplier of a major raw material is doing meeting specifications. As an initial look, you decide to take data from Certificates of Analysis received from the supplier and construct histograms. For each histogram in Figure 8, what conclusions would you draw about the material you are being supplied. Write your comments below.

Figure 8a:

Figure 8b:

Figure 8c:

Figure 8d:

Figure 8e:

Figure 8

HISTOGRAM PATTERNS

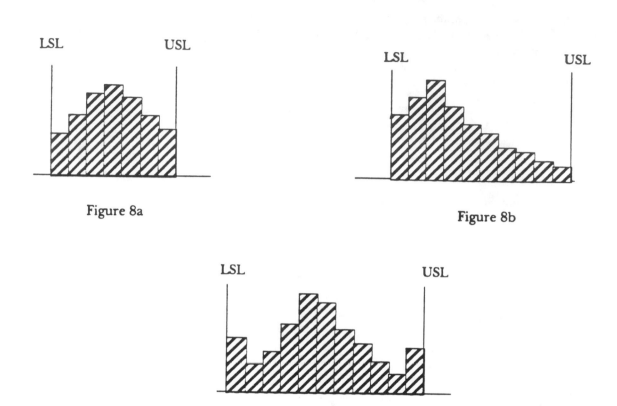

Figure 8a

Figure 8b

Figure 8c

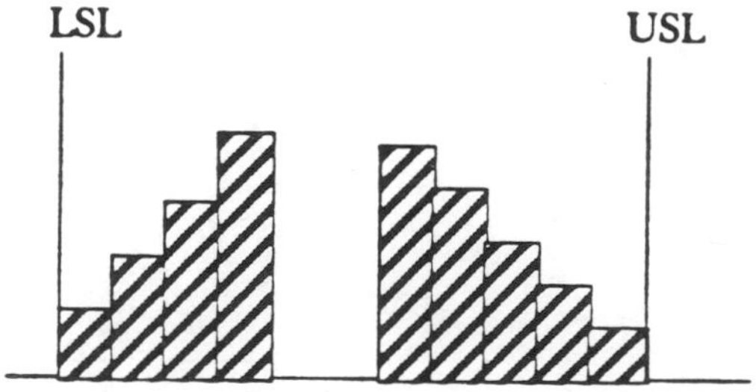

Figure 8d

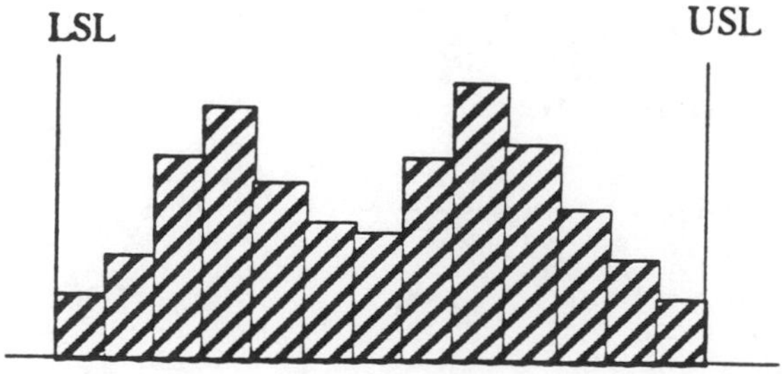

Figure 8e

STEPS IN CONSTRUCTION OF A HISTOGRAM

The steps in construction of a histogram are given below and shown as a process flow diagram in Figure 9.

1. Select the time frame (history) you are interested in and gather the data. The time frame must be sufficiently long so that there are at least 50 data points. It is better to have 75 to 100 points.

2. Select the number of classes (bars) to be used on the histogram using the following guidelines:

Number of data points	Number of classes (K)
50 - 100	5 - 15
100 - 250	15 - 20
over 250	20 - 25

3. Calculate the overall range (OR), where OR equals the maximum value minus the minimum value.

4. Calculate the class width (width of bar on the histogram). The class width, h, is determined by $h = OR/K$.

5. Round h to the nearest convenient number.

6. Select the class boundaries so that data points do not fall on the boundary between two classes. In other words, select the boundaries so the data must fall into one class only. Making the boundary value on 1/2 the unit of measurement will accomplish this.

7. Record the number of data points in each class.

8. Plot the histogram.

    a. The x axis represents the classes
    b. The y axis represents the frequency of occurrence.
    c. The height of the bar represents how often each class occurred.

9. Label the histogram and include any other important information ,such as the time period covered.

Figure 9

STEPS IN CONSTRUCTION OF HISTOGRAMS

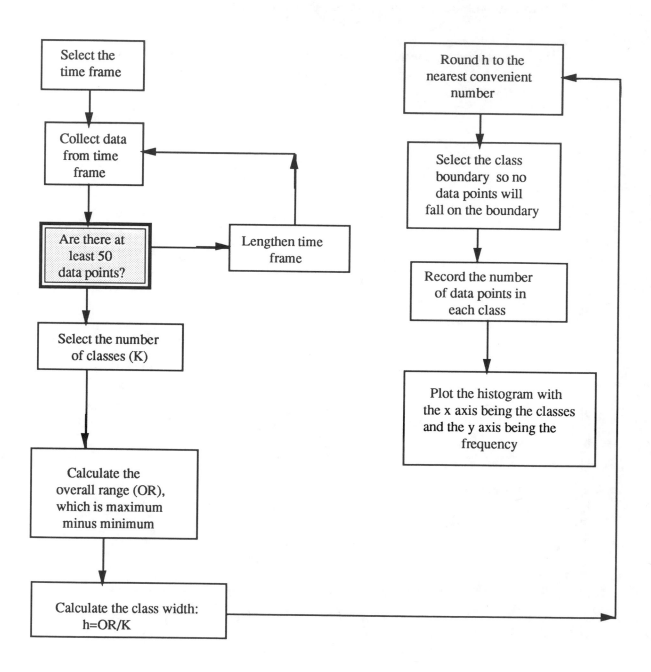

HISTOGRAM EXAMPLE

Reaction Yield

A team in one operating unit is interested in improving the yield from a given batch reaction. Five batches of material are made per day. The team wants to use a histogram to provide them with a snapshot in time of yields over the past month. Data for percent yield from the last 25 days are given below.

Day Number	Sample Number 1	2	3	4	5
1	81.3	80.4	78.6	83.1	81.8
2	74.3	76.4	82.4	77.8	82.5
3	78.7	77.4	79.4	81.6	81.0
4	80.4	81.7	81.4	79.7	80.2
5	79.4	75.6	80.3	80.2	77.4
6	85.0	75.4	73.8	75.8	78.6
7	78.5	86.2	77.1	73.3	76.4
8	81.7	84.0	80.2	78.6	80.9
9	84.5	82.4	78.8	83.2	83.0
10	82.7	80.5	85.9	82.7	84.0
11	78.4	83.1	80.1	78.5	86.6
12	82.9	82.4	78.9	78.2	78.4
13	75.6	80.1	81.1	78.3	80.4
14	78.2	76.4	82.3	81.7	85.1
15	81.8	80.6	79.1	79.3	83.6
16	75.2	82.2	79.6	83.6	81.9
17	78.6	80.1	80.6	79.3	80.4
18	82.3	80.8	79.7	76.5	85.6
19	83.0	83.6	75.2	83.3	81.3
20	77.6	79.1	78.7	80.8	80.2
21	75.0	81.0	82.9	80.0	81.9
22	82.7	78.8	81.2	74.8	81.7
23	76.9	82.5	82.5	81.4	84.4
24	78.1	82.9	73.7	81.5	75.9
25	79.9	78.7	81.3	80.0	78.5

Use these data to construct a histogram for reaction yield (p. 89). After constructing the histogram, answer the following questions.

1. What is the most common value (the mode)?

2. How much variation is there in reaction yield?

3. Describe the pattern of variation.

4. Is there any evidence that something unusual is occurring in the process?

Histogram for Reaction Yield

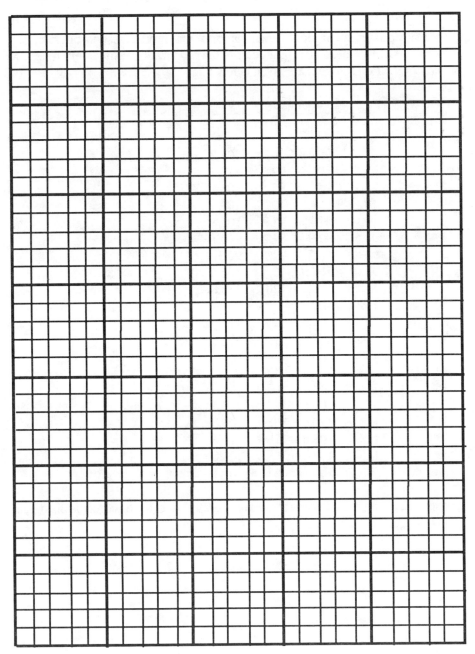

Chapter 8

HISTOGRAM EXAMPLE

Product Brightness

A company produces a product in granular form. One product response of interest to a customer is the product brightness. The brightness of the product is measured four times each hour. The results for the last 20 hours are given below.

Hour Number	Sample 1	2	3	4
1	87.4	93.4	103.4	107.2
2	89.5	105.5	99.6	117.7
3	76.7	96.7	86.6	105.1
4	102.1	103.5	105.7	103.9
5	103.0	87.4	95.9	105.8
6	94.5	92.9	84.2	110.1
7	90.8	98.9	92.5	90.3
8	93.0	106.8	107.0	106.6
9	116.8	118.1	120.0	116.8
10	111.0	109.8	107.4	114.2
11	123.2	108.6	92.4	122.0
12	104.2	127.9	117.5	107.3
13	109.0	106.2	126.3	114.1
14	103.2	105.4	94.6	93.8
15	106.7	114.1	109.0	105.3
16	95.6	103.6	123.0	113.3
17	121.9	123.0	108.4	102.0
18	113.6	103.6	128.9	111.6
19	118.3	102.4	108.3	101.0
20	109.9	100.8	86.0	107.1

Construct a histogram based on these data (p.92). After constructing the histogram, answer the following questions.

1. What is the most common value (the mode)?

2. How much variation is there in product brightness?

3. Describe the pattern of variation.

4. Is there any evidence that something unusual is occurring in the process?

Histogram for Product Brightness

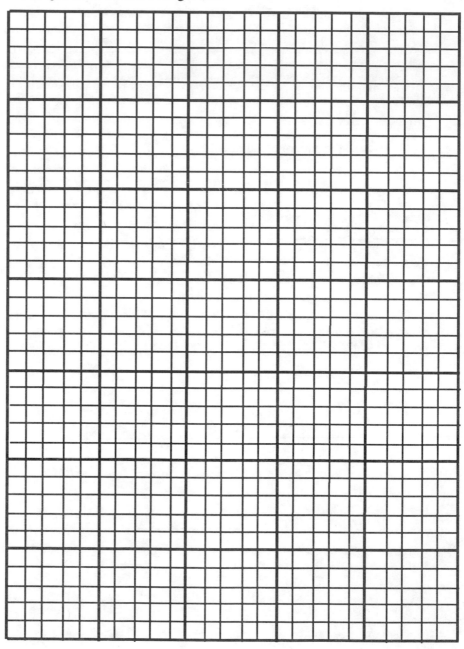

HISTOGRAM EXAMPLE

Reaction Run Time

An operating unit has selected reaction run time as a parameter to monitor. The personnel in the unit feel that run time may affect certain product parameters. Four batches of material are made each day. Data for the last 20 day are given below. The run time is in minutes.

Day Number	Sample 1	2	3	4
1	316	366	361	381
2	334	336	384	339
3	334	364	369	339
4	346	397	335	343
5	384	323	363	388
6	370	372	369	356
7	332	386	351	366
8	383	380	366	366
9	399	405	366	416
10	371	372	373	370
11	359	344	317	364
12	395	345	362	344
13	342	393	377	386
14	364	372	385	360
15	353	365	361	367
16	362	369	366	345
17	402	378	388	401
18	371	381	353	347
19	350	364	374	330
20	348	345	351	347

Construct a histogram based on these data (p. 95). After constructing the histogram, answer the following questions.

1. What is the most common value (the mode)?

2. How much variation is there in reaction run time?

3. Describe the pattern of variation.

4. Is there any evidence that something unusual is occurring in the process?

Histogram for Reaction Run Time

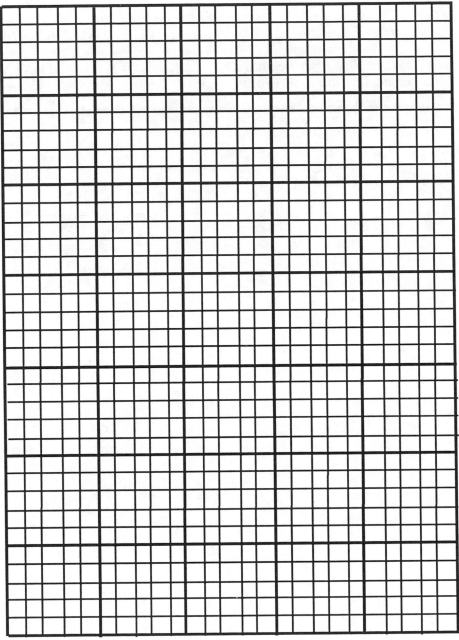

SUMMARY

Histograms represent a picture of the past history of your process. They show how many times a certain data value or range of data values occurred in a given time frame. Histograms are useful in identifying the most common value, the amount of variation and the shape of the distribution, and for comparison with specifications. Histograms arc valuable tools for determining how a process is behaving. They are often used in conjunction with control charts. Histograms also point out problems that might exist.

APPLICATIONS OF HISTOGRAMS

Think about your own job or work area. What are some possible applications of histograms? List three possible applications below.

1.

2.

3.

9

DATA AND DISTRIBUTIONS

Control charts give us a picture of our process over time. This picture tells us when to leave our process alone (i.e., the process is in control) or when to look for a problem (i.e., an assignable cause is present). There are many different types of control charts. However, we can group control charts into two major categories. These two categories are distinguished by the type of data being charted. There are two types of data we can have: attributes data and variables data. Both these types of data are introduced in this chapter. Depending on the type of data, there are some assumptions made concerning the population being sampled. These assumptions involve the type of distribution present. Distributions are introduced in this chapter. With attributes data, there is a need to develop specific descriptions. These descriptions are called operational definitions. Operational definitions are also introduced in this chapter.

OBJECTIVES

In this chapter you will learn:

1. What attributes data are
2. What an operational definition is
3. What the binomial distribution is
4. What the Poisson distribution is
5. What variables data are
6. What the normal distribution is
7. What the central limit theorem is

It is important to know what type of data you will collect so you can determine what type of control chart to construct. Different charts will give different information. Attributes charts include p, np, c and u charts. Variables charts include $\overline{X}$-R charts, $\overline{X}$-s charts, individuals charts and moving average/moving range charts.

INTRODUCTION TO DATA AND DISTRIBUTIONS

This chapter begins to develop the foundation for the construction and use of control charts to monitor processes. Processes are variable. You won't always get the same result each time. Variation can be divided into two types: common cause variation and assignable cause variation. Common cause variation is due to the normal way the process is managed. To reduce common cause variation, the process must be changed. This is management's responsibility. Assignable cause variation is due to non-random influences -- things that don't normally occur. The front line personnel have the responsibility to reduce assignable cause variation.

Statistical control means that all the variation is due to common causes. If a process is in statistical control, it is stable, consistent, and predictable in the near future. Control charts are used to determine if a process is in statistical control. This is accomplished by computing the range of output you would expect from the process if only common cause variation is present. The largest number you would expect is called the upper control limit. The lowest number you would expect is called the lower control limit. As shown below, the process is in control if there are no points outside these control limits and no evidence of non-random patterns. An example of a non-random pattern is a long run of points above the average. When a process is in statistical control, it means that there is only one average and one standard deviation associated with the process. In addition, the shape of the variation about the average does not change.

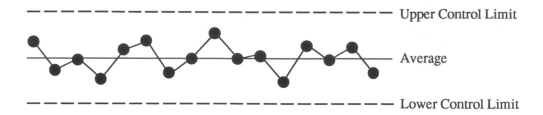

Suppose you are in charge of an operating process that produces a liquid product. One product parameter of interest to a major customer is the purity of the product. The customer wants to know what the average purity is, how much variation there is about that average, and what is the shape of the variation. How can you get this information for the customer?

Assume that the process is in statistical control. One method of getting this information would be to take 300 million samples from the process and test these samples for purity. These 300 million samples represent every possible outcome (purity result) from the process. This collection of every possible outcome is called the "population." The average and standard deviation of this population could be calculated. The shape of the population (called a distribution) could be determined by constructing a histogram. One possible distribution is given below.

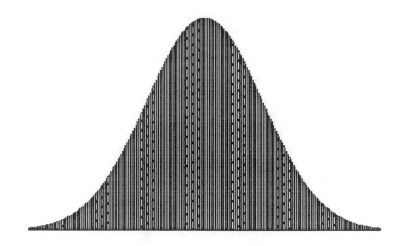

In real life, it is not possible to take 300 million samples of the product. The alternative is to take samples on some basis and use the results to estimate the average, standard deviation, and the shape of the distribution. The average and standard deviation can be determined by using control charts based on the sample results. The shape of the distribution can be estimated by constructing a histogram. Since the process is in control, the average, the standard deviation, and the shape of the distribution do not change over time.

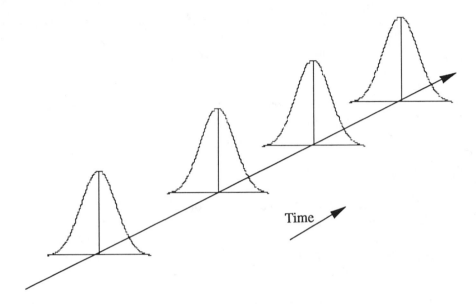

Time

Even though the process is in control, you won't get the same result each time you sample the process. The reason is that you are sampling a population of values. There is a certain probability associated with obtaining a certain value each time you sample. This probability is based on the shape of the population distribution. For example, consider the population of results generated by rolling one die. Each face of the die has the same probability of occurring (1 in 6). You don't know what the next roll will be but you do know the probability is 1 in 6 that it will be a one, for example.

Control limits are based on probability. Control limits are selected so that, if the process remains in control, there is only a small possibility of obtaining a point beyond the control limits. Thus, whenever a point is beyond the control limits, you are fairly certain that something has changed in the process. The control charts and control limit calculations depend on the assumed distribution of the population being sampled.

Control charts are based on the assumption that you are sampling a population governed by one of three distributions: the binomial distribution, the Poisson distribution or the normal distribution. These three distributions are discussed below along with the type of data described by each distribution. There are two types of data: attributes and variables data. It is important to determine what type of data you have because it influences the type of control chart you select to use.

ATTRIBUTES DATA

Attributes control charts are based on attributes data. This type of data is often referred to as discrete data. There are two kinds of attributes data: yes/no data and counting data.

1. Yes/No Data

   The yes/no kind of data has only two possible outcomes: either it passes or it fails some preset specification. Each item inspected is either defective (i.e., it does not meet the specifications) or is not defective (i.e., it meets specifications). Examples of the yes/no kind of attributes data are:

   > mail delivery: is it on time or not on time?
   > phone answered: is it answered or not answered?
   > invoice correct: is it correct or is it not correct?

2. Counting Data

   With counting data, you count the number of defects. A defect occurs when something does not meet a preset specification. It does not mean that the item itself is defective. For example, a television set can have a scratched cabinet (a defect) but still work properly. When looking at counting data, you end up with whole numbers such as 0, 1, 2, 3; you can't have half of a defect. Examples of counting data are:

   > number of injuries in a plant
   > number of surface imperfections on a plastic sheet
   > number of burnt particles in a polymer resin

OPERATIONAL DEFINITIONS

When working with attributes data, you have to have a clear understanding of whether the item you are looking at is defective or not (yes/no data) or whether it should be counted as a defect (counting data). In order to know whether a shipment was on time or to count the number of on-time shipments, you have to have a definition of what "on time" means. Is "on time" anywhere from 1:55 p.m. to 2:05 p.m., anytime before 2:00 p.m., or anytime between 2:00 p.m. and 2:15 p.m.? This clear understanding of a quality expectation is called an operational definition. According to Dr. Deming, an operational definition includes (Deming, 1982):

a. a written statement (and/or a series of examples) of criteria or guidelines to be applied to an object or to a group,

b. a test of the object or group for conformance with the guidelines, which includes specifics such as how to sample, how to test, and how to measure,

c. a decision: yes , the object or the group did meet the guidelines; no, the object or group did not meet the guidelines; or the number of times the object or group did not meet the guidelines.

Using the invoice error example, the written statement may read "An invoice error is an incorrect shipping amount or a wrong price." The test could be to

1. compare every invoice to bills of lading to check for incorrect shipping amounts and,

2. compare every invoice to a price schedule to check for wrong prices.

Based on these guidelines and a test for conformance with these guidelines, you could make a decision as to whether an invoice is defective or how many defects an invoice contains.

OPERATIONAL DEFINITION EXAMPLES

1. Suppose you are interested in helping your teenage son keep his room clean. It is evident from the appearance of the room that you and your son don't have the same operational definition of what constitutes a clean room. In the space below, develop an operational definition of a clean room.

2. Think about the processes at work. What are three areas where you feel operational definitions would be helpful and why? List your responses below.

Chapter 9

BINOMIAL DISTRIBUTION: YES/NO DATA

Whenever you have the yes/no type of attributes data, you will use either a p or np control chart. The basic probability distribution for calculation of control limits on these charts is the binomial distribution. For example, consider the bead bowl experiment. In that experiment, the bowl contained a large number of white beads and a smaller number of yellow beads. The yellow beads were considered to be defective. A customer wanted only white beads. The bowl represented the process. Since all the beads were contained in the bowl, this was the population of all possible results. How can you determine the contents of the bowl? You could count all the yellow beads and white beads in the bowl. However, in real processes, that is not possible. You don't know the contents of the bowl.

To find out what the population (contents of the bowl) is, samples are pulled at random from the bowl. The number of yellow beads in each sample is counted. This is the yes/no type of data. Either the bead is yellow (defective) or it is not. Each sample is returned to the bowl, and the contents are thoroughly mixed. Will each sample contain the same number of yellow beads? No. As a matter of chance variation, the number of yellow beads will vary from one sample to another even though the population (the contents of the bowl) is not changing.

Suppose the bowl contains 10,000 beads. There are 8,500 white beads and 1,500 yellow beads. The fraction nonconforming to specifications is 1,500/10,000 = 0.15. This represents the population (true) average fraction defective, p'. Primes, ', are used to denote population parameters such as the population average. Now suppose you withdraw a random sample of 100 beads from the bowl. How many yellow beads will the sample contain? On average, the sample will contain (100)(0.15) = 15 yellow beads. Will each sample contain 15 yellow beads? No. Chance variation will cause differences. However, if the 100 beads are returned to the bowl after each drawing, the theory of probability will allow us to calculate the probability of obtaining a sample containing 0, 1, 2, 3, etc. yellow beads. The binomial distribution can be used to estimate these probabilities (Grant and Leavenworth, 1980).

The value p' represents the probability of obtaining a yellow bead. Let np be the number of yellow beads present in a sample of size n. The probability of exactly np yellow beads occurring in a sample size of n for a population with a probability of p' is given by:

$$P(np|n, p') = [n!/(np!(n - np)!)](1 - p')^{n-np}(p')^{np}$$

P(np|n,p') has to be greater than or equal to zero and less than or equal to one.

For example, suppose you are only going to draw one bead at a time from the bowl. The bead will either be a white bead or a yellow bead. The above equation can be used to calculate the probability of obtaining a yellow bead with a sample size of 1:

$$P(1|1, 0.15) = [1!/(1!(0)!)](1 - 0.15')^0(0.15)^1 = 0.15$$

The probability of obtaining a white bead can be calculated also. This probability is 0.85. This means that if you draw 100 samples of size 1 from the bowl, you will on average have selected 85 white beads and 15 yellow beads. Figure 1a shows the resulting binomial distribution for this case. Note that the sum of the two probabilities above is $0.15 + 0.85 = 1$.

As sample size increases, the probability of obtaining a sample with a given number of yellow beads can be calculated. For example, for a sample size of 10, what is the probability of drawing a sample with 2 yellow beads? This probability can be calculated as shown below:

$$P(2|10, 0.15) = [10!/(2!(10 - 2)!)](1 - 0.15)^{10-2}(0.15)^2 = 0.2759$$

Similar calculations can be done to determine the probability of obtaining samples with 0, 1, etc. yellow beads. These probabilities are given below.

Number of Yellow Beads (np)	P(np|10, 0.15)	Cumulative Probability
0	0.1969	0.1969
1	0.3474	0.5443
2	0.2759	0.8202
3	0.1298	0.9500
4	0.0401	0.9901
5	0.0085	0.9986
6	0.0012	0.9998
7	0.0001	0.9999

The numbers under cumulative probability represent the probability of obtaining that number of yellow beads or less. For example, the probability of obtaining a sample with two or fewer yellow beads is 0.8202. The cumulative probability is the sum of all previous probabilities. The distribution for this situation is shown in Figure 1b.

The shape of the binomial distribution depends on the sample size, n, and the average fraction defective, p'. Figures 1a through 1e show how the shape of the distribution changes as sample size increases for constant p'. As n becomes larger, the distribution becomes more symmetrical. Figures 2a through 2e show how the shape of the distribution changes as the fraction defective changes for a constant sample size.

The average of the binomial distribution is np'. The average depends on the sample size and the average fraction defective. For example, in Figure 1b, the average of the binomial is np' = 10(.15) = 1.5. For Figure 1c, the average of the binomial is np' = 25(.15) = 3.75. This is the expected value for the number of defective items in a sample size of n that has p' fraction defective. The standard deviation of the binomial is given by:

$$\sigma'_{np} = \sqrt{np'(1 - p')}$$

Figure 1

EXAMPLES OF THE BINOMIAL DISTRIBUTION -- CONSTANT p'

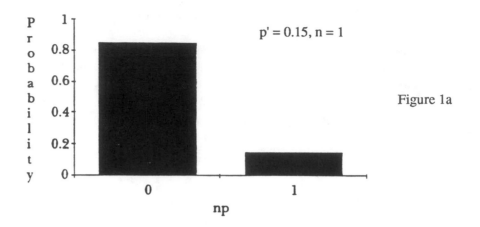

p' = 0.15, n = 1

Figure 1a

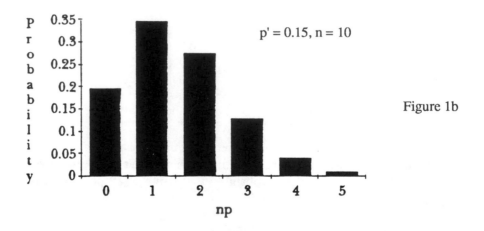

p' = 0.15, n = 10

Figure 1b

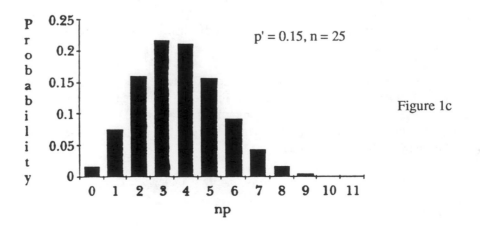

p' = 0.15, n = 25

Figure 1c

Figure 1 (Continued)

EXAMPLES OF THE BINOMIAL DISTRIBUTION -- CONSTANT p'

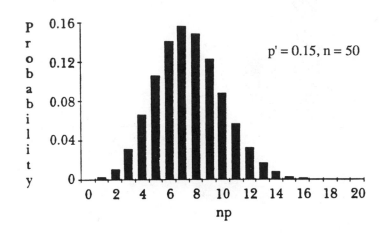

p' = 0.15, n = 50

Figure 1d

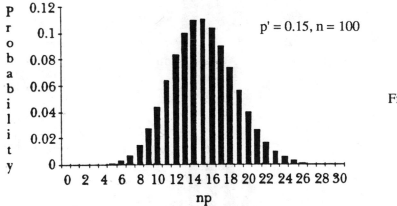

p' = 0.15, n = 100

Figure 1e

Figure 2

EXAMPLES OF THE BINOMIAL DISTRIBUTION -- CONSTANT n

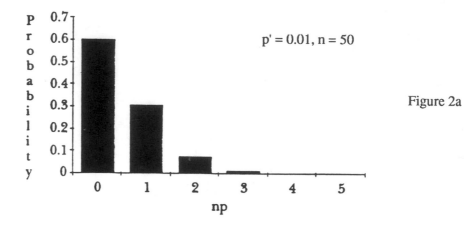

Figure 2a

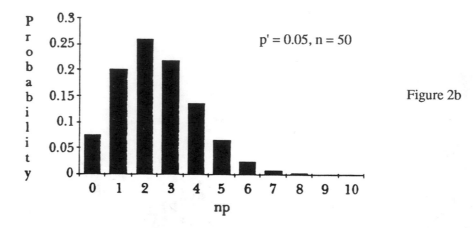

Figure 2b

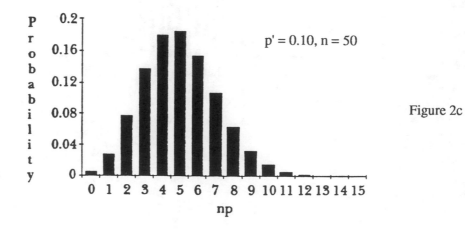

Figure 2c

Figure 2 (continued)

EXAMPLES OF BINOMIAL THE DISTRIBUTION -- CONSTANT n

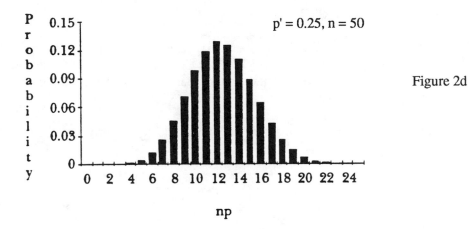

Figure 2d

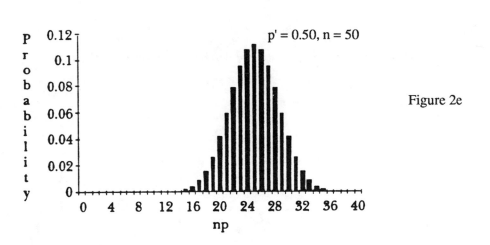

Figure 2e

The standard deviation is a measure of the spread or variation in the data. For example, in Figure 1b and 1c the standard deviations are given by:

$$\sigma'_{np} = \sqrt{np'(1 - p')} = \sqrt{10(.15)(1 - .15)} = 1.13 \quad \text{for Figure 1b}$$

$$\sigma'_{np} = \sqrt{np'(1 - p')} = \sqrt{(25)(.15)(1 - .15)} = 1.47 \quad \text{for Figure 1c}$$

BINOMIAL DISTRIBUTION EXAMPLE

Your company is monitoring the number of on-time deliveries to a customer's plant. The operational definition for an on-time delivery is a delivery that arrives at the customer's gate within 30 minutes of the scheduled time. Assume that it is known that $p' = 0.08$. Each week there are 20 deliveries made to the customer.

1. What is the probability that there will be no late deliveries in a week?

2. What is the probability that there will be 2 late deliveries in a week?

3. What is the probability that there will be 10 late deliveries in a week?

4. What is the average number of late deliveries per week?

5. What is the standard deviation of this process?

POISSON DISTRIBUTION: COUNTING DATA

Whenever you have the counting type of attributes data, you will use either a c or a u control chart. The basic probability distribution used for calculating control limits on these charts is the Poisson distribution. Usually, you will be counting some sort of defect. For example, you might be interested in counting the number of surface imperfections on an extruded piece of plastic. You might be interested in counting the number of first aid cases in a manufacturing plant. In these situations, there are many opportunities for defects to occurs. However, the number of defects that actually occur is small. Under these conditions, the situation can be approximated by using the Poisson distribution.

In these cases, the sample size, n, is very large. It may not be possible to even calculate a value for n. For example, what is the sample size for the extruded plastic sheet? What is the sample size for a manufacturing plant? This leads to the concept of "area of opportunity" for defects to occur. For example, the area of opportunity for first aid injuries to occur is the manufacturing plant itself. The area of opportunity for surface imperfections to occur on an extruded piece of plastic might be 10 square inches.

The Poisson distribution is actually the binomial distribution with n approaching infinity. Let c represent the number of defects. The Poisson distribution represents the probability of c occurrences of a certain event given that the population average for the number of occurrences is c'. This probability is given by (Grant and Leavenworth, 1980):

$$P(c|c') = (c'^c/c!)e^{-c'}$$

For example, suppose you are looking at the number of at-fault accidents each month involving the truck drivers in your distribution system. Suppose you know that the population average, c', is 1.0. The above equation can be used to determine the probability of truck drivers having 0, 1, 2, 3, etc. at-fault accidents each month. The calculations are shown below for this situation.

Number of Accidents	Probability	Cumulative Probability
0	$(1^0/0!)e^{-1} = 0.368$	0.368
1	$(1^1/1!)e^{-1} = 0.368$	0.736
2	$(1^2/2!)e^{-1} = 0.184$	0.920
3	$(1^3/3!)e^{-1} = 0.061$	0.981
4	$(1^4/4!)e^{-1} = 0.015$	0.996
5	$(1^5/5!)e^{-1} = 0.003$	0.999
6	$(1^6/6!)e^{-1} = 0.001$	1.000

The distribution is shown in Figure 3a. Note that even though the population average is small (1.0), the probability of having a month where there are three or more accidents is significant.

Figure 3

EXAMPLES OF THE POISSON DISTRIBUTION

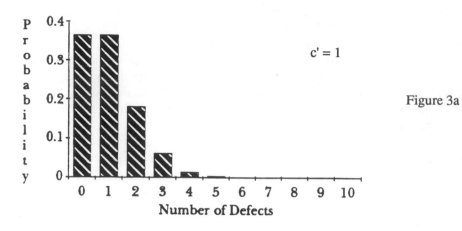

c' = 1

Figure 3a

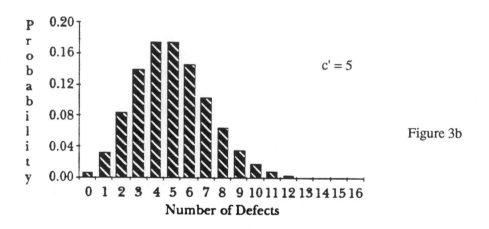

c' = 5

Figure 3b

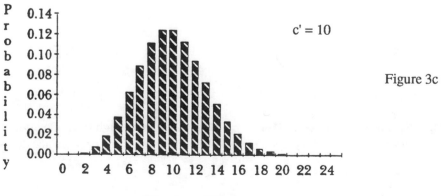

c' = 10

Figure 3c

The average of the Poisson distribution is c'. The standard deviation of the Poisson distribution is $\sqrt{c'}$. The shape of the distribution is determined by c'. Figures 3a through 3c illustrate how the distribution changes shape for different values of c'. As c' becomes larger, the distribution becomes more symmetrical.

POISSON DISTRIBUTION EXAMPLE

Your manufacturing plant is monitoring the number of valves that leak each month. The plant is extremely large and, due to poor record keeping over the years, no one knows how many valves there are in the plant. The operational definition for a valve that leaks is one that fails a pressure test or one that has been discovered to be leaking while in service. Assume that it is known that c' = 7.5. Answer the questions below.

1. What is the probability that no valves will leak in a month?

2. What is the probability that 5 valves will leak in a month?

3. What is the probability that 20 valves will leak in a month?

4. What is the population average for the number of valves that leak?

5. What is the standard deviation of this process?

VARIABLES DATA

Variables control charts are based on variables data. Variables data consist of observations made from a continuum. That is, the observation can be measured to any decimal place you want if your measurement system allows it. Some examples of variables data are contact time with a customer, ratings on a survey, amount of time to make a delivery and various product responses (bulk density, purity, color, etc.).

NORMAL DISTRIBUTION: VARIABLES DATA

There are many types of control charts that are based on variables data. These control charts include the $\overline{X}$-R, $\overline{X}$-s, and individuals control chart. The basic probability distribution that governs the calculation of the control limits on these charts is the normal distribution. The normal distribution is the familiar bell-shaped curve shown in Figure 4.

The normal distribution has several interesting characteristics. The shape of the distribution is determined by the average, $\overline{X}'$, and the standard deviation, σ'. The highest point on the curve is the average. The distribution is symmetrical about the average. Most of the area under the curve (99.7%) lies between $-3\sigma'$ and $+3\sigma'$ of the average. This is shown in Figure 4. The curve is generated by the equation below (Himmelblau, 1970):

$$P(x) = (1/(\sigma')(\sqrt{2\pi}))\exp[-(x - \overline{X}')^2/2\sigma'^2)]$$

where x is a value between $-\infty$ to $+\infty$. While this equation is useful, it is more beneficial to convert the above equation using the standard randomized variable, z :

$$z = (x - \overline{X}')/\sigma'$$

where z measures how many standard deviations some value x is from the average. Using this definition of z, it can be shown that:

$$P(z) = (1/\sqrt{2\pi})\exp(-z^2/2)$$

This equation gives a normal distribution with an average of zero and a standard deviation of one. For this distribution, the area under the curve from $-\infty$ to $+\infty$ is equal to 1.0. The area under the curve is proportional to the fraction of measurements that fall in that region. These two facts will be used to help determine the fraction of measurements that fall above some value (such as a specification limit), below some value or between two values. The relationship between z, $\overline{X}'$ and σ' is also shown in Figure 4.

Figure 5 shows the effect that varying standard deviation has on the shape of the normal distribution. The average in all cases is zero. As the standard deviation becomes larger, the normal distribution becomes wider.

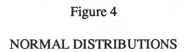

Figure 4

NORMAL DISTRIBUTIONS

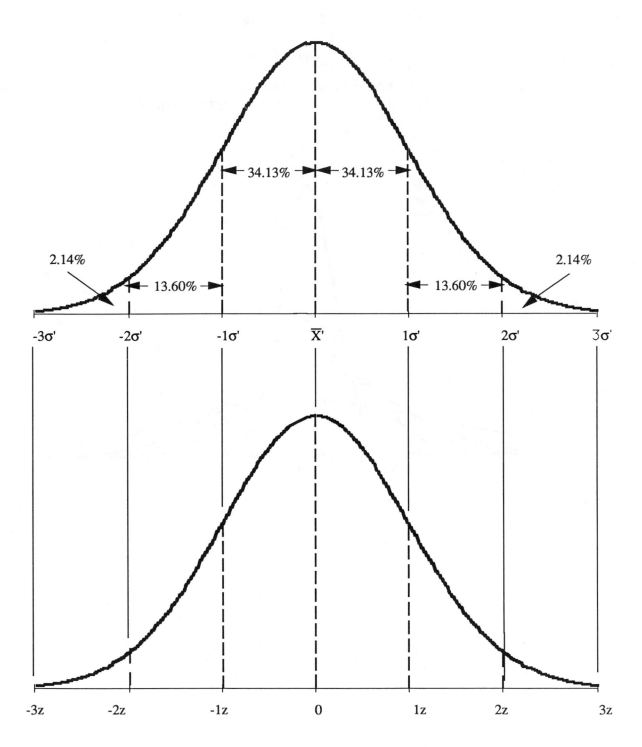

Figure 5

EFFECT OF VARYING STANDARD DEVIATION
ON THE SHAPE OF THE NORMAL DISTRIBUTION

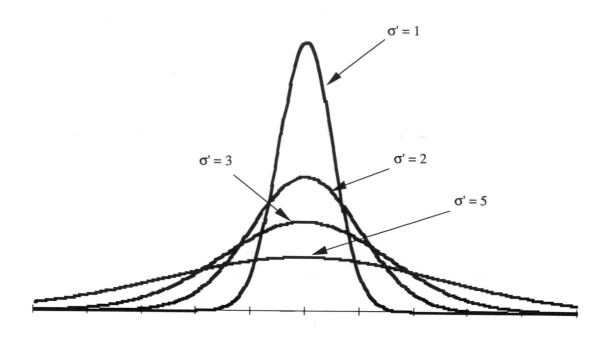

Consider the examples shown in Figure 6. Each distribution in Figure 6 is a normal distribution. We want to find the area of the shaded portion under each curve. In Figure 6a, we want to find the area beyond 2 standard deviations above the average. In this case, z = 2 since z represents the number of standard deviations some value is away from the average. To determine the area of the shaded portion under the curve, we will use the z table (Appendix Table 7, p. 491). The z table gives the percentage of process output that lies beyond a single value that is z standard deviations away from the mean. In Figure 6a, the value of z is 2. The table will then give us the percentage of process output beyond z = 2. Looking up the value of z = 2.00 in the table gives an area of 0.0228. This means that 0.0228 or 2.28% of the process output is above z = 2.

In Figure 6b, we want the area beyond a value 1.53 standard deviations below the average. The z table can be used again. In this case, z = -1.53. It is negative since the value is below the average. To find the shaded area, the value of |z| is looked up in the table. The value for +z or -z will always be the same since the normal distribution is symmetrical. The value for |z| = 1.53 is .0630. This means that 0.0630 or 6.3% of the process output is less than z = -1.53.

The z table gives the percentage of process output above a certain value of z if z is positive or the fraction of process output below a certain value of z if z is negative. In Figure 6c, we want to find the shaded area between z = -1.53 and z = 2. This area is not directly available from the z table. To find this area, we make use of the fact that the total area under the curve is equal to 1.0. We know that the area above z = 2.0 is 0.0228 and that the area below z = -1.53 is 0.0630. The area in between is 1.0 minus the sum of those two areas or 1 - (0.0228 + 0.0630) = 0.9142. Thus 0.9142 or 91.42% of the process output lies between z = -1.53 and z = 2.0.

Suppose that a process has an average, $\overline{X}'$, of 100, a standard deviation, σ', of 10 and is normally distributed. The distribution can be easily drawn as is shown in Figure 7. Suppose you want to know the percentage of process output that is above 118. You can use the z table to determine this. The first step is to calculate a z value based on 118. The calculation is given below.

$$z = (x - \overline{X}')/\sigma' = (118 - 100)/10 = 1.8$$

For x = 118, the z value is 1.8. This means that 118 is 1.8 standard deviations above the mean. The area under the curve above z = 1.8 can be determined from the z table. The area is 0.0359. This means that 3.59% of the process output is above x = 118.

Figure 6

EXAMPLES OF USING THE z TABLE

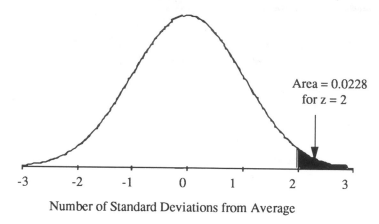

Area = 0.0228
for z = 2
-3
-2
-1
0
1
2
3
Number of Standard Deviations from Average

Figure 6a

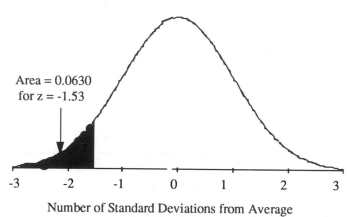

Area = 0.0630
for z = -1.53
-3
-2
-1
0
1
2
3
Number of Standard Deviations from Average

Figure 6b

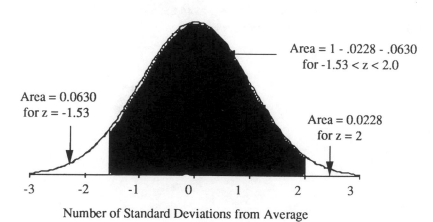

Area = 1 - .0228 - .0630
for -1.53 < z < 2.0
Area = 0.0630
for z = -1.53
Area = 0.0228
for z = 2
-3
-2
-1
0
1
2
3
Number of Standard Deviations from Average

Figure 6c

Figure 7

PROCESS WITH NORMAL DISTRIBUTION

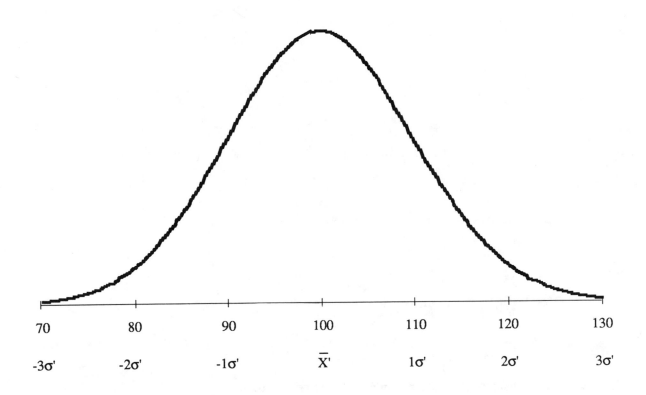

NORMAL DISTRIBUTION EXAMPLE

Reaction yield is one parameter being monitored by a plant on a regular basis. Assume that it is known that the population average is 91%, the population standard deviation is 2.5 and the population is normally distributed. Answer the questions below.

1. Draw a picture of the distribution.

2. What percent of the reaction yields are below 87?

3. What percent of the reaction yields are above 95?

4. What percent of the reaction yields are below 90?

5. What percent of the reaction yields are between 87 and 95?

6. What percent of the reaction yields are between 87 and 90?

DETERMINING IF A POPULATION IS NORMALLY DISTRIBUTED

There are many naturally occurring distributions. How can you determine if the population is normally distributed? There are two simple methods that can be used. One method is to construct a histogram of the individual measurements and see if it resembles a normal distribution. The other method is to construct a normal probability plot. This method is covered below.

1. Divide and group at least 50 individual measurements into intervals (just as in constructing a histogram).
2. For these data, determine the frequency distribution, the cumulative frequency distribution and the cumulative percentage corresponding to each interval.
3. On normal probability paper (Figure 9, p. 121), plot the cumulative percentage on the vertical axis against the upper limit of the corresponding interval on the horizontal axis.
4. If the data came from a normal distribution, the points should lie roughly on a straight line.

The following example will illustrate the concepts. Suppose an operating unit has been monitoring reaction run time. A histogram of the individual reaction run times has been constructed and is shown in Figure 8.

Figure 8

REACTION RUN TIME HISTOGRAM

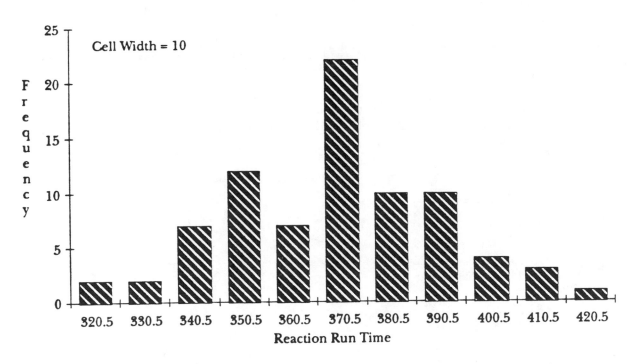

Is the population of individual reaction run times normal? The histogram appears to be normally distributed. To double check this, a normal probability plot will be used. The intervals used in the construction of the histogram are shown below along with the frequency with which individual measurements fall in the intervals.

Intervals	Frequency	Cumulative Frequency	Cumulative Percentage
315.5 - 325.5	2	2	2.50
325.5 - 335.5	2	4	5.00
335.5 - 345.5	7	11	13.75
345.5 - 355.5	12	23	28.75
355.5 - 365.5	7	30	37.50
365.5 - 375.5	22	52	65.00
375.5 - 385.5	10	62	77.50
385.5 - 395.5	10	72	90.00
395.5 - 405.5	4	76	95.00
405.5 - 415.5	3	79	98.75
415.5 - 425.5	1	80	100.0

Also included in the table is the cumulative frequency. The cumulative frequency is the number of measurements that fall in that interval or below it. The cumulative percentage is also shown. The cumulative percentage is determined by dividing the cumulative frequency by the total number of measurements (in this case 80) and multiplying by 100.

The cumulative percentage is plotted against the upper limit of the interval on normal probability paper (Figure 9). The points appear to line in a straight line. Thus, the population of individual reaction times is normally distributed.

EXAMPLE OF DETERMINING IF A DISTRIBUTION IS NORMALLY DISTRIBUTED

A team has been monitoring reaction yield. Is the population of individual reaction yields normal? Use a normal probability plot (p. 122) to determine this. The intervals and frequency distribution are given in the table below.

Intervals	Frequency	Cumulative Frequency	Cumulative Percentage
73.25 - 74.75	4		
74.75 - 76.25	9		
76.25 - 77.75	9		
77.75 - 79.25	22		
79.25 - 80.75	25		
80.75 - 82.25	23		
82.25 - 83.75	23		
83.75 - 85.25	6		
85.25 - 86.75	4		

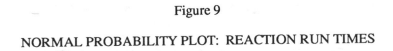

Figure 9

NORMAL PROBABILITY PLOT: REACTION RUN TIMES

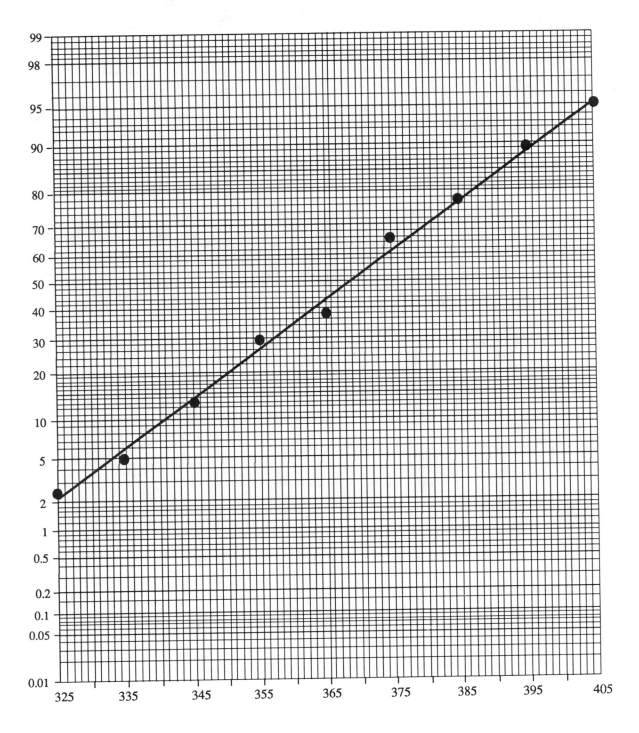

NORMAL PROBABILITY PLOT: REACTION YIELDS

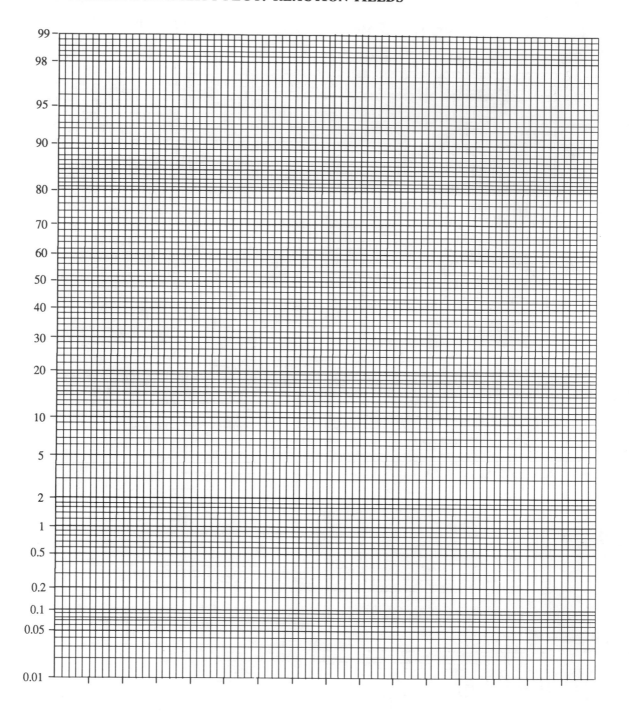

CENTRAL LIMIT THEOREM

As discussed previously, most control charts are based on one of three basic probability distributions: the binomial, the Poisson or the normal distribution. There are many naturally occurring distributions that do not fit any of these three distributions. For example, skewed distributions are common in processes where a parameter is being maximized or minimized. Does this mean that there are no control charts for these situations? There are control charts for these situations because of the Central Limit Theorem. This very important and powerful theorem can be stated as:

- Regardless of the shape of the distribution of a population, the distribution of average values, $\overline{X}$'s, of subgroups of size n drawn from that population will tend toward a normal distribution as the subgroup size n becomes large.

In most cases, the size of n does not have to be very large. Usually subgroup sizes of four to five are sufficient to allow the averages to be normally distributed. Note that the binomial and Poisson distributions tend toward a normal distribution as np' and c' become large (see Figures 1-3).

The standard deviation ($\sigma_{\overline{x}}'$) of the distribution of average values is related to the standard deviation (σ') of the individual values by the following:

$$\sigma_{\overline{x}}' = \sigma'/\sqrt{n}$$

CENTRAL LIMIT THEOREM EXAMPLE

Consider the distribution that would exist because of a toss of a die. The distribution will be rectangular in shape if the die is true. The die will read either 1, 2, 3, 4, 5, or 6. All have an equal opportunity of appearing.

1. Throw the die 50 times and construct a histogram based on the individual results.

2. Throw two dice and calculate the average. Do this 50 times and construct a histogram based on the averages.

3. Throw four dice and calculate an average. Do this 50 times and construct a histogram based on the averages.

4. What are the differences among the three histograms? Is the central limit theorem working?

CENTRAL LIMIT THEOREM EXAMPLE -- DICE

One Die Result

Frequency

Average of Two Dice Results

Frequency

Average of Four Dice Results

Frequency

SUMMARY

This chapter began to lay the foundation for the construction and use of control charts. Control charts can be divided into two major categories: attribute control charts and variables control charts. Attributes control charts are based on attributes data. There are two types of attributes data: the yes/no type and the counting type of data. Yes/no attributes data have only two possible outcomes: either the item is defective or it is not defective. The basic probability distribution underlying the calculation of control limits for yes/no data is the binomial distribution. With counting type attributes data, the number of defects is counted. The basic probability distribution underlying the calculation of control limits for counting data is the Poisson distribution. With attributes data, there is the need for operational definitions. Operational definitions are used to determine what constitutes a defective item or a defect. Variable control charts are based on variables data. Variables data are data from a continuum. The basic probability distribution underlying the calculation of control limits for variables data is the normal distribution. The Central Limit Theorem states that, regardless of the shape of a population's distribution, the distribution of average values of sample size n will tend to be normally distributed as n becomes large.

APPLICATIONS OF DATA AND DISTRIBUTIONS

Think about the material covered in this chapter. What are the three most important things you learned and why? List your responses below.

1.

2.

3.

10
INTERPRETATION OF CONTROL CHARTS

Control charts are being used more and more to monitor the variation in many types of processes. The control chart represents a picture of a process over time. To effectively use control charts, one must be able to interpret the picture. What is this control chart telling me about my process? Is this picture telling me that everything is all right and I can relax? Is this picture telling me that something is wrong and I should get up and find out what has happened? This chapter discusses various methods of interpreting a control chart.

OBJECTIVES

In this chapter you will learn:

1. The general model used for control charts
2. How control charts relate to a process
3. How to recognize a process in statistical control
4. The various tests for out-of-control situations

The control chart is a powerful tool for monitoring variation in a process. The chart allows you to determine when variation is simply due to random variation or when the variation is due to assignable causes.

INTRODUCTION TO CONTROL CHART INTERPRETATION

A control chart tells you if your process is in statistical control (i.e., only random variation is present) or if your process is out of statistical control (i.e., assignable cause variation is also present). To be able to determine which situation is present, you must be able to interpret the control chart. The most important factor in finding the reason for an assignable cause is time. The faster the out-of-control situation is detected, the better the odds of finding out what happened. This is why control charts are very powerful tools for front line personnel. If the front line personnel are keeping the charts, they can immediately begin to look for the reasons for out-of-control points when they appear on the chart.

Control charts are based on a general model. The general model is that the control limits are ±3 standard deviations from the average. The control limits are sometimes called 3 sigma limits. The model is shown below for a process, Y, with an average of $\overline{Y}$.

$$\text{Upper Control Limit} = UCLy = \overline{Y} + 3\sigma_y$$

$$\text{Lower Control Limit} = LCLy = \overline{Y} - 3\sigma_y$$

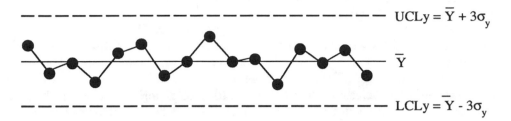

$\overline{Y}$ can either be the average of a normal distribution, the binomial distribution or the Poisson distribution. For example, suppose the population you are sampling can be represented by the Poisson distribution. Samples from the process indicate that the process average number of defects is $\overline{c}$. Since the standard deviation of the Poisson distribution is $\sqrt{\overline{c}}$, the control limits are given by:

$$UCLc = \overline{c} + 3\sqrt{\overline{c}}$$

$$LCLc = \overline{c} - 3\sqrt{\overline{c}}$$

The ± 3 sigma limits have been used because they represent a balance between the costs of looking for assignable causes when they are not present and the costs of not hunting for them when they are present. This can intuitively been seen in the following example. Suppose you are charting a product response that is normally distributed. The control limits on this variable, X, are given by:

$$UCLx = \overline{X} + 3\sigma_x$$

$$LCLx = \overline{X} - 3\sigma_x$$

The distribution is shown below.

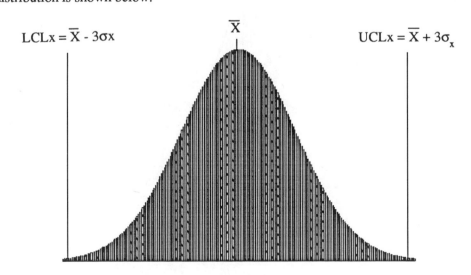

The area beyond the 3 sigma limits can be determined from the z table (Appendix Table 7, p. 491). The area beyond each limit is 0.00135. This represents the fraction of process output we would expect beyond the 3 sigma limits. The total of both areas is 0.0027. This also represents the probability that a sample will be obtained beyond the 3 sigma limits. Since this probability is low (about 3 times out of 1,000), the occurrence of such a sample would be rare for a normally distributed product response. Similar probabilities hold for the other distributions. The point is that the control limits are selected so that the probability of obtaining a point beyond the control limits is very small. This means that when a point falls beyond the control limits, you are fairly certain that something has changed in the process, and you should begin to look for an assignable cause. Since this takes time (and thus money), we want to be sure that there is really something wrong in the process before you begin to look for assignable causes. The 3 sigma limits help us accomplish this.

CONTROL CHARTS AND THE PROCESS

When determining whether a process is in statistical control, you must determine *in control with respect to what*. A process can be in control (or out of control) with respect to the average value, the variation in individual measurements, or both. If a process is in control, it means that (statistically) the process has only one average and one variation. These do not change over time. This is shown below.

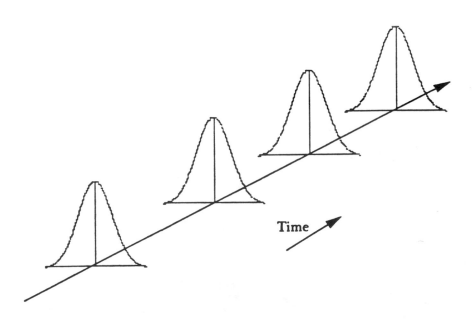

Time

A process can be out of control with respect to the average but in control with respect to the variation. At a given time, you do not know what the average of the process is. This situation is shown below.

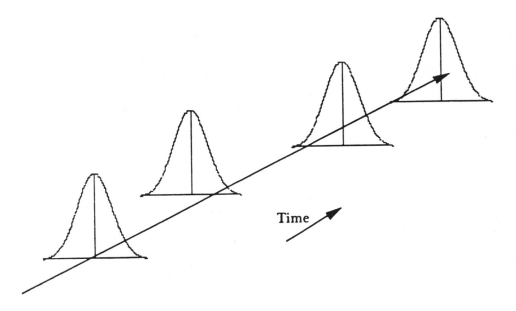

A process can be out of control with respect to the variation but in control with respect to the average. At a given time, you do not know how much variation there is about the average. This situation is shown below.

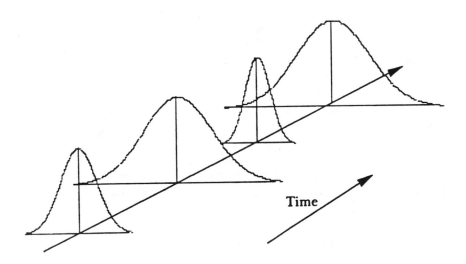

Chapter 10

A process can be out of control with respect to both the average and the variation. In this case, you do not know what the average is or how much variation there is at any given time. This situation is shown below.

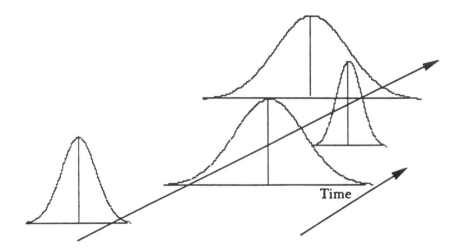

To be able to determine which of these situations exist in a process, one must be able to interpret a control chart. Tests for interpreting a control chart are given below.

TESTS FOR OUT-OF-CONTROL SITUATIONS

A control chart tells you if your process is in statistical control. Figure 1 is an example of a stable or in-statistical-control process. The pattern is typical of processes that are stable. Three characteristics of a process that is in control are:

- Most points are near the average
- A few points are near the control limits
- No points are beyond the control limits

If a control chart does not look similar to Figure 1, there is probably an assignable cause present. Various tests for determining if an assignable cause is present are given below (AT&T, 1956). Each test is described, i.e., what kind of pattern on a chart indicates that an out-of-control situation is present. The probability of obtaining this pattern if the process is stable is also given. The purpose of doing this is to demonstrate that these patterns are really indications of out-of-control situations. The probabilities of obtaining these patterns from a process that is in statistical control are very small. For calculations of the probabilities, a normal distribution is assumed.

Not all tests can be applied to all control charts. No differentiation is made of this fact in this chapter. The individual chapters for each control chart discuss which tests are applicable.

Figure 1

RANDOM PATTERNS FOR A STABLE PROCESS

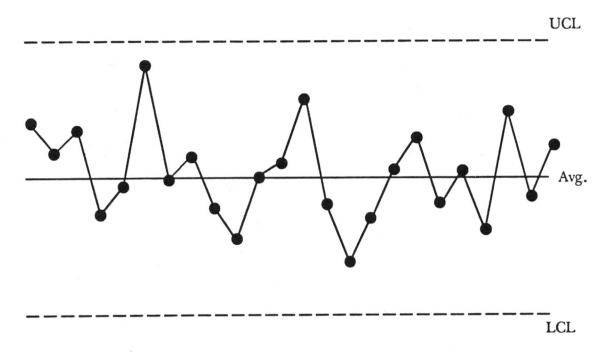

Characteristics of a random pattern:

- Most points near the average (centerline)
- A few points near the control limits
- No points beyond the control limits

These tests assume that the data have been plotted in "production run order." This means that the data are plotted in the order in which they are generated over time. That the data must be time ordered is a very important item to remember.

POINTS BEYOND THE CONTROL LIMITS

An assignable cause is present in the process if any points fall above the upper control limit or below the lower control limit. This type of situation is shown in Figure 2. If your process is in control, the probability of getting a point above the upper control limit is 0.00135 (or one out of 740). The same probability exists for a point below the lower control limit. The probability of getting a point beyond either control is then $0.00135 + 0.00135 = 0.0027$ (one out 370). Therefore, if you have a point beyond either control limit, the probability that your process is in control is very small. It is very likely that an assignable cause is present in the process. Action should be taken to find the assignable cause and permanently remove it from the process.

If there is a point beyond the control limit, there is no need to apply the other tests for out-of-control situations. Points on the control limits are considered to be out of control.

131

Figure 2

POINTS BEYOND THE CONTROL LIMITS

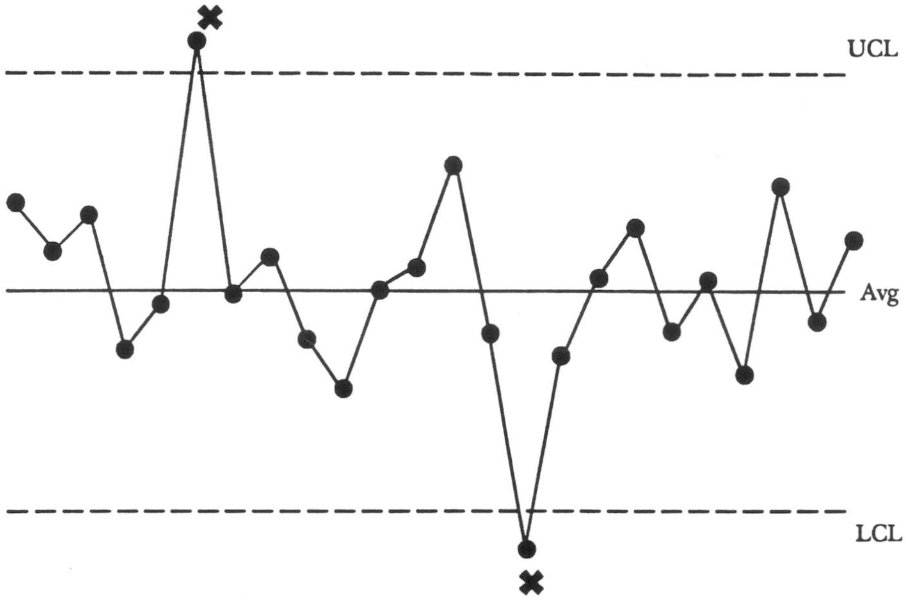

An assignable cause exists if any points fall above the upper control limit or below the lower control limit. ✖ = Out-of-control point.

LENGTH-OF-RUN TEST

A stable process has random variation about the average. A run is defined as consecutive points on one side of the average. The number of points in the run is called the length of the run. If the length of a run becomes too large, it indicates an out-of-control situation. Table 1 lists the critical values for the length-of-run test. The table lists the number of points on the graph and the critical value for run length. If the length of run on a chart is larger than this critical value, the process is out of control. Figure 3 shows an example of this test.

When the length of a run is too long, it means that the average of the process has changed. If the run is above the average, something has caused the average to increase. If the run is below the average, something has caused the average to decrease.

NUMBER-OF-RUNS TEST

The number of runs on a control chart can be an indication of an-out-of control situation. The number of runs can be determined by counting the runs on each side of the average. An easier method is to count the number of times the average is crossed and add one.

How many runs do you expect there to be on a chart? Usually the number of runs will be about one-half the number of points on the chart. If there are too few runs (i.e., the average is not crossed frequently enough), something is causing the process to cycle slowly. If there are too many runs, something is causing the process to cycle too quickly (e.g., overcontrol).

Table 1

LENGTH-OF-RUN TEST

Number of Total Points on Chart	Critical Value (.05 Probability)	Critical Value (.01 Probability)
20 - 29	7	8
30 - 39	8	9
40 - 49	9	10
50 or over	10	11

Runs that are larger than the critical values listed above represents an out of control situation.

Figure 3

LENGTH-OF-RUN TEST

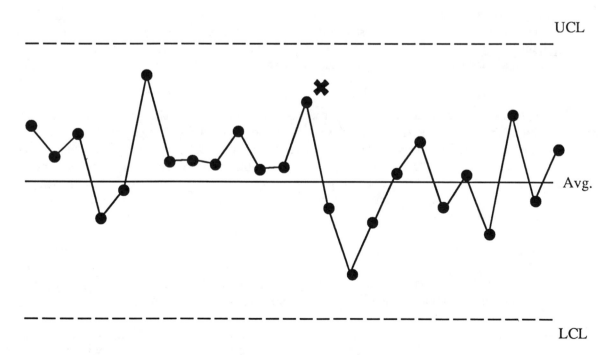

A run length of 8 or more is an indication that an out of control situation exists with these 24 points shown on the chart. ✖ = Out-of-control point.

Table 2 lists the critical values for the number of runs tests. To apply the test, count the number of runs on the chart and the number of points. Look up the number of points listed in the table. If the number of runs is less than the small critical value, there are too few runs present. Figure 4 shows an example of this test. If the number of runs is greater than the large critical value, there are too many runs. Figure 5 shows an example of this test.

If a point lies on the average line, move the point either up or down so that the number of runs becomes larger. An example of this is shown below. This approach helps avoid looking for assignable causes until you are sure that one is present.

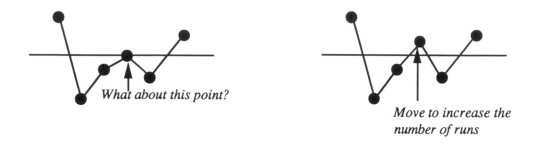

What about this point?

Move to increase the number of runs

ZONE TESTS

Zone tests are valuable for enhancing the ability of control charts to detect small shifts quickly. The first step in using the zone test is to divide the control chart into zones. As shown in Figure 6, this is done by dividing the area between the average and the upper control limit into three equally spaced areas. This is then repeated for the area between the average and the lower control limit.

The zones are called zones A, B and C. There is a zone A for the top half of the chart and a zone A for the bottom half of the chart. The same is true for zones B and C. Remember that control charts are based on 3 sigma limits. Thus, each zone is one standard deviation in width. For example, considering the top half of the chart, zone C is the region from the average to the average plus one standard deviation. Zone B is the region between the average plus one standard deviation to the average plus two standard deviations. Zone A is the region between the average plus two standard deviations and the average plus three standard deviations.

Assuming that the process is in control and normally distributed, one can calculate the probability of obtaining a point in each zone. These probabilities are given in Figure 7. For example, suppose 100 points are plotted on a control chart. There will be about 34 points in each zone C, 14 points in each zone B, 2 points in each zone A and no points beyond the control limits.

There are tests for each zone to determine if the process is out of control. These tests are described below. The tests are applied for the area above the average and then for the area below the average.

Table 2

NUMBER-OF-RUNS TEST

Total Number of Points on Chart	Expected Number of Runs	Small Critical Value	Large Critical Value
10 - 11	5	2	9
12 - 13	6	2	11
14 - 15	7	3	12
16 - 17	8	4	13
18 - 19	9	4	15
20 - 21	10	5	16
22 - 23	11	6	17
24 - 25	12	7	18
26 - 27	13	7	20
28 - 29	14	8	21
30 - 31	15	9	22
32 - 33	16	10	23
34 - 35	17	10	25
36 - 37	18	11	26
38 - 39	19	12	27
40 - 41	20	13	28
42 - 43	21	14	29
44 - 45	22	14	31
46 - 47	23	15	32
48 - 49	24	16	33
50 - 59	25	17	34
60 - 69	30	21	40
70 - 79	35	25	46
80 - 89	40	30	51
90 - 99	45	34	57
100 - 109	50	38	63
110 - 119	55	43	68
120 - 129	60	47	74

If the number of runs is $\leq$ the small critical value or $\geq$ the large critical value, there is evidence of an assignable cause. The critical values are based on 0.01 probability. The number of runs is the number of times the centerline is crossed plus one.

Figure 4

TOO FEW RUNS

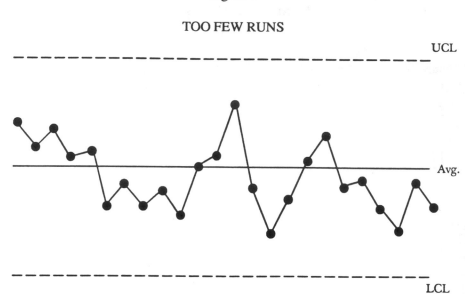

The small critical value for 24 points is 7 runs. This chart has only 6 runs. Thus, there is evidence of an assignable cause.

Figure 5

TOO MANY RUNS

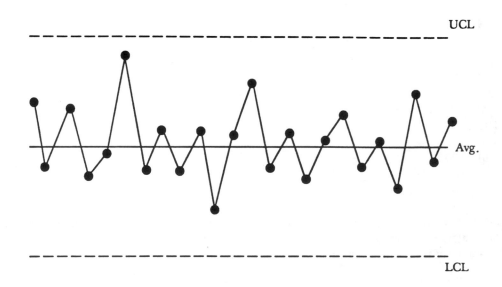

The large critical value for 24 points is 18 runs. This chart has 21 runs. Thus, there is evidence of an assignable cause.

Figure 6

DIVIDING A CHART INTO ZONES

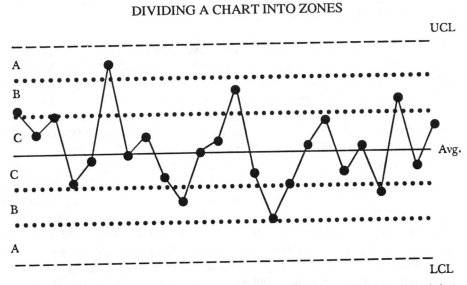

To divide a chart into zones, consider only the area between one control limit and the average. Divide this area into three equal zones. Repeat this fro the other half of the chart. The zones are called zone A, zone B and zone C.

Figure 7

PROBABILITIES FOR ZONE TESTS

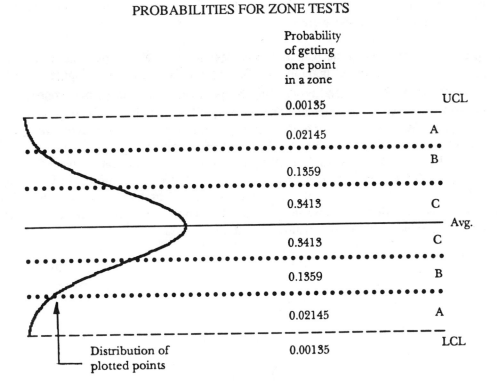

137

Test for Zone A: An assignable cause exists if two out of three consecutive points fall in zone A or beyond. Figure 8 shows an example of this test. The test is applied for the zone A above the average and then for the zone A below the average.

The probability of getting two out of three consecutive points in zone A or beyond can be determined. The probability of getting one point in zone A or beyond is 0.0228 (from Figure 7, 0.00135 + 0.02145). The probability of getting two points in a row in zone A or beyond is then (0.0228)(0.0228) = 0.00052. Note that this probability is smaller than the probability of getting one point beyond one of the control limits. Thus, if two points in a row fall in zone A or beyond, it is a stronger indication of an out-of- control situation than if a point is beyond the control limits. This situation is shown in Figure 8. Since this probability is so small, the requirement can be loosened somewhat by requiring that only two out of three consecutive points fall in zone A or beyond. The probability of getting a point somewhere else on the chart besides zone A or beyond is 1 - 0.0228 = 0.9772. The probability of getting two out of three consecutive points in zone A or beyond is then (0.0228)(0.0228)(0.9972)(3) = 0.00156 (or one out of 640). You multiply by 3 because the point not in zone A could be the first, second or third point. The probability of obtaining this pattern for a process that is in control is then 0.00156, a small number. Thus, if this pattern occurs, it is very likely that an assignable cause is present in the process. Action should be taken to discover the assignable cause and permanently remove it from the process.

These tests are applied to both halves of the chart. However, only one half is considered at a time. For example, if one point falls in zone A above the average and the next point falls in zone A below the average, this is not two out of three consecutive points in zone A or beyond. The two points in zone A must be on the same side of the average.

Test for Zone B: An assignable cause exists if four out five consecutive points fall in zone B or beyond. Figure 9 shows an example of this test. This test is applied for zone B above the average and then for zone B below the average.

The probability of obtaining this pattern for a process in statistical control is easily determined. The probability of getting one point in zone B or beyond is 0.1587 (from Figure 7, 0.1359 + 0.02145 + 0.00135). The probability of getting four points in a row in zone B or beyond is 0.1587^4 = 0.000634. This probability is smaller than the probability of getting a point beyond one of the control limits. The probability can be increased by requiring that only four out of five points fall in zone B or beyond. The probability of getting a point somewhere else on the graph besides zone B or beyond is 1 - 0.1587 = 0.8413. Thus, the probability of getting four out of five points in zone B or beyond is (0.0006234)(0.8413)(5) = 0.0027 (or one out of 370). The probability of obtaining this pattern for a process that is in statistical control is small. Thus, if this pattern occurs, it is very likely there is an assignable cause in the process. Action should be taken to discover the assignable cause and permanently remove it from the process.

Figure 8

TEST FOR ZONE A

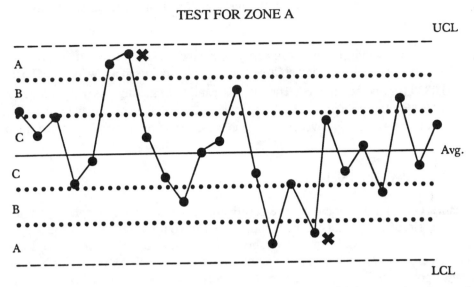

An assignable cause exists if two out of three consecutive points fall in zone A or beyond. ✖ =
Out of control point.

Figure 9

TEST FOR ZONE B

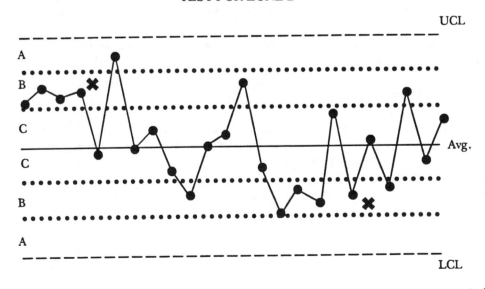

An assignable cause exists if four out of five consecutive points fall in zone B or beyond. ✖ =
Out-of-control point.

Test for Zone C: An assignable cause exists if eight consecutive points fall in zone C or beyond. An example of this test is shown in Figure 10. The test should be applied for the zone C above the average and then for the zone C below the average. Note that this test is similar to the test for the length of a run.

The probability of obtaining this pattern is easily determined. The probability of getting one point in zone C or beyond is 0.5. The probability of getting eight points in a row in zone C or beyond is $0.5^8 = 0.0039$ (or one out of 256). Thus, the probability of getting this pattern if the process is in control is small. If this pattern occurs, it is very likely that an assignable cause is present. Action should be taken to discover the assignable cause and permanently remove it from the process.

TEST FOR STRATIFICATION

Stratification occurs if two or more processes (distributions) are being sampled systematically. For example, stratification can occur if samples are taken once a shift and a subgroup of size 3 is formed, based on the results from three shifts. It is possible that the shifts are operating at a different average or variability. Stratification (an assignable cause) exists if fifteen or more consecutive points fall in zone C either above or below the average. Figure 11 is an example of this test. Note that the points tend to hug the center line. This test involves the use of the zones but is applied to the entire chart and not one-half of the chart at a time.

The probability of obtaining this pattern if the process is in control can be determined. The probability of getting a point in zone C (above or below the average) is 0.6826 (from Figure 7, 0.3413 + 0.3413). The probability of getting 15 points in zone C (above or below the average) is $0.6826^{15} = 0.0032$ (or one out of 307). The probability of obtaining this pattern from a process that is in control is low. Thus, if this pattern occurs, it is likely that an assignable cause is present. Action should be taken to find the assignable cause and remove it from the process.

If stratification is occurring, a histogram of the individual measurements will probably be bimodal.

TEST FOR MIXTURES

A mixture exists when there is more than one process present but sampling is done for each process separately. For example, suppose you take three samples per shift and form a subgroup based on these three samples. If different shifts are operating at different averages, a mixture can occur. A mixture (an assignable cause) is present if eight or more consecutive points lie on both sides of the average with none of the points in zone C. Figure 12 shows an example of this test. Note the absence of points in zone C. This test is applied to the entire chart.

Figure 10

TEST FOR ZONE C

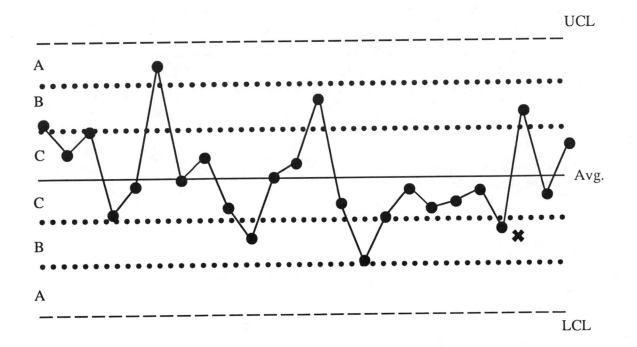

An assignable cause exists if eight consecutive points fall in zone C or beyond. ✖ = Out-of-control point.

Figure 11

TEST FOR STRATIFICATION

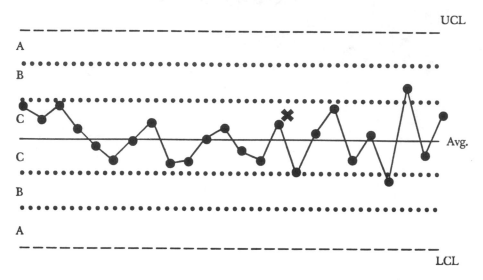

An assignable cause exists if fifteen or more consecutive points fall in zone C (either above or below the average). ✖ = Out-of-control point.

Figure 12

TEST FOR MIXTURES

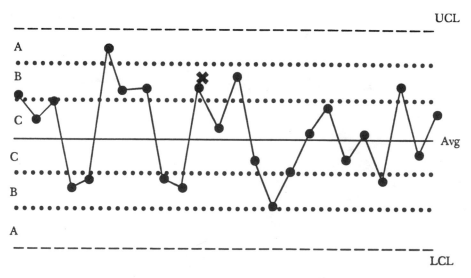

An assignable cause exists if eight or more consecutive points lie on both sides of the average with none of the points in zone C. ✖ = Out-of-control point.

The probability of obtaining this pattern if the process is in control can be determined. The probability of obtaining a point not in zone C is $1 - 0.6826 = 0.3174$. The probability of obtaining eight consecutive points on either side of the average but not in zone C is $0.3174^8 = 0.0001$. The probability of obtaining this pattern from a process in control is small. If this pattern occurs, it is likely that an assignable cause is present. Action should be taken to find the assignable cause and remove it from the process.

RULE-OF-SEVEN TEST

This test is a combination of several tests described above. This test is often taught to front line personnel as the method for interpreting control charts (along with points beyond the limits). The test states that an out- of-control situation is present if one of the following conditions is true:

1. Seven points in a row above the average
2. Seven points in a row below the average
3. Seven points in a row trending up
4. Seven points in a row trending down

These four conditions are shown below.

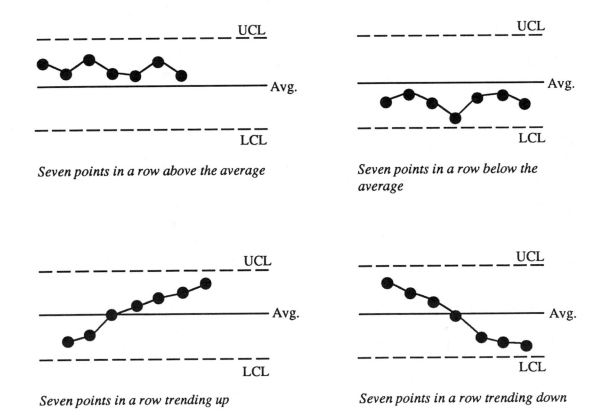

Seven points in a row above the average

Seven points in a row below the average

Seven points in a row trending up

Seven points in a row trending down

SUMMARY

To effectively use control charts, one must be able to interpret the chart. What is this chart telling me about my process? The control chart provides a method of determining if only common cause variation is present or if assignable cause variation is also present. Various tests to determine if the process is out of control were introduced in this chapter. An-out-of control situation is present if one of the following conditions holds:

1. Points beyond the control limits
2. A run length is too large
3. Too many or too few runs
4. 2 out of 3 consecutive points are in zone A or beyond
5. 4 out of 5 consecutive points are in zone B or beyond
6. 8 consecutive points lie in zone C or beyond
7. 15 consecutive points lie in zone C
8. Eight consecutive points lie on both sides of the average with none in zone C

A simple test, called the Rule of Seven, was also introduced.

APPLICATIONS OF CONTROL CHART INTERPRETATION

Do you feel that all these tests should be taught to everyone? If yes, why? If no, which tests should be taught and why? Is there an order in which the tests should be taught? Record your responses below.

11
p CHARTS

Many customers today are examining quality from many different aspects. Not only do customers want a product that meets their expectations, but they also want quality in items associated with the product. These items include things such as accurate paperwork that might accompany a shipment, on-time arrival of shipments and prompt answering of the customer's phone calls. How can we monitor these types of situations over time? Attributes chart can be used to monitor variation over time in attributes data. The p chart is useful for determining the variation in the yes/no type of data, e.g., the paperwork is right or it is not. The p chart is introduced in this chapter.

OBJECTIVES

In this chapter you will learn:

      1. What a p chart is
      2. When to use a p chart
      3. How to construct a p chart

The p chart is a type of control chart that looks at variation in attributes data.

INTRODUCTION TO p CHARTS

p charts are used in the second step of the problem solving model. This step involves analyzing how the process is behaving. The p chart will tell us if the process is in statistical control.

A p chart is used to look at variation in the yes/no type of attributes data. It is used to determine the percentage of defective items in a group of items. A product or service is defective if it fails to conform to specifications or a standard in some respect. For example, consider the case of a customer calling the plant to place an order. The customer would probably not like to have the phone ring 10 to 15 times before it is answered. Suppose you have determined that the operational definition for answering the phone in a timely fashion is "to answer the phone on three or fewer rings." Using this definition, you could monitor the percentage of phone calls answered or not answered in a timely fashion. If a phone call is answered on or before the third ring, the item (answering the phone call) is not defective. If the phone call is not answered on or before the third ring, the item is defective.

Figure 1 is an example of a p chart. In this example, the percentage of late shipments received at a customer's plant is being monitored. The operational definition of a late shipment (a defective item) is any shipment that is over 1 hour late arriving at the customer's plant. Twenty shipments are made each week to the customer. The percentage late each week (p) is simply the number of late shipments that week divided by the total number of shipments. The percentage for each week

Figure 1

p CHART: PERCENTAGE OF LATE SHIPMENTS

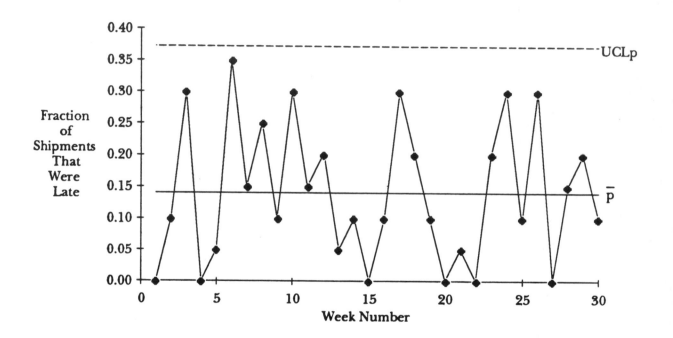

is calculated and plotted. For example, the percentage late in week one was 0; the percentage late in week two was 0.10. Data for the past thirty weeks are shown. The overall average and the control limits have also been calculated and plotted.

The chart in Figure 1 is in statistical control. What does it mean when the p chart is in statistical control? It means that the percentage of deliveries that are late to the customer's plant will remain consistent over time. We can predict what will happen to deliveries in the near future. The average percentage of late shipments per week will be 0.14. Some weeks the percentage may be as high as 0.37. Other weeks it may be as low as 0. Although the p chart is in control, it does not mean that the results are acceptable. Indeed, the goal should be to have no late deliveries. Since the chart is in control, the system must be changed to decrease the percentage of late deliveries.

Operational definitions, as used above, are required to determine if an item is defective. The operational definition must be clear enough so that there is no doubt that an item is or is not defective.

STEPS IN CONSTRUCTION OF A p CHART

The steps in constructing the p chart are given below. A process flow diagram of these steps is shown in Figure 2.

1. Gather the data

    a. Select the subgroup size (n). Attributes data often require large subgroup sizes (50 - 200). The subgroup size should be large enough to have several defective items. If possible, subgroup sizes should be constant. If not, the size should not vary by more than +/- 25%.

    b. Select the frequency with which the data will be collected. Data should be collected in the order in which they are generated.

    c. Select the number of subgroups (k) to be collected before control limits are calculated (at least 25).

    d. Inspect each item in the subgroup and record the item as either defective or non-defective. If an item has several defects, it is still counted as one defective item.

    e. Determine the following for each subgroup:

        n = number of items inspected
        np = number of defective items found

    f. Record the data.

---

g. Calculate the percentage defective (p) for each subgroup and record the results:

$$p = (np)/n$$

2. Plot the data

   a. Select the scales for the control chart.

   b. Plot the values of p for each subgroup on the control chart.

   c. Connect consecutive points with straight lines.

3. Calculate the process average and control limits.

   a. Calculate the process average percentage defective ($\bar{p}$):

   $$\bar{p} = \Sigma np/\Sigma n = (np_1 + np_2 + ... + np_k)/(n_1 + n_2 + ... + n_k)$$

   where np_1, np_2, etc. are the number of defective items in subgroups 1, 2, etc.; n_1, n_2, etc. are the number of items in subgroups 1, 2, etc; and k is the number of subgroups.

   Note: If the subgroup size is constant, $\bar{p}$ can be calculated as:

   $$\bar{p} = \Sigma p/k = (p_1 + p_2 + ... + p_k)/k$$

   where p_1, p_2, etc. are the percentage of defective items in subgroups 1, 2, etc.

   b. Draw the process average percentage defective, $\bar{p}$, on the control chart as a solid line and label.

   c. Calculate the average subgroup size:

   $$\bar{n} = \Sigma n/k = (n_1 + n_2 + ... + n_k)/k$$

   Note: If the subgroup size is constant, $\bar{n} = n$.

   d. Calculate the control limits for the p chart. The upper control limit is given by UCLp. The lower control limit is given by LCLp.

   $$UCLp = \bar{p} + 3\sqrt{\bar{p}(1 - \bar{p})/\bar{n}}$$

   $$LCLp = \bar{p} - 3\sqrt{\bar{p}(1 - \bar{p})/\bar{n}}$$

e. Draw the control limits on the control chart as dashed lines and label.

Note: Theoretically, control limits should be calculated for each different subgroup size. The value of $\bar{n}$ in the control limit equations would be replaced by the value of n for that subgroup. For practical purposes, this is not necessary as long as the subgroup sizes do not vary by more than +/-25%.

4. Interpret the chart for statistical control.

a. The following tests for statistical control are valid:

Points beyond the control limits
Length-of-run test
Number-of-runs test

Figure 2

STEPS IN CONSTRUCTION OF A p CHART

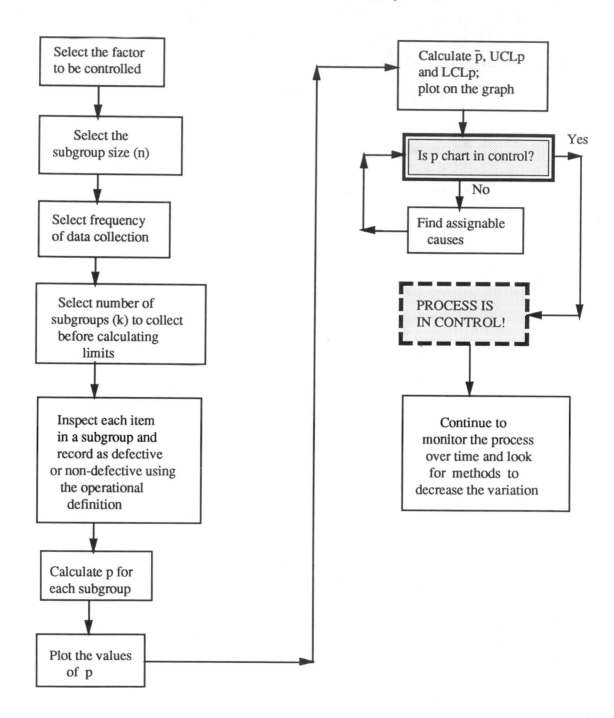

p CHART EXAMPLE

Percentage of Invoices with Errors

A team in a purchasing group has been working on improving the processing of invoices. The group processes 100,000 invoices each year. The cost to process an invoice averages $35 per invoice. The company spends $3.5 million per year just processing invoices. The team is trying to reduce this cost by decreasing the percentage of invoices with errors. The team developed the following operational definition for what constitutes a defective invoice: An invoice is defective if it has incorrect price, incorrect quantity, incorrect coding, incorrect address or incorrect name. The team decided to pull a random sample of 100 invoices per day. The data from the last 25 days are given below.

Day Number	Invoices Inspected (n)	Number Defective (np)	Percentage Defective (p)
1	100	22	0.22
2	100	33	0.33
3	100	24	0.24
4	100	20	0.20
5	100	18	0.18
6	100	24	0.24
7	100	24	0.24
8	100	29	0.29
9	100	18	0.18
10	100	27	0.27
11	100	31	0.31
12	100	26	0.26
13	100	31	0.31
14	100	24	0.24
15	100	22	0.22
16	100	22	
17	100	29	
18	100	31	
19	100	21	
20	100	26	
21	100	24	
22	100	32	
23	100	17	
24	100	25	
25	100	21	

Use these data to construct a p chart. The calculations for the first 15 weeks have been done for you. The first 15 points have also been plotted for you (p. 153). Once you have completed the p chart, answer the following questions.

1. What variation is the p chart examining?

2. Is the process in statistical control? What does this mean?

3. What should be done next to work on improving this process?

4. What other type of statistical tool could be used in conjunction with this p chart and why?

5. What other sampling schemes might be considered instead of sampling 100 invoices at random each day?

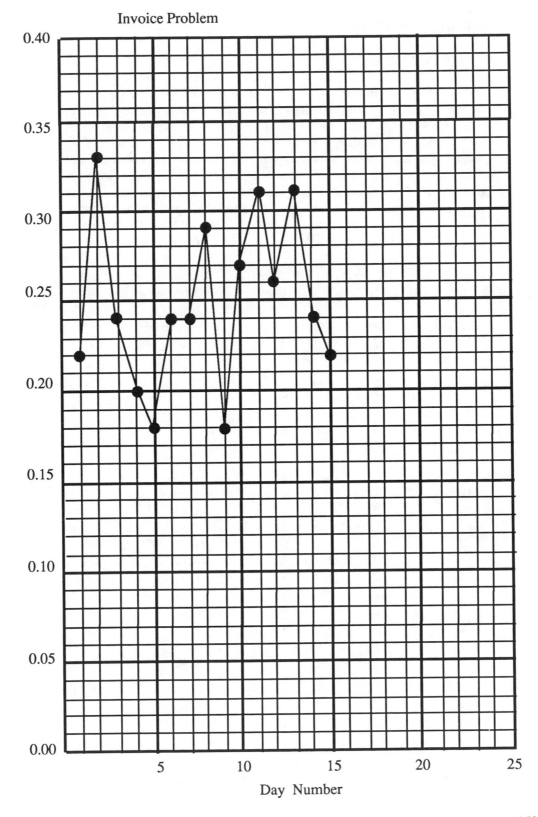

Invoice Problem

p

Day Number

Chapter 11

p CHART EXAMPLE

Product Samples Below Detectable Limits

A manufacturing plant produces a liquid product for a major customer. The customer requires that one component of this product be below 50 ppm. The composition of the liquid is measured using a gas chromatograph. The plant produces the product with the component often being below the detectable limit (<10 ppm) of the chromatograph and always being less than 30 ppm. Three samples are taken each day and analyzed. Operating personnel decided to use a p chart to monitor the percentage of samples above the detectable limit. The objective was to decrease the frequency of occurrence of such samples. The operational definition of a defective sample was a sample that contained above 10 ppm (the detectable limit) of the component. The operating personnel decided to form a subgroup based on one week's worth of samples. Data for the past 25 weeks are shown below.

Week Number	Subgroup Size (n)	Samples > 10 ppm (np)	Percentage Defective (p)
1	21	3	0.14
2	21	5	0.24
3	21	2	0.10
4	21	3	0.14
5	21	4	0.19
6	21	2	0.10
7	21	4	0.19
8	21	10	0.48
9	21	8	0.38
10	21	6	0.29
11	21	3	0.14
12	21	7	0.33
13	21	8	0.38
14	21	2	0.10
15	21	7	0.33
16	21	6	0.29
17	21	6	0.29
18	21	9	0.43
19	21	4	0.19
20	21	10	0.48
21	21	3	
22	21	5	
23	21	7	
24	21	4	
25	21	8	

Use these data to construct a p chart. The calculations for the first 20 weeks have been done for you. The first 20 points have also been plotted for you (p. 156). Once you have completed the p chart, answer the following questions.

1. What variation is the p chart examining?

2. Is the process in statistical control? What does this mean?

3. What should be done next to work on improving this process?

4. What other alternatives are available for process improvement in situations like this?

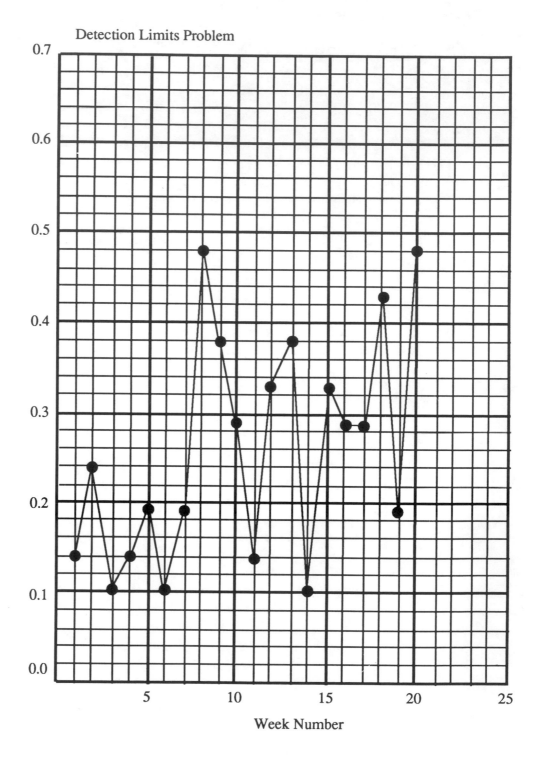

Detection Limits Problem

p CHARTS WITH VARYING SUBGROUP SIZE

In the examples above, the subgroup size used for the p chart was constant ($\bar{n} = n$). At times this may not be possible. This may occur when there are not enough items to form a large subgroup. For example, the number of invoices received may vary from only 10 to 30 a day. In this case, it makes more sense to use all the invoices received instead of taking a random sample. The only other alternative is to wait several days to get enough invoices to form a larger subgroup. The problem with this approach is that it takes too long to obtain a point to plot on the p chart. The longer the time period between points, the harder it is to find reasons for assignable causes.

In the steps of construction of a p chart, it was said that subgroup sizes should not vary by more than +/- 25%. The average subgroup size, $\bar{n}$, was used in calculating the control limits. Theoretically, control limits should be recalculated each time n changes. As n becomes larger, the control limits become tighter (closer to the center line). To avoid calculating control limits for each different n, $\bar{n}$ was used. However, points near the control limits need to examined closely in this situation. It may be necessary to recalculate the limits for these points.

Figure 3 illustrates the two situations. In Figure 3a, $n > \bar{n}$. This means that the control limits are tighter for n than for $\bar{n}$. Points beyond the control limits based on n will be out of control for n since the control limits would just become tighter. The points just inside the control limits may or may not be out of control. The control limits should be recalculated for these points. In Figure 3b, $n < \bar{n}$. The situation is just the opposite of Figure 3a. The control limits are wider for n than for $\bar{n}$. Points inside the control limits will be in control. Points outside the control limits may or may not be out of control. The control limits should be recalculated for these points.

Figure 3

RECALCULATION OF CONTROL LIMITS FOR p CHARTS WITH VARYING
SUBGROUP SIZES

For $n > \bar{n}$:

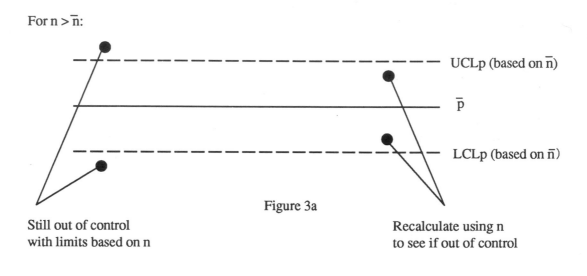

UCLp (based on $\bar{n}$)

$\bar{p}$

LCLp (based on $\bar{n}$)

Figure 3a

Still out of control
with limits based on n

Recalculate using n
to see if out of control

For $n < \bar{n}$:

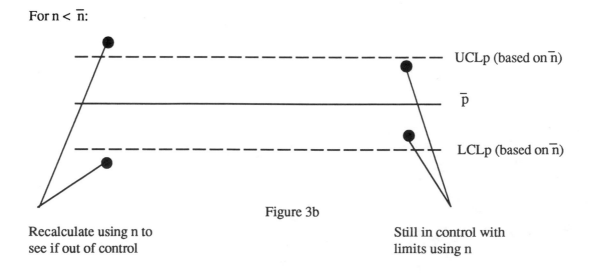

UCLp (based on $\bar{n}$)

$\bar{p}$

LCLp (based on $\bar{n}$)

Figure 3b

Recalculate using n to
see if out of control

Still in control with
limits using n

p CHART EXAMPLE

Expense Accounts

A company was interested in improving the accuracy of expense accounts employees turned into Accounts Payable for reimbursement. When mistakes were made on an expense, it had to be returned to the employee for correction. A team decided to determine how frequently this was occurring and to determine if the problem was a system problem. For an operational definition, the team decided that a defective expense account would be one with one or more of the following: incorrect addition or subtraction, incorrect coding, or missing receipts. The team decided to use all expense accounts received each day. The number received each day varied. Data from the last 25 days are given below.

Day Number	Expense Accounts (n)	Number Defective (np)	Percentage Defective (p)	UCLp	LCLp
1	25	5	0.20	0.528	-
2	34	7	0.21	0.491	0.037
3	56	14	0.25	0.441	0.087
4	43	7	0.16	0.466	0.062
5	36	13	0.36	0.484	0.044
6	42	17	0.40	0.468	0.060
7	21	3	0.14	0.553	-
8	24	8	0.33	0.534	-
9	36	5	0.14	0.484	0.044
10	29	9	0.31	0.510	0.018
11	41	15	0.37	0.471	0.057
12	35	12	0.34	0.488	0.040
13	34	17	0.50	0.491	0.037
14	37	6	0.16	0.481	0.047
15	41	9	0.22	0.471	0.057
16	42	17	0.40	0.468	0.060
17	43	5	0.12	0.466	0.062
18	23	6	0.26	0.540	-
19	38	16	0.42	0.479	0.049
20	47	14	0.30	0.457	0.071
21	30	7	0.23	0.505	0.023
22	42	3	0.07	0.468	0.060
23	33	14	0.42		
24	31	3	0.10		
25	35	5	0.14		

Use the data above to construct a p chart. Since the subgroup size varies so greatly, control limits should be calculated for each different subgroup size. Most of the calculations have been done for you. The points have also been plotted for you (p. 161). Determine control limits for the last three subgroups based on n for that subgroup. Determine control limits based on $\bar{n}$ and plot

159

those limits on the p chart. When finished, answer the following questions.

1. What variation is the chart examining?

2. Which points should be examined further on the chart by recalculating control limits?

3. Is the process in statistical control? What does this mean?

4. What should be done next to improve the process?

5. What different sampling scheme might you recommend for this process and why?

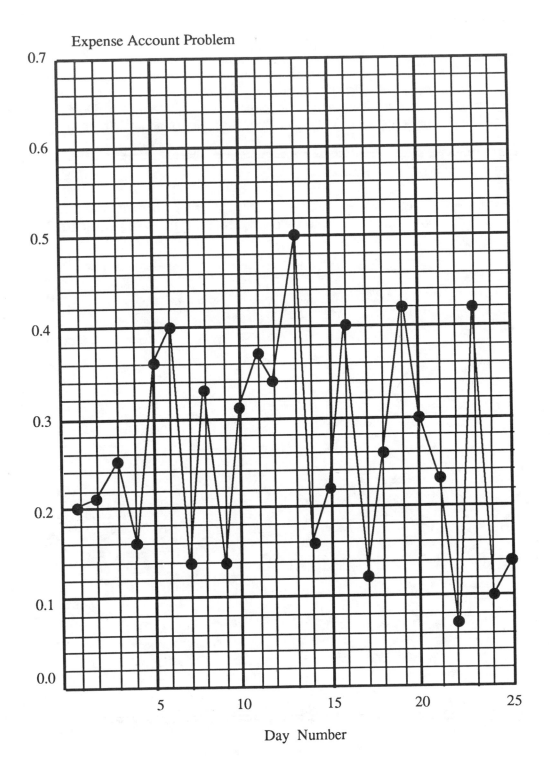

Expense Account Problem

Day Number

SUMMARY

p charts are control charts used with the yes/no type of attributes data. They are used to determine if a process is in statistical control. p charts look at the variation in the percentage of defective items in a subgroup. Operational definitions are required to determine what makes an item defective. Subgroup sizes do not have to be constant to construct a p chart. Depending on the situation, it may be necessary to recalculate control limits for certain points.

APPLICATIONS OF p CHARTS

Think about the processes at work. What are possible applications for p charts. Write down three possible applications and describe what the objective of using the p chart would be.

1.

2.

3.

12

np CHARTS

Customers continue to demand that suppliers track various quality parameters such as on-time arrival and accuracy of paperwork. Many of these situations require the use of attributes data --particularly the yes/no type. p and np charts are used to monitor the variation in yes/no data. The p chart can always be used with yes/no data. The np chart can be used in special situations with yes/no data. The np chart is introduced in this chapter.

OBJECTIVES

In this chapter you will learn:

1. What the np chart is
2. When to use the np chart
3. How to construct the np chart

The np chart is a type of control chart that looks at variation in yes/no data when the subgroup size remains constant.

INTRODUCTION TO np CHARTS

The np chart is used in the second step of the problem solving model. This step involves analyzing how the process is behaving. The np chart will tell us if the process is in statistical control.

The np chart, like the p chart, is used to look at variation in yes/no type attributes data. np charts are used to determine the number of defective items in a group of items. The p chart looked at the fraction of defective items in a group of items. The major difference between the np and p chart is that the subgroup size has to be constant for the np chart. This is not true for the p chart.

The np chart monitors the number of defective items over time. Operational definitions must be used in conjunction with np charts to determine what constitutes a defective item. A defective item is one that fails to conform to specifications or to a standard in some respect.

Figure 1 is an example of the np chart. In this example, the number of late shipments received at a customer's plant is being monitored. The operational definition of a late shipment (a defective item) is any shipment that is over 1 hour late arriving at the customer's plant. Twenty shipments are made each week to the customer. The number of late shipments each week is plotted on the np chart. For example, in week one there were no late shipments; in week two there were 2 late shipments. The overall average and control limits based on 30 weeks of data have been calculated and plotted in Figure 1.

Figure 1

np CHART: NUMBER OF LATE SHIPMENTS

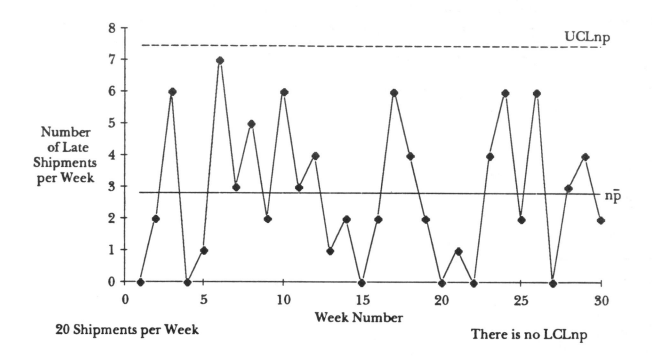

20 Shipments per Week

There is no LCLnp

The chart in Figure 1 is in statistical control. What does it mean when the np chart is in statistical control? It means that the variation in the number of late shipments each week will remain consistent over time. The average number of late shipments will be about 3 per week. Some weeks it may be as high as 7 or as low as 0. The only way to improve this process (decrease the number of late shipments) is to change the process.

The data used in Figure 1 are the same data as used to construct the p chart on late shipments in the previous chapter. Any data that can be plotted on the np chart can be plotted on the p chart. Subgroup sizes must be constant when using the np chart. Thus, the np chart provides the same information as using a p chart. The choice of which chart to use when the subgroup size is constant is a matter of preference. The choice depends on which allows you to communicate best.

STEPS IN CONSTRUCTION OF AN np CHART

The steps in constructing the np chart are given below. A process flow diagram of these steps is shown in Figure 2.

1. Gather the data

    a. Select the subgroup size (n). Attributes data often require large subgroup sizes (50 - 200). The subgroup size should be large enough to have several defective items. The subgroup size must be constant.

    b. Select the frequency with which the data will be collected. Data should be collected in the order in which it is generated.

    c. Select the number of subgroups (k) to be collected before control limits are calculated (at least 25).

    d. Inspect each item in the subgroup and record the item as either defective or non-defective. If an item has several defects, it is still counted as one defective item.

    e. Determine np for each subgroup:

    np = number of defective items found

    f. Record the data.

2. Plot the data

    a. Select the scales for the control chart.

    b. Plot the values of np for each subgroup on the control chart.

    c. Connect consecutive points with straight lines.

3. Calculate the process average and control limits.

    a. Calculate the process average number defective ($\bar{np}$):

$$\bar{np} = Snp/k = (np_1 + np_2 + ... + np_k)/k$$

    where np_1, np_2, etc. are the number of defective items in subgroups 1, 2, etc. and k is the number of subgroups.

    b. Draw the process average number defective, $\bar{np}$, on the control chart as a solid line and label.

    c. Calculate the control limits for the np chart. The upper control limit is given by UCLnp. The lower control limit is given by LCLnp.

$$UCLnp = \bar{np} + 3\sqrt{\bar{np}(1 - (\bar{np}/n))}$$

$$LCLnp = \bar{np} - 3\sqrt{\bar{np}(1 - (\bar{np}/n))}$$

    d. Draw the control limits on the control chart as dashed lines and label.

4. Interpret the chart for statistical control.

    a. The following tests for statistical control are valid:

    Points beyond the control limits
    Length-of-run test
    Number-of-runs test

Figure 2

STEPS IN CONSTRUCTION OF THE np CHART

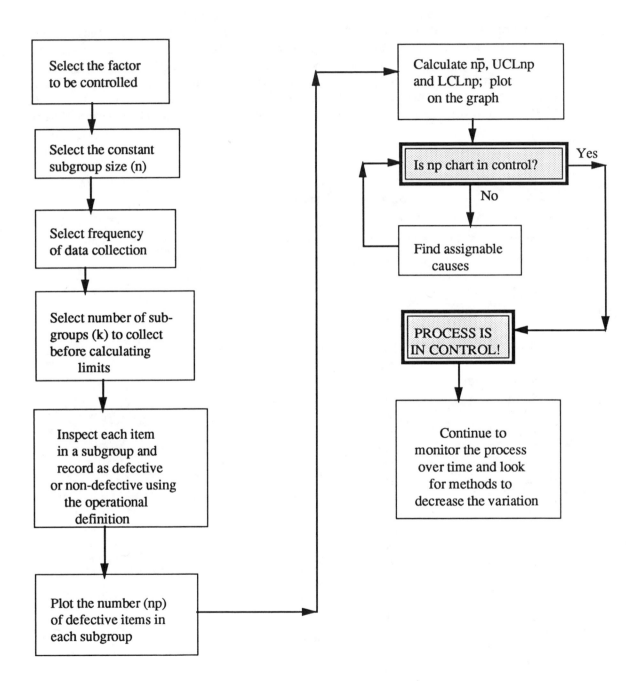

np CHART EXAMPLE

Number of Invoices with Errors

A team in a purchasing group has been working on improving the processing of invoices. The group processes 100,000 invoices each year. The cost to process an invoice averages $35 per invoice. The company spends $3.5 million per year just processing invoices. The team is trying to reduce this cost by decreasing the number of invoices with errors. The team developed the following operational definition for what constitutes a defective invoice: An invoice is defective if it has incorrect price, incorrect quantity, incorrect coding, incorrect address or incorrect name. The team decided to pull a random sample of 100 invoices per day. The data from the last 25 days are given below.

Day Number	Invoices Inspected (n)	Number Defective (np)
1	100	22
2	100	33
3	100	24
4	100	20
5	100	18
6	100	24
7	100	24
8	100	29
9	100	18
10	100	27
11	100	31
12	100	26
13	100	31
14	100	24
15	100	22
16	100	22
17	100	29
18	100	31
19	100	21
20	100	26
21	100	24
22	100	32
23	100	17
24	100	25
25	100	21

Use these data to construct an np chart. The first 15 points have been plotted for you (p. 170). Once you have completed the np chart, answer the following questions.

1. What variation is the np chart examining?

2. Is the process in statistical control? What does this mean?

3. What should be done next to work on improving this process?

4. What other type of statistical tool could be used in conjunction with this np chart and why?

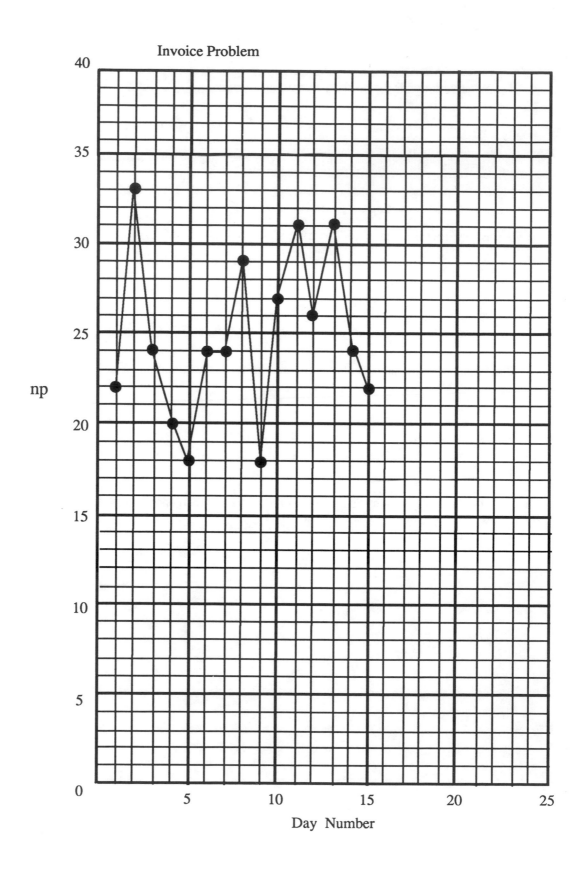

Invoice Problem

np

Day Number

np CHART EXAMPLE

Number of Defective Calculators

A company has been involved in training its employees in how to use statistical techniques. Since these statistical techniques often include calculations, the company purchased a calculator for each course participant. The supplier of the calculator indicated that less than 1% of the calculators would be defective. The course instructor decided to monitor the defective calculators. Since each course had 25 participants, the instructor always ordered 25 calculators. She decided to use an np chart to monitor the variation in the number of defective calculators. The operational definition for a defective calculator was any calculator that would not turn on or one with any key that did not perform the function indicated. Data were collected for 25 courses. The data are given below.

Course Number	Number Inspected (n)	Number Defective (np)
1	25	0
2	25	2
3	25	1
4	25	4
5	25	0
6	25	0
7	25	8
8	25	2
9	25	1
10	25	2
11	25	0
12	25	7
13	25	2
14	25	2
15	25	0
16	25	0
17	25	2
18	25	0
19	25	0
20	25	1
21	25	0
22	25	6
23	25	2
24	25	3
25	25	1

Use these data to construct an np chart. The first 20 points have been plotted for you (p. 173). When you have completed the np chart, answer the following questions.

1. What variation is being examined in the np chart?

2. Is the np chart in statistical control? What does this mean?

3. Is the supplier's claim of less than 1% of the calculators being defective valid?

4. What should be done next?

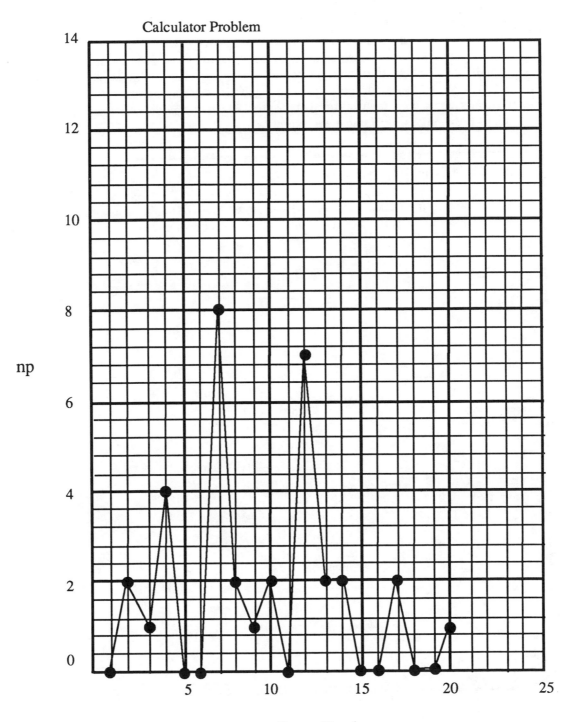

Calculator Problem

np

Course Number

SUMMARY

np charts are control charts that are used with the yes/no type of attributes data when the subgroup size remains constant and the number of defective items in a group of items is more meaningful than the fraction defective. np charts look at the variation in the number of defective items in a group of items. Operational definitions are required to determine what makes an item defective. np charts are used in the second step of the problem solving model (Analyzing the Process) and tell us if the process is in statistical control.

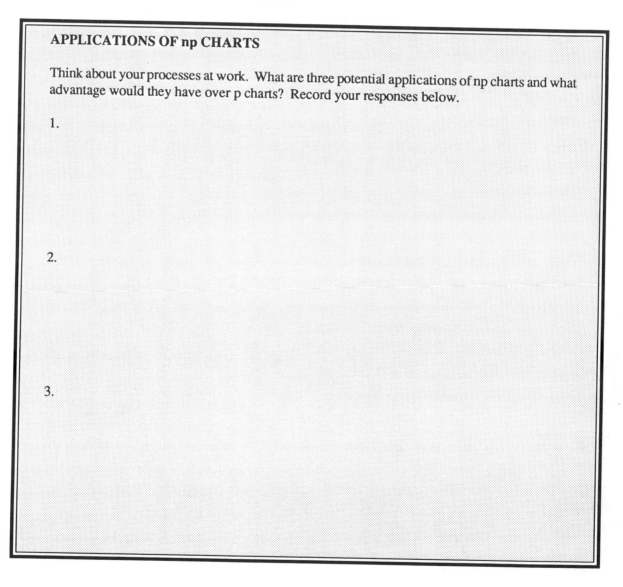

APPLICATIONS OF np CHARTS

Think about your processes at work. What are three potential applications of np charts and what advantage would they have over p charts? Record your responses below.

1.

2.

3.

13
c CHARTS

There are thousands of valves in this plant. Some are going to leak over time. How can you monitor the number of valves that fail over time? A customer uses one product to make a plastic sheet. The sheet cannot have too many surface imperfections. How can the customer monitor the number of surface imperfections? Both these situations are examining the counting type of attributes data. Each "count" is considered a defect. For example, each valve that leaks or each surface imperfection is a defect. "c" represents the number of defects. The c chart is one technique to use for examining variation in counting-type attributes data over time. The c chart is introduced in this chapter.

OBJECTIVES

In this chapter you will learn:

      1. What a c chart is
      2. When to use a c chart
      3. How to construct a chart

The c chart is a type of control chart that looks at variation in counting-type attributes data.

INTRODUCTION TO c CHARTS

c charts are used in the second step of the problem solving model. This step involves analyzing how the process is behaving. The c chart will tell us if the process is in statistical control.

A c chart is used to look at variation in counting-type attributes data. It is used to determine the variation in the number of defects in a constant subgroup size. Subgroup size refers to the area being examined. For example, a c chart can be used to monitor the number of injuries in a plant. In this case, the plant is the subgroup. Since the plant doesn't change size, it is a subgroup of constant size. A c chart can be used to monitor the number of surface imperfections on a plastic sheet. The area examined must be the same size each time (e.g., 10 square inches).

To use the c chart, the opportunities for defects to occur in the subgroup must be very large, but the number that actually occur must be small. For example, the opportunity for injuries to occur in a plant is very large but the number that actually occurs is small.

The area examined can be one inspection unit. For example, you may be interested in determining the number of errors per invoice. In this case, the invoice represents the subgroup. Each error on the invoice represents a defect.

The fact that there are defects present does not mean that the item is defective. For example, a plastic sheet can contain some surface imperfections without being defective. "Defective" implies that the product or service is not usable.

Operational definitions must be used to determine what constitutes a defect. A subgroup can contain different types of defects. On a plastic sheet, surface imperfections may be air bubbles or streaks. An operational definition should exist for each type of defect.

Figure 1 is an example of a c chart. In this example, the variation in the number of valve failures per week is being monitored. The subgroup size in this case is the manufacturing plant where the valves are located. The operational definition for a valve failure is any valve that leaks. The opportunity for defects (valve failures) to occur is very large since there are many valves in the plant. However, the number of defects (valve failures) that actually occur is small. A c chart is appropriate in this situation. To consider the number of valve failures per week, a time frame of one week was selected. However, the c chart could be used for valve failures per day, per month, etc.

The number of valve failures per week was plotted for 25 consecutive weeks. For example, in week one there were 21 valves that failed; for week two there were 34 valves that failed. The overall average and control limits have been calculated and plotted.

The chart in Figure 1 is in statistical control. What does it mean when the c chart is in statistical control? It means that the number of valve failures per week will remain consistent in the near future. The average number per week will be about 26. Some weeks it may be as high as 41 or as low as 11. Suppose the objective is to decrease the number of valve failures per week. Since the process is in control, the system must be changed to decrease the number.

STEPS IN CONSTRUCTION OF A c CHART

The steps in constructing a c chart are given below. A process flow diagram of these steps is given in Figure 2.

1. Gather the data.

    a. Select the subgroup size. The subgroup size is the area where defects have the opportunity to occur. It must be constant from subgroup to subgroup.

    b. Select the frequency with which the data will be collected. Data should be collected in the order in which they are generated.

    c. Select the number of subgroups (k) to be collected before control limits will be calculated (at least 25).

    d. Count the number of defects (c) in each subgroup.

    e. Record the data.

2. Plot the data.

    a. Select the scales for the control chart.

    b. Plot the values of c for each subgroup on the control chart.

    c. Connect consecutive points with straight lines.

3. Calculate the process average and control limits.

    a. Calculate the process average number of defects ($\bar{c}$):

    $$\bar{c} = Sc/k = (c_1 + c_2 + ... + c_k)/k$$

    where c_1, c_2, etc. are the number of defects in subgroups 1, 2, etc. and k is the number of subgroups.

    b. Draw the process average number of defects, $\bar{c}$, on the control chart as a solid line and label.

    c. Calculate the control limits for the c chart. The upper control limit is given by UCLc. The lower control limit is given by LCLc.

    $$UCLc = \bar{c} + 3\sqrt{\bar{c}}$$

    $$LCLc = \bar{c} - 3\sqrt{\bar{c}}$$

    d. Draw the control limits on the control chart as dashed lines and label.

4. Interpret the chart for statistical control.

    a. The following tests for statistical control are valid:

    Points beyond the control limits
    Length-of-run test
    Number-of-runs test

Figure 1

c CHART: NUMBER OF VALVE FAILURES PER WEEK

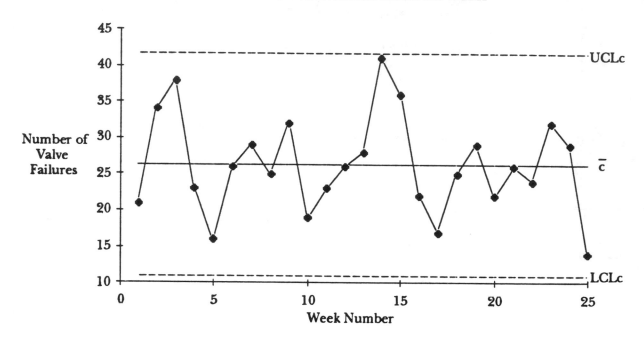

Figure 2

STEPS IN CONSTRUCTION OF A c CHART

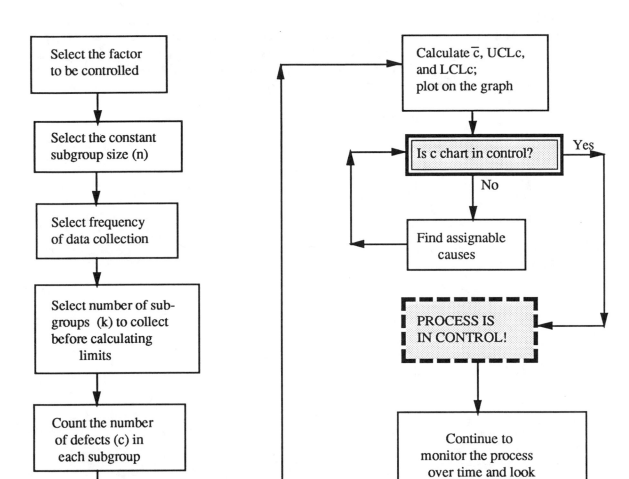

Chapter 13

c CHART EXAMPLE

Plant Safety

A plant manager was interested in determining if a new safety program had helped reduce injuries in the plant. The new safety program had been in effect for four months. The plant manager decided to look at data for the past two years. She elected to use a c chart. The operational definition for a plant injury was any injury that required first aid treatment or more. The data from the past 24 months are given below.

Month Number	Number of Injuries
1	5
2	3
3	2
4	4
5	0
6	3
7	4
8	1
9	4
10	8
11	3
12	1
13	0
14	4
15	6
16	2
17	4
18	5
19	3
20	6
21	2
22	3
23	1
24	5

Use the above data to construct a c chart. The first 15 points have been plotted for you (p. 182). Once you have completed the c chart, answer the following questions.

1. Is the above situation applicable to a c chart? Why?

2. What variation is the c chart examining?

3. Is the process in control? What does this mean?

4. Has the plant manager's new safety program had any impact? Why?

5. What other statistical tool could be used in conjunction with this c chart and why?

6. What should be done next to improve the process?

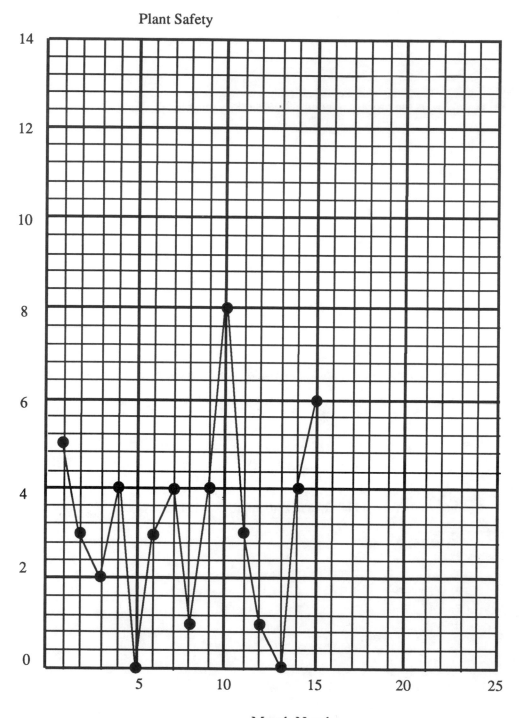

Plant Safety

Month Number

c CHART EXAMPLE

Emergency Jobs in Maintenance

A maintenance team in a large manufacturing facility was working on planning its work in a more organized and efficient manner. One barrier to planning its work was the number of emergencies work orders that came in each day. The team decided to determine what the variation was in the number of emergency jobs each day. The operational definition for emergency job was any work order that came in marked with an "E" for emergency. They decided to use a c chart to examine the data. Data from the last 25 days were obtained from records. These data are given below.

Day Number	Number of "E" Jobs
1	10
2	14
3	17
4	12
5	20
6	8
7	6
8	16
9	17
10	13
11	11
12	15
13	17
14	15
15	9
16	13
17	17
18	15
19	20
20	16
21	17
22	17
23	18
24	18
25	15

Use these data to construct a c chart. The first 20 points have been plotted for you (p. 185). Once you have completed the c chart, answer the following questions.

1. Why is the c chart applicable in this situation?

2. What variation is the c chart examining?

3. Is the process in statistical control? What does this mean?

4. What should be done next?

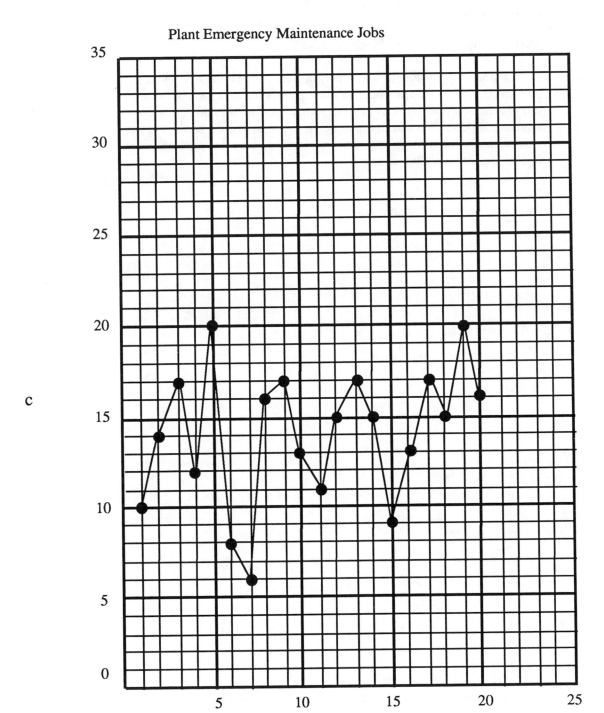

Plant Emergency Maintenance Jobs

c

Day Number

Chapter 13

c CHART EXAMPLE

Surface Imperfections on Plastic Bottles

A manufacturer produces plastic bottles that hold various beverages. Too many surface imperfection makes the product contained in the bottle look less appealing. The specification for this process was no more than 14 surface imperfections per bottle. The operational definition for a surface imperfection was any "pimple" that appeared on the bottle. For each sample, one bottle was examined out of each 100 bottles made. The bottle was randomly selected. Data from the last 25 samples are given below.

Sample Number	Number of Imperfections
1	5
2	3
3	8
4	6
5	9
6	7
7	4
8	14
9	19
10	3
11	4
12	8
13	10
14	3
15	5
16	7
17	2
18	6
19	4
20	8
21	3
22	3
23	6
24	3
25	6

Use the data above to construct a c chart. The first 20 points have been plotted for you (p. 188). Once you have completed the chart, answer the following questions.

1. Is the c chart applicable in this situation and why?

2. What variation is the c chart examining?

3. Is the process in statistical control? What does this mean?

4. Is the process meeting the specification of no more than 14 surface imperfections per bottle?

5. What should be done next to improve the process?

c

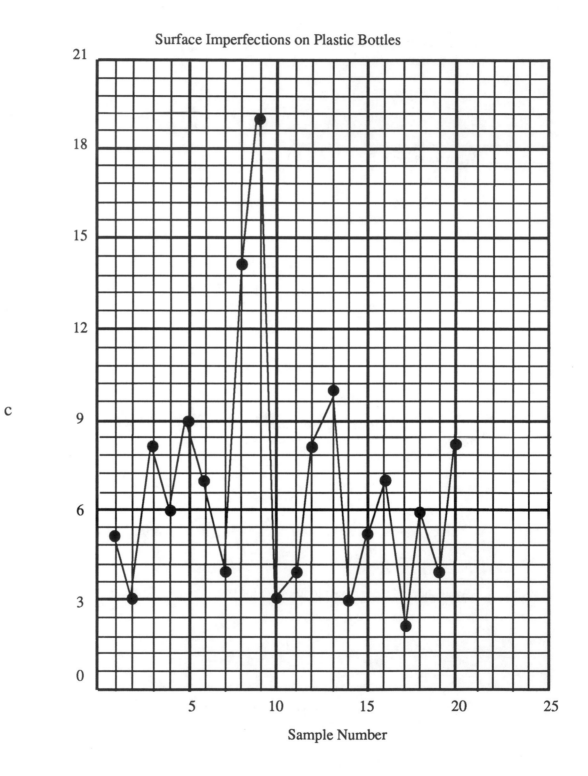

SUMMARY

c charts are control charts that look at variation in counting-type attributes data when the area of opportunity for defects remains constant. The c chart examines the variation in the number of defects in this area. Operational definitions are required to determine exactly what constitutes a defect in the subgroup area. c charts are used in the second step of the problem solving model (Analyzing the Process) and tell us if the process is in statistical control.

APPLICATIONS OF c CHARTS

Think about your processes at work. What are three potential applications of c charts? Record your responses below.

1.

2.

3.

14
u CHARTS

Sometimes defects are seldom found in one item. For example, suppose you are monitoring the number of errors or defects on Certificates of Analysis sent to customers. Errors occur very rarely. There may be only 1 error per 100 certificates sent. This gives an average of 0.01 errors per certificate. This number does not have a lot of meaning to most people. To make the reporting more meaningful, it helps to look at the results on a basis of errors per 10 certificates or errors per 50 certificates. The u chart is one method of doing this. u is the number of defects per inspection unit. Like the c chart, the u chart monitors the variation in counting-type attributes data. The u chart can be used at all times with counting-type attributes data. The u chart is introduced in this chapter.

OBJECTIVES

In this chapter you will learn:

1. What a u chart is
2. When to use a u chart
3. How to construct a u chart

The u chart is a type of control chart that looks at variation in counting-type attributes data.

INTRODUCTION TO u CHARTS

u charts are used in the second step of the problem solving model. This step involves analyzing how the process is behaving. The u chart will tell us if the process is in statistical control.

A u chart is used to examine the variation in counting-type attributes data. It is used to determine the variation in the number of defects in a subgroup. The subgroup refers to the area being examined and does not have to be constant. The subgroup size, n, for a u chart is expressed in terms of the number of inspection units. An inspection unit can be one item, such as one Certificate of Analysis, or it can be multiple items, such as 10 Certificates of Analysis. If the inspection unit is 10 Certificates of Analysis, the subgroup size, n, indicates how many tens were inspected. For example, if 42 Certificates of Analysis were inspected, the subgroup size, n, would be 4.2 inspection units.

The u chart is very similar to the c chart. The only difference is that the subgroup size for the c chart must be constant. The subgroup size for the u chart does not have to be constant. To use the u chart, the opportunities for defects to occur in the subgroup must be very large, but the number that actually occur must be small.

Operational definitions must be used to determine what constitutes a defect. A subgroup can contain different types of defects. An operational definition should exist for each type of defect.

Figure 1 is an example of a u chart. In this example, the variation in the number of pump defects per inspection unit is being charted. A large manufacturing facility regularly examines pumps as part of its preventative maintenance program. A record is kept of the number of pumps inspected and the number of defects found. The operational definition for a pump defect is one of the following conditions: low oil level in reservoir, any leakage, case vibration above a given value or parts worn beyond a given tolerance. The number of defects found on each pump was usually zero. To make the results easier to interpret, it was decided to monitor the number of defects per 10 pumps. A different number of pumps were inspected each week. Since the subgroup size is not constant, the u chart is appropriate.

The number of pump defects per inspection unit is plotted for 25 consecutive weeks. In week one, there were 0.44 pump defects per inspection unit. In week two, there were 0.63 pump defects per inspection unit. The overall average has been plotted. The control limits (based on the average subgroup size) have also been plotted.

The chart in Figure 1 is in statistical control. What does it mean when the u chart is in statistical control? It means that the average number of defects per inspection unit will remain consistent in the near future. The average number will be about 0.54 defects per inspection unit (10 pumps). Some weeks it may be as high as 1.51 or as low as 0 defects per inspection unit. Since the process is in control, it is predictable. The only way to improve the process is to change the system.

In Figure 1, the inspection unit used was 10 pumps. The inspection unit is arbitrary. It can be selected to be number of pumps (e.g., 20 pumps, 30 pumps, 5 pumps, etc.) The key to selecting the size of an inspection unit is using whatever will help get the information into a meaningful form.

STEPS IN CONSTRUCTION OF A u CHART

The steps in construction of the u chart are given below. A process flow diagram of these steps is given in Figure 2.

1. Gather the data.

    a. Select the size of one inspection unit (IU). The inspection unit should be large enough so defects have an opportunity to occur. An inspection can be one item (such as 1 invoice) or area (such as 1 square foot). It can be multiple items (such as 100 invoices) or larger areas (such as 10 square feet).

    b. Select the frequency with which to collect data. Data should be collected in the order in which they are generated.

    c. Select the number of subgroups (k) to be collected before control limits will be calculated (at least 25).

    d.   The number of samples in a subgroup does not have to be constant. However, the number of samples should not very by more than +/- 25% or the calculation of limits becomes more difficult.

    e.   For each subgroup, determine the following:

c = number of defects found in the subgroup
n = subgroup size = Number of samples/IU

    f.   For each subgroup, calculate the number of defects per inspection unit (u):

$$u = c/n$$

    g.   Record the data.

2.   Plot the data.

    a.   Select the scales for the control chart.

    b.   Plot the values of u for each subgroup on the control chart.

    c.   Connect consecutive points with straight lines.

3.   Calculate the process average and control limits.

    a.   Calculate the process average number of defects per inspection unit ($\bar{u}$):

$$\bar{u} = \Sigma c/\Sigma n = (c_1 + c_2 + ... + c_k)/(n_1 + n_2 + ... + n_k)$$

where c_1, c_2, etc. are the number of defects in subgroups 1, 2, etc.; n_1, n_2, etc. are the number of inspection units in subgroups 1, 2, etc.; and k is the number of subgroups.

    b.   Draw the process average number of defects per inspection unit ($\bar{u}$) on the chart as a solid line and label.

    c.   Calculate the average subgroup size:

$$\bar{n} = \Sigma n/k = (n_1 + n_2 + ... + n_k)/k$$

d. Calculate the control limits for the u chart. The upper control limit is given by UCLu. The lower control limit is given by LCLu.

$$UCLu = \bar{u} + 3\sqrt{(\bar{u}/\bar{n})}$$

$$LCLu = \bar{u} - 3\sqrt{(\bar{u}/\bar{n})}$$

e. Draw the control limits on the control chart as dashed lines and label.

Note: Theoretically, control limits should be calculated for each different subgroup size. The value of $\bar{n}$ in the control limit equations would be replaced by the value of n for that subgroup. For practical purposes, this is not necessary as long as the subgroup sizes do vary by more than +/-25%.

4. Interpret the chart for statistical control.

a. The following tests for statistical control are valid:

Points beyond the control limits
Length-of-run test
Number-of-runs test

Figure 1

u CHART: NUMBER OF PUMP DEFECTS PER INSPECTION UNIT

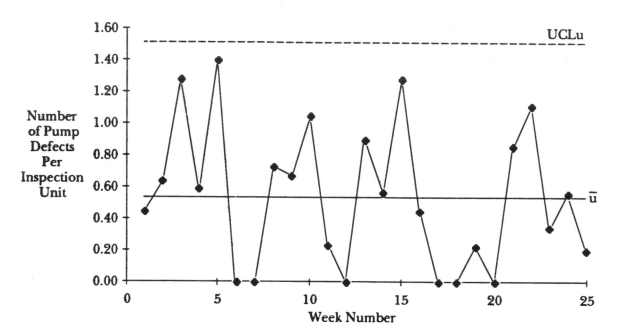

One inspection unit = 10 pumps. Control limits are based on average subgroup size (n). There is no LCLu.

Figure 2

STEPS IN CONSTRUCTION OF THE u CHART

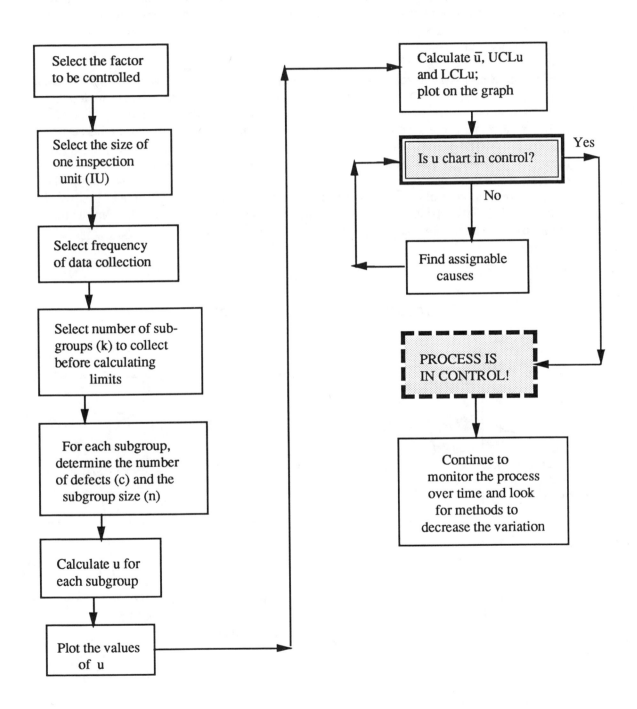

u CHARTS WITH VARYING SUBGROUP SIZE

In Figure 1, the average subgroup size ($\bar{n}$) was used to calculate the control limits. This is adequate as long as the control limits do not vary by more than +/- 25%. Theoretically, control limits should be recalculated each time the subgroup size changes. This has been done in Figure 3 for the pump defect data used in Figure 1.

It may not be necessary to recalculate control limits for each different subgroup size. As n becomes larger, the control limits become tighter. Depending on where the points are located, it may or may not be necessary to recalculate the limits. Figure 4 illustrates the two situations. In Figure 4a, $n > \bar{n}$. These means that the control limits are tighter for n than for n. Points beyond the control limits based on $\bar{n}$ will be out of control for n since the control limits would just become tighter. The points just inside the control limits may or may not be out of control. The control limits should be recalculated for these points. In Figure 4b, $n < \bar{n}$. The situation is just the opposite of Figure 4b. The control limits are wider for n than for $\bar{n}$. Points inside the control limits will be in control. Points outside the control limits may or may not be out control. The control limits should be recalculated for these points.

Figure 3

u CHART WITH VARYING SUBGROUP SIZE

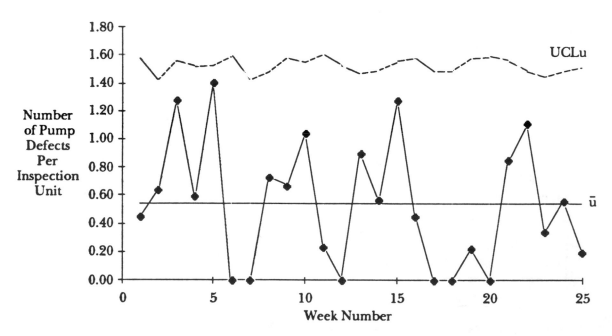

One inspection unit = 10 pumps. Control limits are based on actual subgroup size. There were no LCLu.

Figure 4

RECALCULATION OF CONTROL LIMITS FOR u CHARTS WITH VARYING
SUBGROUP SIZES

For n > n̄:

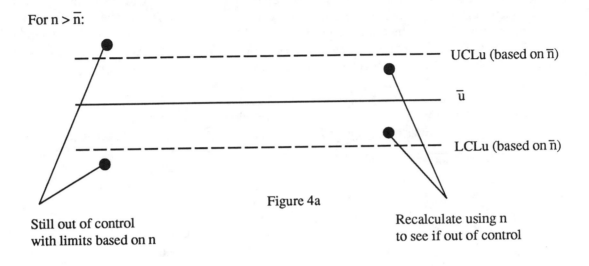

UCLu (based on n̄)

ū

LCLu (based on n̄)

Figure 4a

Still out of control
with limits based on n

Recalculate using n
to see if out of control

For n < n̄:

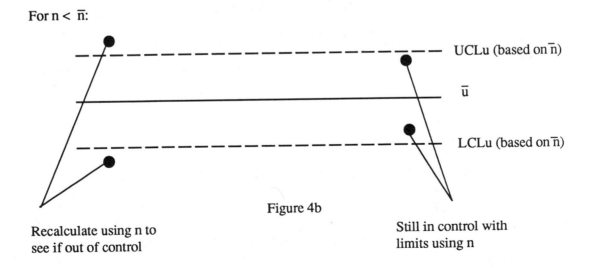

UCLu (based on n̄)

ū

LCLu (based on n̄)

Figure 4b

Recalculate using n to
see if out of control

Still in control with
limits using n

u CHART EXAMPLE

Number of Revisions to Engineering Drawings

The Process Engineering Department of a company was trying to decrease the number of revisions made to P&I drawings for new processes. Continual changes in these drawings cause the department to spend over 25% of its time revising the P&I drawings. The department manager decided to monitor the number of revisions to P&I drawings. He decided to use all the drawings completed each week. A system was set up to collect the data. The operational definition for a revision to a P&I diagram was any time an engineer had to make one change in a P&I diagram (after the P&I has been completed by the engineer). He decided that an inspection unit (IU) would be 1 P&I drawing. Data were collected for 25 weeks. The data are given below.

Week Number	Number of P&I Drawings	No. of Inspection Units (n)	No. of Revisions (c)	Revisions per Unit (u)
1	20	20	56	2.8
2	15	15	43	2.9
3	16	16	65	4.1
4	18	18	39	2.2
5	32	32	60	1.9
6	27	27	55	2.0
7	21	21	53	2.5
8	17	17	46	2.7
9	22	22	38	1.7
10	26	26	45	1.7
11	27	27	62	2.3
12	30	30	35	1.2
13	20	20	44	2.2
14	16	16	34	2.1
15	13	13	40	3.1
16	17		65	
17	24		43	
18	26		55	
19	23		57	
20	29		53	
21	25		56	
22	17		43	
23	20		38	
24	24		69	
25	27		44	

Use the data above to construct a u chart. The calculations for the first 15 weeks have been done for you. The points for the first 15 weeks have also been plotted (p. 198). Use control limits based on the average subgroup size. When you have completed the u chart, answer the following questions.

1. What variation is the u chart examining?

2. Is the process in statistical control? What does this mean?

3. Which points should be examined further on the chart by recalculating control limits?

4. What other statistical tool could be used in conjunction with this chart and why?

5. What should be done next to improve the process?

6. Could a c chart have been used instead of a u chart? Why?

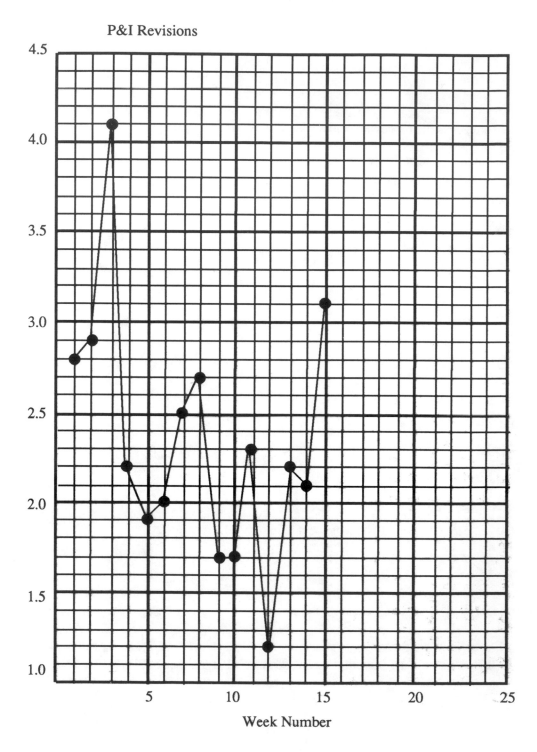

P&I Revisions

u

Week Number

u CHART EXAMPLE

Maintenance Rework Jobs

A team in a maintenance department of a large manufacturing plant had been criticized about the number of jobs that had been completed and then had to be redone for one reason or another. The team decided to determine how to decrease the number of these rework jobs. One of the first steps the team decided to take was to determine if the process was in statistical control. It collected historical data from the past 25 weeks. The team decided to use a u chart with an inspection unit equal to 20 maintenance jobs. A rework job was defined as any maintenance job that required maintenance personnel to return and redo one or more parts of the job after it had been turned over to the operating unit. A defect was the reason the job had to be reworked. There could be more than one reason per job. Data from the past 25 weeks are given below.

Week Number	No. of Maintenance Jobs	No. of Inspection Units (n)	No. of Rework Defects (c)	Defects Per Unit (u)
1	105	5.25	5	0.95
2	95	4.75	7	1.47
3	101	5.05	4	0.79
4	97	4.85	9	1.86
5	95	4.75	7	1.47
6	93	4.65	4	0.86
7	110	5.50	6	1.09
8	108	5.40	7	1.30
9	105	5.25	8	1.52
10	94	4.70	4	0.85
11	99	4.95	7	1.41
12	102	5.10	5	0.98
13	93	4.65	6	1.29
14	96	4.80	5	1.04
15	111	5.55	8	1.44
16	97	4.85	4	0.82
17	91	4.55	4	0.88
18	90	4.50	7	1.56
19	101	5.05	6	1.19
20	103	5.15	9	1.75
21	99		5	
22	98		6	
23	108		4	
24	93		6	
25	96		7	

Use the data above to construct a u chart. The calculations for the first 20 points have been completed for you. The first 20 points have also been plotted (p. 203). Use the average subgroup size to calculate control limits. When you have completed the u chart, answer the following questions.

1. What variation is the u chart examining?

2. Is the process in statistical control? What does this mean?

3. What chart could be used instead of a u chart and why?

4. What should be done next to improve the process?

5. What would you have used for an inspection unit? What effect would this have on the appearance of the control chart?

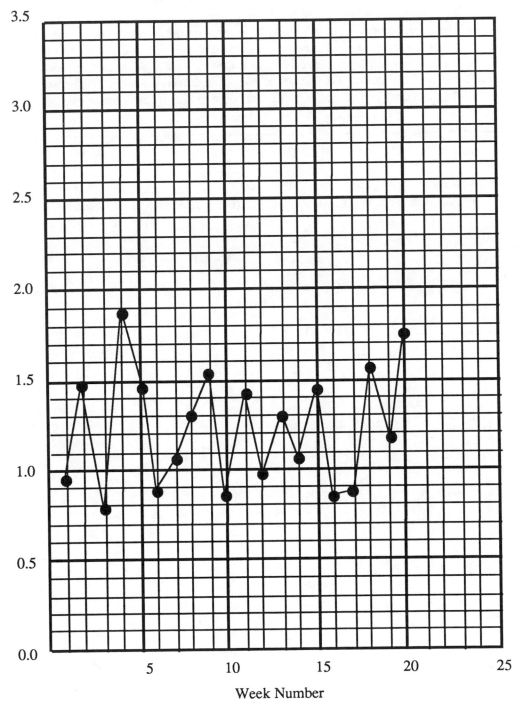

Maintenance Rework Jobs

SUMMARY

u charts are control charts used with counting-type data. They are used to determine if a process is in statistical control. u charts look at the variation in the number of defects in a subgroup. The subgroup size refers to the area being inspected and does not have to be constant. The subgroup size for a u chart is defined as the number of inspection units. An inspection unit can be one item or it can be many items. The size of the inspection unit is set so that the data are meaningful. It may be necessary to recalculate limits from time to time if the subgroup size is not constant.

APPLICATIONS OF u CHARTS

Think about the processes at work. What are possible applications for u charts? Write down three possible applications and describe what the objective of each chart would be.

1.

2.

3.

15
SPECIAL TOPICS IN ATTRIBUTES CONTROL CHARTS

The four types of attributes control charts have been covered in detail. Sometimes it is difficult to determine which type of chart should be used in a given situation. This chapter presents a method of determining which type of chart to select. The control limit equations for the attributes charts are valid only if $n\bar{p}$ or $\bar{c}$ is significantly large. This chapter covers what to do when np or c is small. There will also be times when you may want to change the probability associated with the control limits. This chapter presents a method for changing these probabilities.

OBJECTIVES

In this chapter you will learn:

1. How to select the correct attribute control chart
2. How to handle the situation when $n\bar{p}$ or $\bar{c}$ is small
3. How to determine limits other than three sigma limits

It is important to be able to select the correct type of control chart in a given situation. It is also important to know how to handle the situation when $n\bar{p}$ or $\bar{c}$ is small or to adjust the limits based on probabilities. These situations will occur on occasion. Knowledge of how to handle these situations increases the power and flexibility of the techniques.

SELECTION OF ATTRIBUTES CONTROL CHARTS

The four types of attributes control charts are the p, np, c and u charts. To determine which chart to use in a given situation, the two questions below must be answered.

1. Are the data yes/no or counting type?
2. Is the subgroup size constant?

If the data are of the yes/no type data, then either the p or np chart is used. If the subgroup size is constant, the np chart is used. If the subgroup size is not constant, the p chart is used. Note that there are times when the p chart will be used even if the subgroup size is constant simply because the fraction defective is more meaningful than the number defective. If the data are of the counting type data, either the c or u chart is used. If the subgroup size is constant, the c chart is used. If the subgroup size is not constant, the u chart is used. Figure 1 can be used to determine the type of attribute control chart to use. It is based on the two questions listed above.

Figure 1

SELECTION OF ATTRIBUTES CONTROL CHARTS

Type of Attributes Data

	Yes/No	Counting
Constant Subgroup Size	np Chart or p Chart	c Chart
Changing Subgroup Size	p Chart	u Chart

EXAMPLES OF SELECTING ATTRIBUTES CONTROL CHARTS

What kind of attributes chart would you select to use in each of the situations given below? You may have to make some assumptions about the objective of using a particular attributes chart. Record you answers below.

1. Late shipments each month to a customer's plant

2. Surface imperfections on a extruded sheet

3. Contamination in a powdered product

4. Revisions to P&I diagrams

5. Invoice errors

6. Errors on bills of lading

7. Injuries in a plant

8. Rework jobs in maintenance

9. Overtime in a plant department

10. Errors on certificates of analysis

SMALL-SAMPLE CASE FOR p AND np CHARTS

Suppose you have a process that is in statistical control with an average fraction defective of $\bar{p}$. The control limits for this process are then:

$$UCLp = \bar{p} + 3\sqrt{\bar{p}(1 - \bar{p})/\bar{n}}$$

$$LCLp = \bar{p} - 3\sqrt{\bar{p}(1 - \bar{p})/\bar{n}}$$

Since the process is in control, any p values obtained should fall between the control limits in a random fashion. The chance that p will fall outside the control limits is approximately 3 out of 1,000. These control limits are valid as long as $n\bar{p}$ is sufficiently large. In these cases, the binomial distribution is symmetrical and can be approximated by a normal distribution. The equations provide good estimates of the control limits.

If $n\bar{p}$ is not sufficiently large, the binomial distribution is not symmetrical. In these cases, the control limit equations are no longer valid. $n\bar{p}$ is not sufficiently large if $n\bar{p} < 5$ or if $n(1 - \bar{p}) < 5$. This is referred to as a small- sample case for np and p charts. Figure 2 demonstrates how the shape of the binomial distribution changes as $n\bar{p}$ changes from 0.5 to 5.0. As can be seen, the binomial distribution becomes more symmetrical and approaches the shape of a normal distribution as $n\bar{p}$ becomes larger.

If $n\bar{p} < 5$ or if $n(1 - \bar{p}) < 5$, Appendix Table 12 (p. 497) must be used to determine the control limits. The table gives the upper and lower control limits for various values of $\bar{p}$ from 0.001 to 0.5 and for values of n from 5 to 50. These control limits are exact solutions of the equation governing the binomial distribution, with the assumption that the probability of obtaining a point beyond the control limits is less than or equal to 0.003:

$$P(p \leq LCLp) + P(p \geq UCLp) \leq 0.003$$

The limits given in the table are for np charts for various values of $\bar{p}$ and n. To obtain the limits for a p chart or convert np to p, use the following relationships:

$$UCLp = UCLnp/n$$

$$LCLp = LCLnp/n$$

$$\bar{p} = n\bar{p}/n$$

To understand how to use Appendix Table 12 and how it was developed, consider the following example. Suppose you are sampling 10 items (such as invoices or expense accounts) on a regular basis. The average percentage defective has been determined to be 0.01. Thus:

$$n\bar{p} = (n)(\bar{p}) = (10)(0.01) = 0.1$$

Since $n\bar{p} < 5$, Appendix Table 12 must be used to determine the control limits. If the control limit equations were used, the control limits would be:

$$UCLp = \bar{p} + 3\sqrt{\bar{p}(1 - \bar{p})/n} = 0.01 + 3\sqrt{0.01(1 - 0.01)/10} = 0.10$$

$$LCLp = \bar{p} - 3\sqrt{\bar{p}(1 - \bar{p})/n} \ \ 0.01 - 3\sqrt{0.01(1 - 0.01)/10} = None$$

The control limits from Table 1 for $\bar{p} = 0.01$ and n = 10 are:

$$UCLnp = 3$$

$$LCLnp = None$$

Figure 2

EFFECT OF n$\bar{\text{p}}$ ON THE SHAPE OF THE BINOMIAL DISTRIBUTION

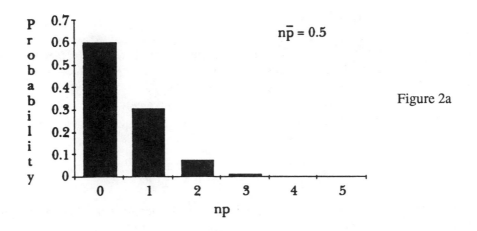

Figure 2a

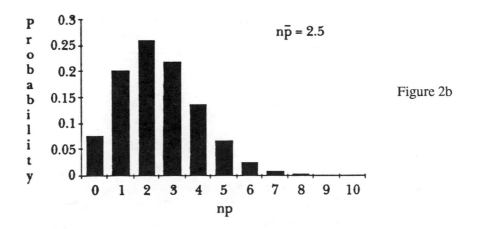

Figure 2b

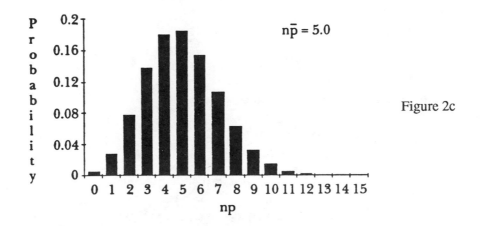

Figure 2c

The control limits are converted from an np chart to a p chart by dividing by n:

$$UCLp = UCLnp/n = 3/10 = 0.3$$

$$LCLp = LCLnp/n = None/10 = None$$

Note the difference between the UCLp calculated using the equations (UCLp = 0.1) and that obtained from the table (UCLp = 0.3). This difference is simply due to the fact that when $n\bar{p} < 5$, the binomial distribution is no longer symmetrical. The control limit equations no longer provide the same probability as when $n\bar{p} > 5$.

The control limits in Appendix Table 12 were obtained from the equation governing the binomial distribution ($n\bar{p}$ is an estimate of np'):

$$P(np|n, p') = [n!/(np!(n - np))!](1 - p')^{n-np}p'^{np}$$

For this example, the probability of finding 0, 1, 2, and 3 defective items in a sample size of 10 can be calculated. The calculation results are summarized below.

Number of Defective Items	Probability of Getting This Number	Cumulative Probability
0	0.904	0.904
1	0.091	0.995
2	0.004	0.999
3	0.0001	0.9991

The probability of the sample containing 0 defective items is 904 out of 1,000. The probability of the sample containing 0 or 1 defective item is 995 out of 1,000. The probability of the sample containing 0, 1, or 2 defective items is 999 out of 1,000. The control limits in Appendix Table 12 are determined so that the probability of obtaining a point beyond the control limits is less than or equal to 0.003 (or 3 out of 1,000). For this case, the probability becomes less than 0.003 when the number of defective items is 3 or more. Thus, the upper control limit for this example is 3 (which agrees with Appendix Table 12).

EXAMPLES OF SMALL-SAMPLE CASE FOR p AND np CHARTS

Determine the control limits for the examples below.

1.   Suppose an np chart has $n\bar{p} = 4$, with n = 20. What are the control limits?

2.   Suppose a p chart has $\bar{p} = 0.14$, with n = 50. What are the control limits?

3.   Suppose a p chart has $\bar{p} = 0.04$, with n = 30. What are the control limits?

4.   Suppose a np chart has $n\bar{p} = 3.5$, with n = 20. What are the control limits?

SMALL-SAMPLE CASE FOR c AND u CHARTS

The c and u charts are based on or approximated by the Poisson distribution. If $\bar{c}$ is sufficiently large, the Poisson distribution is symmetrical and approaches the shape of a normal distribution. In these cases, the equations for the control limits on the c and u charts are valid. However, if $\bar{c}$ is small, the Poisson distribution is not symmetrical and the equations are no longer valid. Figure 3 demonstrates how the shape of the Poisson distribution changes as $\bar{c}$ increases.

If $\bar{c} < 2$, the control limit equations are not valid. In this case, Appendix Table 13 (p. 305) must be used to determine the control limits. Appendix Table 13 lists the cumulative probabilities (multiplied by 1,000) of obtaining up through a certain number of defects for given values of $\bar{c}$. The control limits are determined with the assumption that the probability of obtaining a point beyond the control limit is less than or equal to 0.003:

$$P(c \leq LCLc) + P(c \geq UCLc) \leq 0.003$$

To understand how to use Appendix Table 13, consider the following example. Suppose you are charting the number of injuries in a plant and that $\bar{c} = 0.20$. To determine the control limits, look up $\bar{c} = 0.20$ in Appendix Table 13. Read across that line until the probability is larger than 997. The number of defects for this column plus one is the upper control limit. In this case, the upper control limit is 3.

The probabilities in Appendix Table 13 were obtained using the equation that governs the Poisson distribution ($\bar{c}$ is an estimate of c'):

$$P(c|c') = (c'^c/c!)e^{-c'}$$

Figure 3

EFFECT OF c̄ ON THE SHAPE OF THE POISSON DISTRIBUTION

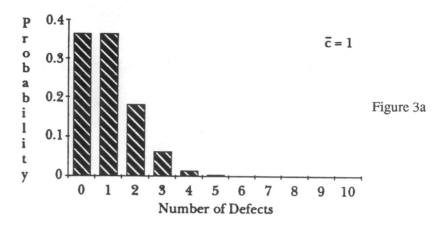

Figure 3a

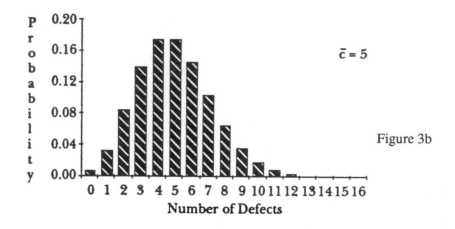

Figure 3b

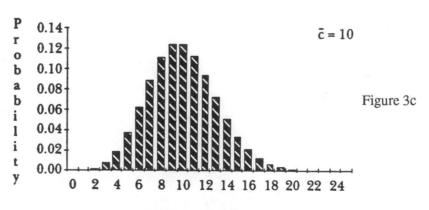

Figure 3c

For this example, the probabilities have been calculated and are shown below.

Number of Defects	Probability	Cumulative Probability
0	0.8187	0.8187
1	0.1637	0.9824
2	0.0163	0.9988
3	0.0011	0.9999

Thus, the probability of obtaining 0 defects is 819 out of 1,000. The probability of obtaining 0 or 1 defect is 982 out of 1,000. The probability of obtaining 0, 1, or 2 defects is 999 out of 1,000. Note that even though this is larger than 997 out of 1,000, the upper control limit is 3. This is because the probability of obtaining 2 defects is 16 out of 1,000. The control limits are set so that the probability of obtaining of point beyond the control is less than or equal to 3 out of 1,000.

Appendix Table 13 is for the c chart. To convert the numbers to a u chart format, use the following relationships:

$$UCLu = UCLc/n$$

$$LCLu = LCLu/n$$

$$\bar{u} = \bar{c}/n$$

where n is the number of inspection units.

EXAMPLES OF SMALL-SAMPLE CASE FOR c AND u CHARTS

Determine the control limits for the examples below.

1. Suppose a c chart has $\bar{c} = 0.45$. What are the control limits?

2. Suppose a u chart has $\bar{u} = 2.1$, with n = 10. What are the control limits?

3. Suppose a u chart has $\bar{u} = 0.05$, with n = 40. What are the control limits?

4. Suppose a c chart has $\bar{c} = 0.95$. What are the control limits?

DETERMINING LIMITS OTHER THAN THREE-SIGMA LIMITS

The three-sigma limits on control charts really represent a probability. The limits are set so that the probability of getting a point beyond the control limits is less than or equal to 3 out of 1,000 if a process is in statistical control. These limits represent a tradeoff between the cost of looking for an assignable when one is not present and the cost of not looking for one when it is present. In some cases, you may be willing to take more of a risk. For example, if you are charting injuries in a plant, you may be more willing to look for assignable causes when they may not be present. You may want to use two-sigma limits. In this case, the probability of obtaining a point beyond the control limits is about 5 in 100. Control limits with lower probabilities will give more false signals.

To determine control limits other than three-sigma limits when $n\bar{p} > 5$ or $\bar{c} > 2$ (i.e., $n\bar{p}$ or $\bar{c}$ is significantly large), simply use the control limit equations with the "3" replaced by the number of sigma limits you desire. For example, the p chart control chart equations become:

$$UCLp = \bar{p} + a\sqrt{\bar{p}(1 - \bar{p})/\bar{n}}$$

$$LCLp = \bar{p} - a\sqrt{\bar{p}(1 - \bar{p})/\bar{n}}$$

If two-sigma limits are desired, a is set equal to 2. If 2.5-sigma limits are desired, a is set equal to 2.5.

If $n\bar{p} < 5$, the equation for the binomial distribution must be used. For example, consider the case where $n\bar{p} = 0.1$, with n = 10. The probability of obtaining samples of size 10 with 0, 1, 2, or 3 defects is given below.

Number of Defective Items	Probability of Getting This Number	Cumulative Probability
0	0.904	0.904
1	0.091	0.995
2	0.004	0.999
3	0.0001	0.9991

Suppose the probability desired for obtaining a point beyond the control limits for a process that is in statistical control is 5 out of 100 (or 0.95). The upper control limit in this case would be 2. There is no lower control limit.

If $\bar{c} < 2$, Appendix Table 13 can be used to determine the limits. For example, suppose $\bar{c} = 0.80$ and you want 95% probability control limits. To determine this, look up $\bar{c} = 0.80$ in Appendix Table 13. Go across until the number is larger than 950. The number of defects for this column plus one is the upper control limit. In this case, UCLc = 3. There is no lower control limit.

EXAMPLES OF LIMITS OTHER THAN THREE-SIGMA LIMITS

1. Suppose $\bar{c} = 0.50$. What should the control limits be if you want the probability of obtaining a point beyond the control limits (for a process in control) to be 5 out of 100.

2. Suppose $\bar{p} = 0.65$ and $n = 10$. What are the two-sigma limits?

3. Suppose $\bar{c} = 4.2$. What are the two-sigma limits?

4. Suppose $\bar{p} = 0.20$ and $n = 10$. What should the control limits be if you want the probability of obtaining a point beyond the control limits (for a process in control) to be 5 out of 100?

SUMMARY

Several different topics related to attributes control charts were introduced in this chapter. A procedure for determining which type of control chart to use in a given situation was presented. This procedure involved determining what type of data you have (either yes/no or counting-type attributes data) and determining whether the subgroup size is constant. Once these two questions are answered, the proper control chart can be selected. A procedure for handling the cases when $n\bar{p}$ or $\bar{c}$ is small was introduced. In these cases, tables must be used to determine the control limits since the underlying distributions are not symmetrical. A procedure for determining control limits representing different probabilities than the normal three-sigma limits was introduced.

APPLICATIONS OF SPECIAL TOPICS

Think about your department at work. What are some possible areas for application of the material introduced in this chapter? Record your responses below.

16
$\overline{X}$-R CHARTS

Suppose you are a member of a bowling team. You bowl three games a night, once a week in a bowling league. You are interested in determining if you are improving your bowling game. What are some different approaches you could use? One idea is that you could plot the score from each game. However, you are more interested in what your average score is on a given night. So another idea is to plot the average of the three games each night. You definitely would like to increase that average over time. You are also interested in being more consistent, i.e., not having one great game followed by a poor one. Thus, another idea is to keep track of the range in scores for the three games each night. In situations such as this (when you want to monitor averages over time but still keep track of the variation between individual results), the $\overline{X}$-R chart is very useful. The $\overline{X}$-R chart is introduced in this chapter.

OBJECTIVES

In this chapter you will learn:

      1. What the $\overline{X}$-R chart is
      2. When to use the $\overline{X}$-R chart
      3. How to construct the $\overline{X}$-R chart

The $\overline{X}$-R chart is a type of control chart that can be used with variables data. Like most other variables control charts, it is actually two charts. One chart is for subgroup averages ($\overline{X}$). The other chart is for subgroup ranges (R). These charts are a very powerful tool for monitoring variation in a process and detecting changes in either the average or the amount of variation in the process.

INTRODUCTION TO THE $\overline{X}$-R CHART

The $\overline{X}$-R chart is a method of looking at two different sources of variation. One source is the variation in subgroup averages. The other source is the variation within a subgroup. Consider the bowling example above. You have data available on a fairly frequent basis (three games each week). You can also rationally subgroup the data. The three individual games you bowl on one night can be used to form a subgroup.

Continuing with the bowling example, suppose that one night your three bowling scores are 169, 155 and 189. These three scores form a subgroup. You can calculate the range of this subgroup by subtracting the minimum score from the maximum score. Thus the range is 189 - 155 = 34. You can plot this value on a range (R) chart. This is done for each subgroup (night of bowling three games). The range chart shows how much variation there is within each subgroup, i.e., the amount of variation in your bowling scores on one night. You would like this variation to be small and to be consistent over time.

The chart for averages ($\overline{X}$) presents a different variation from the range chart. Using the three scores above, you can calculate an average score for the night by taking the average of the three individual scores. The subgroup average is $(169 + 155 + 189)/3 = 171$. You can plot this value on the $\overline{X}$ chart. When this is done for each subgroup, the $\overline{X}$ chart shows how much week-to-week variation there is in your weekly average bowling score. You would like this variation to be small and to be consistent over time. This permits you to predict what your average score will be on any night, within certain limits.

Figure 1 is an example of the $\overline{X}$-R chart for this bowling example. The top part of the figure is the $\overline{X}$ chart. Each weekly average bowling score (i.e., the average of the three individual games) is plotted. The overall average $(\overline{\overline{X}})$ has been calculated and plotted as a solid line. $\overline{\overline{X}}$ is the average of all the subgroup averages. Upper and lower control limits have also been calculated and plotted. The $\overline{X}$ chart is in statistical control. The lower part of Figure 1 is the range (R) chart. The range is plotted for each week. The average range and control limits have been calculated and plotted. The range is also in statistical control.

What does it mean when the $\overline{X}$-R chart is in statistical control? It means that the subgroup average is consistent over time and the the variation within a subgroup is consistent over time. We can predict what the process will do in the near future.

WHEN TO USE THE $\overline{X}$-R CHART

$\overline{X}$-R charts are used in the second step of the problem solving model. This step involves analyzing how the process is behaving.

$\overline{X}$-R charts should be used when you have taken data frequently. How often you plot points on the charts depends on your subgroup size. For example, if your subgroup size is four, after taking take four samples you can calculate the average and range and plot the points. If you take only one sample per day, it will be four days before you can plot the points. If the point is out of control, the reason for it could have occurred four days ago. This often makes it difficult to find out what happened.

$\overline{X}$-R charts should be used if the individual measurements are not normally distributed. Subgrouping the data and calculating averages tends to make the average values normally distributed (this is called the Central Limit Theorem). Usually a subgroup size of 4 or 5 is sufficient to make the averages normally distributed.

$\overline{X}$-R charts should be used if you can rationally subgroup the data and are interested in detecting differences between subgroups over time. This means there should be some logical basis for the way the subgroups are formed. They should be formed to examine the variation of interest to you. You might be interested in the variation from day to day. In this case, samples from one day would be used to form a subgroup. The $\overline{X}$ chart would examine the variation from day to day, while the R chart would examine the variation within a day.

The R chart is a measure of the short-term variation in the process. Subgroups should be formed to minimize the amount of variation within a subgroup. This causes the $\overline{X}$ chart to do the work in detecting process changes.

Figure 1

WEEKLY BOWLING SCORES (n = 3)

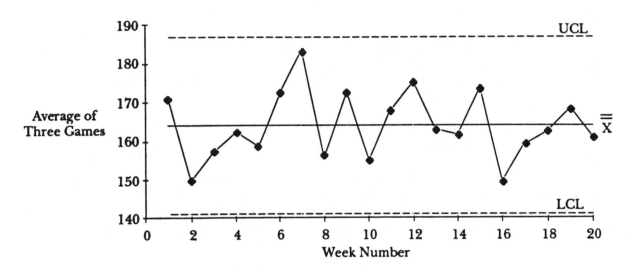

RANGE CHART (n = 3)

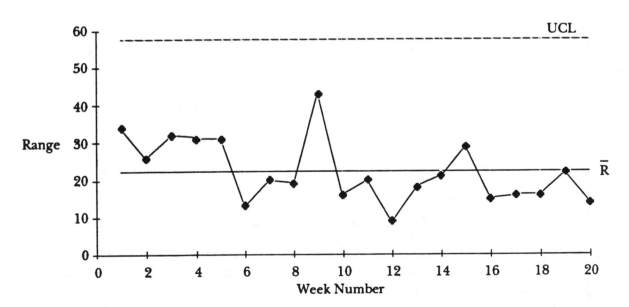

STEPS IN CONSTRUCTION OF THE $\overline{X}$-R CHART

The steps in constructing the $\overline{X}$-R chart are given below. A process flow diagram of these steps is shown in Figure 2 (Ford Motor Company, 1984).

1. Gather the data.

   a. Select the subgroup size (n). Typical subgroup sizes are 4 to 5. The concept of rational subgrouping should be considered. The objective is to minimize the amount of variation within a subgroup. This helps you "see" the variation in the averages chart more easily.

   b. Select the frequency with which the data will be collected. Data should be collected in the order in which it is generated (in most cases).

   c. Select the number of subgroups (k) to be collected before control limits are calculated (at least 20).

   d. For each subgroup, record the individual, independent sample results.

   e. For each subgroup, calculate the subgroup average:

   $$\overline{X} = \Sigma X_i/n = (X_1 + X_2 + ... + X_n)/n$$

   where X_1, X_2, etc. are the individual sample results and n is the subgroup size.

   f. For each subgroup, calculate the subgroup range:

   $$R = X_{max} - X_{min}$$

   where X_{max} is the maximum individual sample result in the subgroup and X_{min} is the minimum individual sample result in the subgroup.

2. Plot the data.

   a. Select the scales for the x and y axes for both the $\overline{X}$ and R charts.

   b. Plot the subgroup ranges on the R chart and connect consecutive points with a straight line.

   c. Plot the subgroup averages on the $\overline{X}$ chart and connect consecutive points with a straight line.

3. Calculate the overall process averages and control limits.

    a. Calculate the average range ($\overline{R}$):

$$\overline{R} = \Sigma R_i/k = (R_1 + R_2 + ... + R_k)/k$$

    where R_1, R_2, etc. are the ranges for subgroups 1, 2, etc. and k is the number of subgroups.

    b. Plot $\overline{R}$ on the range chart as a solid line and label.

    c. Calculate the overall process average ($\overline{\overline{X}}$):

$$\overline{\overline{X}} = \Sigma \overline{X}_i/k = (\overline{X}_1 + \overline{X}_2 + ... + \overline{X}_k)/k$$

    where X_1, X_2, etc. are the subgroup averages for subgroups 1, 2, etc.

    d. Plot $\overline{\overline{X}}$ on the $\overline{X}$ chart as a solid line and label.

    e. Calculate the control limits for the R chart. The upper control limit is given by UCLr. The lower control limit is given by LCLr.

$$UCLr = D_4\overline{R}$$

$$LCLr = D_3\overline{R}$$

    where D_4 and D_3 are control chart constants that depend on subgroup size.

    f. Plot the control limits on the R chart as dashed lines and label.

    g. Calculate the control limits for the $\overline{X}$ chart. The upper control limit is given by UCLx. The lower control limit is given by LCLx.

$$UCLx = \overline{\overline{X}} + A_2\overline{R}$$

$$LCLx = \overline{\overline{X}} - A_2\overline{R}$$

    where A_2 is a control chart constant that depends on subgroup size.

    h. Plot the control limits on the $\overline{X}$ chart as dashed lines and label.

4. Interpret both charts for statistical control.

   a. Always consider variation first. If the R chart is out of control, the control limits on the $\overline{X}$ chart are not valid since you do not have a good estimate of $\overline{R}$. All tests for statistical control apply to the $\overline{X}$ chart. Points beyond the limits, number-of-runs and length-of-run tests apply to the R chart.

5. Calculate the process standard deviation, if appropriate.

   a. If the R chart is in statistical control, the process standard deviation, $\hat{\sigma}'$, can be calculated as:

   $$\hat{\sigma}' = \overline{R}/d_2$$

   where d_2 is a control chart constant that depends on subgroup size.

To calculate control limits and to estimate the process standard deviation, you must use the control chart constants D_4, D_3, A_2, and d_2. These control chart constants depend on the subgroup size (n). These control chart constants are summarized in Table 1. For example, if your subgroup is 4, then $D_4 = 2.282$, $A_2 = 0.729$, and $d_2 = 2.059$. There is no value for D_3. This simply means that the R chart has no lower control limit when the subgroup size is 4.

Figure 2

STEPS IN CONSTRUCTION OF THE $\overline{\text{X}}$-R CHART

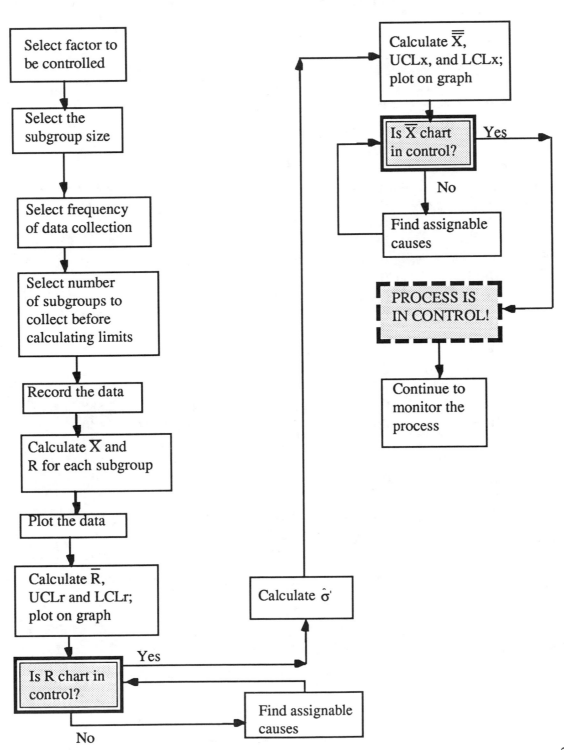

223

Table 1

FACTORS FOR USE WITH $\overline{X}$-R CHARTS

Subgroup Size (n)	A_2	D_3	D_4	d_2
2	1.880		3.267	1.128
3	1.023		2.574	1.693
4	0.729		2.282	2.059
5	0.577		2.114	2.326
6	0.483		2.004	2.534
7	0.419	0.076	1.924	2.704
8	0.373	0.136	1.864	2.847
9	0.337	0.184	1.816	2.970
10	0.308	0.223	1.777	3.078
11	0.285	0.256	1.774	3.173
12	0.266	0.284	1.716	3.258
13	0.249	0.308	1.692	3.336
14	0.235	0.329	1.671	3.407
15	0.223	0.348	1.652	3.472
16	0.212	0.364	1.636	3.532
17	0.203	0.379	1.621	3.588
18	0.194	0.392	1.608	3.640
19	0.187	0.404	1.596	3.689
20	0.180	0.414	1.586	3.735
21	0.173	0.425	1.575	3.778
22	0.167	0.434	1.566	3.819
23	0.162	0.443	1.557	3.858
24	0.157	0.452	1.548	3.895
25	0.153	0.459	1.541	3.931

$\overline{X}$-R CHART EXAMPLE

Reaction Yield

A team in one operating unit is interested in improving the yield from a given batch reaction. The team decides to use a control chart to determine if the process is in statistical control. The team also decides that the control chart will be used in the future to help monitor the process over time. Once any assignable causes are eliminated, the control chart can be used to monitor attempts at process improvement. Five batches of product are made each day. This provides frequent data plus a method of rationally subgrouping the data. The team decides to use a subgroup size of n = 5. Data from the past 25 days are available. The team decides to use this historical data for the control charts. The data for percent yield from the last 25 days are given below.

Subgroup (Day) Number	Sample % Reaction Yield Results (n = 5)					$\overline{X}$	R
	1	2	3	4	5		
1	81.3	80.4	78.6	83.1	81.8	81.04	4.5
2	74.3	76.4	82.4	77.8	82.5	78.68	8.2
3	78.7	77.4	79.4	81.6	81.0	79.62	4.2
4	80.4	81.7	81.4	79.7	80.2	80.68	2.0
5	79.4	75.6	80.3	80.2	77.4	78.58	4.7
6	85.0	75.4	73.8	75.8	78.6	77.72	11.2
7	78.5	86.2	77.1	73.3	76.4	78.30	12.9
8	81.7	84.0	80.2	78.6	80.9	81.08	5.4
9	84.5	82.4	78.8	83.2	83.0	82.38	5.7
10	82.7	80.5	85.9	82.7	84.0	83.16	5.4
11	78.4	83.1	80.1	78.5	86.6	81.34	8.2
12	82.9	82.4	78.9	78.2	78.4	80.16	4.7
13	75.6	80.1	81.1	78.3	80.4	79.10	5.5
14	78.2	76.4	82.3	81.7	85.1	80.74	8.7
15	81.8	80.6	79.1	79.3	83.6	80.88	4.5
16	75.2	82.2	79.6	83.6	81.9		
17	78.6	80.1	80.6	79.3	80.4		
18	82.3	80.8	79.7	76.5	85.6		
19	83.0	83.6	75.2	83.3	81.3		
20	77.6	79.1	78.7	80.8	80.2		
21	75.0	81.0	82.9	80.0	81.9		
22	82.7	78.8	81.2	74.8	81.7		
23	76.9	82.5	82.5	81.4	84.4		
24	78.1	82.9	73.7	81.5	75.9		
25	79.9	78.7	81.3	80.0	78.5		

Use these data to construct the $\overline{X}$-R chart for reaction yield. The first 15 subgroup averages and ranges have been calculated for you. The first 15 points have also been plotted (p. 226). After constructing the $\overline{X}$-R chart, answer the questions on page 227.

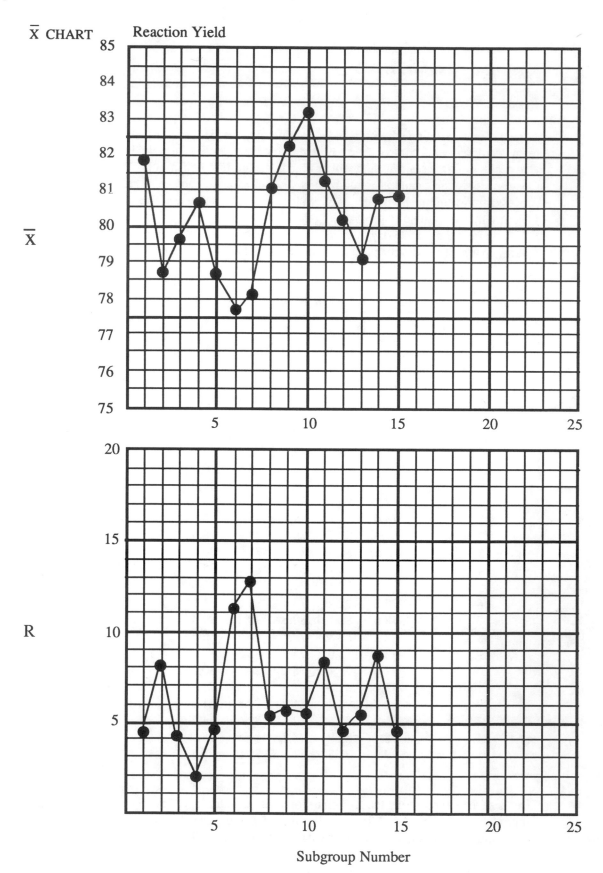

Questions on reaction yield problem:

1. What variation is being examined on the R chart?

2. Is the R chart in or out of statistical control? What does this mean?

3. If the R chart is in statistical control, what is the estimate of the process standard deviation, $\hat{\sigma}'$? What is $\hat{\sigma}'$ measuring?

4. What variation is being examined on the $\overline{X}$ chart?

5. Is the $\overline{X}$ chart in or out of statistical control? What does this mean?

6. What should the team do next, i.e., where should it begin looking for improving the yield of this reaction process?

Chapter 16

X̄-R CHART EXAMPLE

Product Brightness

A company produces a product in granular form. One product response of interest to a customer is the product brightness. The brightness of the product is measured (in process) four times each hour. This provides frequent data and a method of rationally subgrouping the data. A subgroup size of 4 is selected. Data from the last 20 hours are given below.

Subgroup (Hour) Number	Product Brightness Results				X̄	R
	1	2	3	4		
1	87.4	93.4	103.4	107.2	97.85	19.8
2	89.5	105.5	99.6	117.7	103.08	28.2
3	76.7	96.7	86.6	105.1	91.28	28.4
4	102.1	103.5	105.7	103.9	103.80	3.6
5	103.0	87.4	95.9	105.8	98.03	18.4
6	94.5	92.9	84.2	110.1	95.43	25.9
7	90.8	98.9	92.5	90.3	93.13	8.6
8	93.0	106.8	107.0	106.6	103.35	14.0
9	116.8	118.1	120.0	116.8	117.93	3.2
10	111.0	109.8	107.4	114.2	110.60	6.8
11	123.2	108.6	92.4	122.0	111.55	30.8
12	104.2	127.9	117.5	107.3	114.23	23.7
13	109.0	106.2	126.3	114.1	113.90	20.1
14	103.2	105.4	94.6	93.8	99.25	11.6
15	106.7	114.1	109.0	105.3	108.78	8.8
16	95.6	103.6	123.0	113.3		
17	121.9	123.0	108.4	102.0		
18	113.6	103.6	128.9	111.6		
19	118.3	102.4	108.3	101.0		
20	109.9	100.8	86.0	107.1		

Use these data to construct the X̄-R chart for product brightness. The first 15 subgroup averages and ranges have been calculated for you. The first 15 points have also been plotted (p.229). After constructing the X̄-R chart, answer the questions on page 230.

$\overline{X}$ CHART

Product Brightness

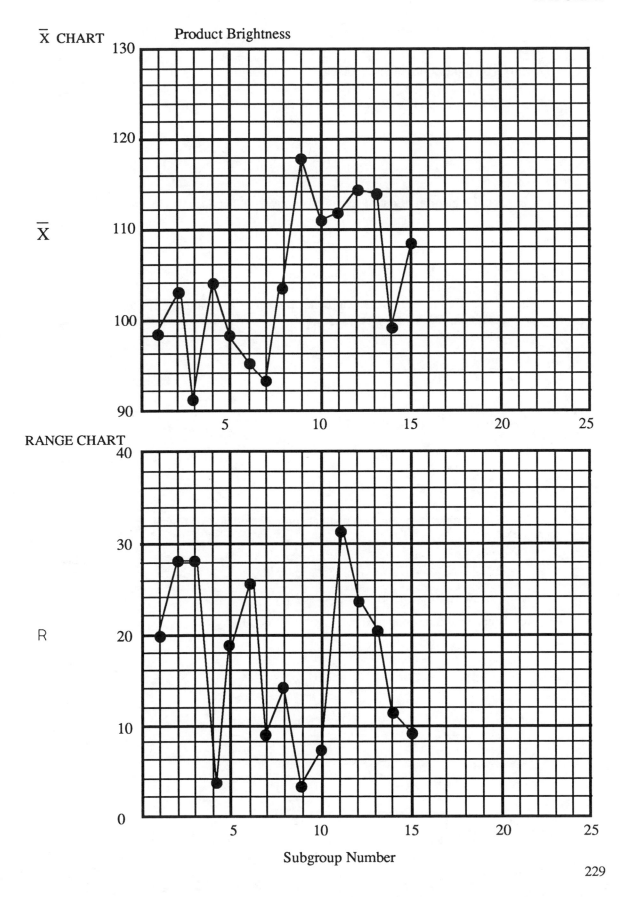

RANGE CHART

Subgroup Number

Questions on product brightness problem:

1. What variation is being examined on the R chart?

2. Is the R chart in or out of statistical control?   What does this mean?

3. If the R chart is in statistical control, what is the estimate of the process standard deviation, $\hat{\sigma}$'?   What is $\hat{\sigma}$ measuring?

4. What variation is being examined on the $\overline{X}$ chart?

5. Is the $\overline{X}$ chart in or out of statistical control?   What does this mean?

6. What should be done next?

$\overline{X}$-R CHART EXAMPLE

Reaction Run Time

Run time for a batch reactor is known to have an effect on several finished product parameters. The operating unit has selected this in-process variable to monitor. Four batches are made each day giving a run time of about 360 minutes for each batch. The operating unit decides to use the $\overline{X}$-R chart with a subgroup size of 4. Data from the last 20 days are given below.

Subgroup (Day) Number	Run Time Sample Results				$\overline{X}$	R
	1	2	3	4		
1	316	366	361	381	356.0	65
2	334	336	384	339	348.3	50
3	334	364	369	339	351.5	35
4	346	397	335	343	355.3	62
5	384	323	363	388	364.5	65
6	370	372	369	356	366.8	16
7	332	386	351	366	358.8	54
8	383	380	366	366	373.8	17
9	399	405	366	416	396.5	50
10	371	372	373	370	371.5	3
11	359	344	317	364	346.0	47
12	395	345	362	344	361.5	51
13	342	393	377	386	374.5	51
14	364	372	385	360	370.3	25
15	353	365	361	367	361.5	14
16	362	369	366	345		
17	402	378	388	401		
18	371	381	353	347		
19	350	364	374	330		
20	348	345	351	347		

Use these data to construct the $\overline{X}$-R chart for run time. The first 15 subgroup averages and ranges have been calculated for you. The first 15 points have also been plotted (p. 232). After constructing the $\overline{X}$-R chart, answer the questions on page 233.

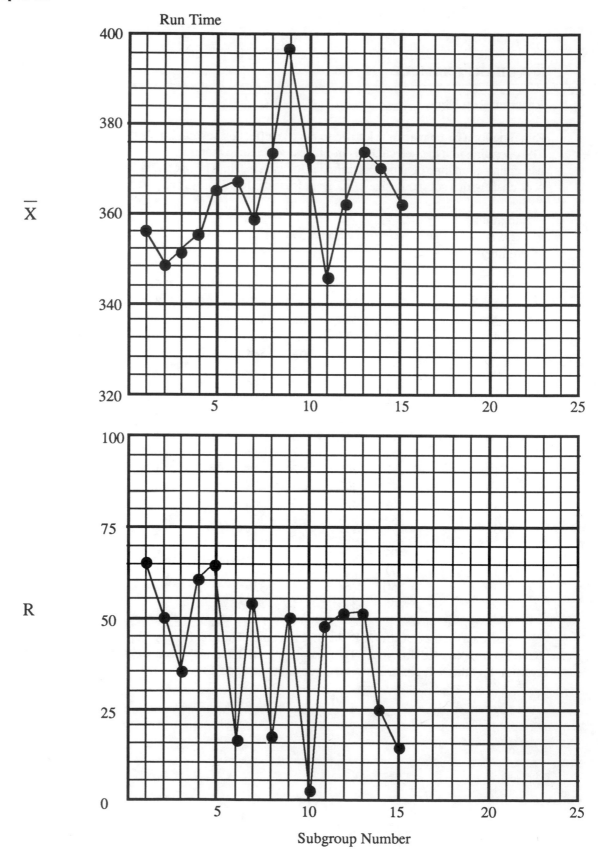

Questions on run time problem:

1. What variation is being examined on the R chart?

2. Is the R chart in or out of statistical control?   What does this mean?

3. If the R chart is in statistical control, what is the estimate of the process standard deviation, $\hat{\sigma}'$? What is $\hat{\sigma}'$ measuring?

4. What variation is being examined on the $\overline{\text{X}}$ chart?

5. Is the $\overline{\text{X}}$ chart in or out of statistical control?   What does this mean?

6. What should be done next?

SUMMARY

The $\overline{X}$-R chart is a useful problem solving tool. It helps us attain and maintain control of our processes. $\overline{X}$-R charts can be used when you have frequently available data, if the data can be subgrouped rationally, and if the individual measurements are not normally distributed. The R chart is usually a measure of the short-term variation in the process. The $\overline{X}$ chart is a measure of the long-term variation in the process. The data are subgrouped to allow you to investigate the variation of interest to you.

APPLICATIONS OF $\overline{X}$-R CHARTS

Think about your own work or work area. What are three possible applications for $\overline{X}$-R charts and why? Write your responses below.

1.

2.

3.

17

INDIVIDUALS CONTROL CHARTS

Suppose your process generates data on a very limit frequency. Maybe you get data only once a day, once a week or once every two weeks. How can you apply control charts to these types of data? If you wait to get several data points to form a subgroup, you won't be able to plot a point very often. Perhaps the test method used to analyze the process is very expensive to run or takes a long time. How can you handle this type of situation? In these instances, individuals control charts are useful. This type of chart is useful when you have only one data point at a time to represent a given situation. The individuals control chart is introduced in this chapter.

OBJECTIVES

In this chapter, you will learn:

1. What the individuals control chart is
2. When to use the individuals control chart
3. How to construct the individuals control chart

The individuals control chart is a type of control chart that can be used with variables data. Like most other variable control charts, it is actually two charts. One chart is for the individual sample result (X). The other chart is for the moving range (R) between successive individual samples. The individuals chart is very useful for monitoring processes where data are not available on a frequent basis.

INTRODUCTION TO THE INDIVIDUALS CONTROL CHART

The individuals control chart is a method of looking at variation. One source of variation is the variation in the individual sample results. This represents "long term" variation in the process. The second source of variation is the variation in the ranges between successive samples. This represents "short term" variation.

Figure 1 is an example of an individuals control chart. In this example, a sample is pulled once a day from a given process stream and tested for a certain contaminant. Since data are not obtained very frequently, an individuals control chart was selected. The top part of the figure is the plot of the individual sample results. For example, on the first day the test method indicated that there were 105 ppm of the contaminant in the product stream. On the second day, the sample result was 93 ppm. The overall process average ($\overline{X}$) has been calculated and plotted as a solid line. The upper and lower control limits have also been calculated and plotted as dashed lines.

The bottom part of Figure 1 is the range chart. This chart represents the range between successive data points. This range is often called a moving range. The range value plotted for the second day is simply the range between day 1 and day 2 (largest minus smallest). This range is 105 - 93 = 12. The average range ($\overline{R}$) has been calculated and plotted as a solid line. The upper control

Figure 1

INDIVIDUALS CONTROL CHART: ppm CONTAMINANT

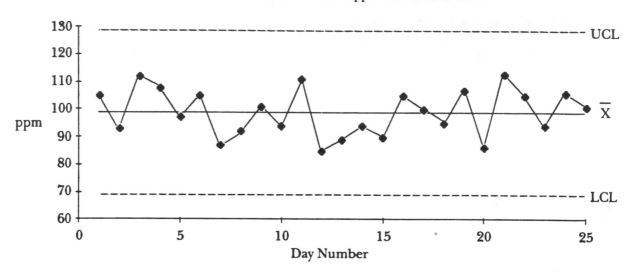

MOVING RANGE CHART (n = 2)

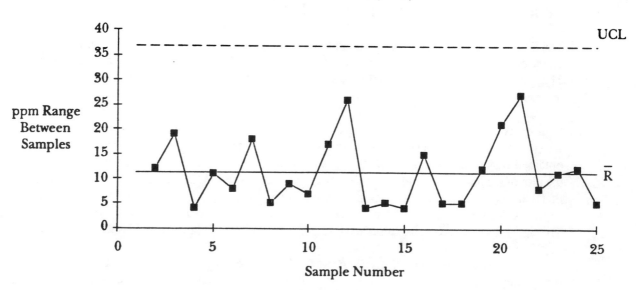

limit has also be calculated and plotted as a dashed line. There is no lower control limit on the range chart for an individuals chart.

The charts in Figure 1 are in statistical control. What does it mean when the individuals control chart is in statistical control? It means that the individual sample results are consistent over time, i.e., they are not significantly different from the process average $(\overline{X})$. It also means that the difference between successive sample results is consistent over time. We can predict what the process will make in the near future.

WHEN TO USE THE INDIVIDUALS CONTROL CHART

Individuals control charts can be used in the second step of the problem solving model. This step involves analyzing how the process is behaving.

Individuals control charts should be used when there is only one data point to represent a situation at a given time. The individuals control chart allows you to plot a point on the chart for each sample taken. This permits you to determine if the process is in statistical control or not for each sample taken. This may seem to imply that you should always use individuals charts. That is not necessarily true, as is discussed below.

To use the individuals control chart, the individual sample results must be normally distributed. If not, the individuals control chart will give more false signals, i.e., will indicate more often that the process is out of control when it actually isn't.

Individual control charts are not as sensitive to changes as $\overline{X}$-R charts. In addition, values of X and R can have significant variation (even though the process is in control) until the number of individual data points reaches 100.

STEPS IN CONSTRUCTION OF THE INDIVIDUALS CONTROL CHART

The steps in constructing the individuals control chart are given below. A process flow diagram of these steps is shown in Figure 2.

1. Gather the data.

    a. Select the frequency with which the data will be collected. Data should be collected in the order in which they are generated.

    b. Select the number of data points (k) to be collected before control limits are calculated (at least 20). These will be tentative limits. Control limits should be recalculated after you have collected 100 data points.

    c. Record the individual sample results. Each sample must be independent.

---

d. Calculate the moving range (R_{i+1}) between consecutive sample results:

$$R_{i+1} = |\ X_{i+1} - X_i\ |$$

where X_{i+1} is the result for sample $i + 1$ and X_i is the result for sample i. The range value is always positive.

2. Plot the data.

a. Select the scales for the x and y axes for both the X and R charts.

b. Plot the ranges on the R chart and connect consecutive points with a straight line.

c. Plot the individual sample results on the X chart and connect consecutive points with a straight line.

3. Calculate the overall process averages and control limits.

a. Calculate the average range $(\overline{R})$:

$$\overline{R} = \Sigma R_i/(k - 1) = (R_2 + R_3 + ... + R_k)/(k - 1)$$

where R_2, R_3, etc. are the ranges between samples 2 and 1, samples 3 and 2, etc. and k is the number of subgroups. Note there is always one less range value than individual sample results for the individuals control chart.

b. Plot $\overline{R}$ on the range chart as a solid line and label.

c. Calculate the overall process average $(\overline{X})$:

$$\overline{X} = \Sigma X_i/k = (X_1 + X_2 + ... + X_k)/k$$

where X_1, X_2, etc. are the individual sample results 1, 2, etc.

d. Plot X on the $\overline{X}$ chart as a solid line and label.

e. Calculate the control limits for the R chart. The upper control limit is given by UCLr. The lower control limit is given by LCLr.

$$UCLr = 3.267\overline{R}$$

$$LCLr = none$$

f. Plot the upper control limit on the R chart as a dashed line and label.

g. Calculate the control limits for the X chart. The upper control limit is given by UCLx. The lower control limit is given by LCLx.

$$UCLx = \overline{X} + 2.66\overline{R}$$

$$LCLx = \overline{X} - 2.66\overline{R}$$

h. Plot the control limits on the X chart as dashed lines and label.

4. Interpret both charts for statistical control.

   a. Always consider variation first. If the $\overline{R}$ chart is out of control, the control limits on the X chart are not valid since you do not have a good estimate of R.

   b. All tests for statistical control apply to the X chart. However, the data on the range chart are not independent. Each data point is used twice. The only test that is valid for the range chart is points beyond the control limits.

5. Calculate the process standard deviation, if appropriate.

   a. If the R chart is in statistical control, the process standard deviation, $\hat{\sigma}'$, can be calculated as:

$$\hat{\sigma}' = \overline{R}/1.128$$

239

Figure 2

STEPS IN CONSTRUCTION OF THE INDIVIDUALS CONTROL CHART

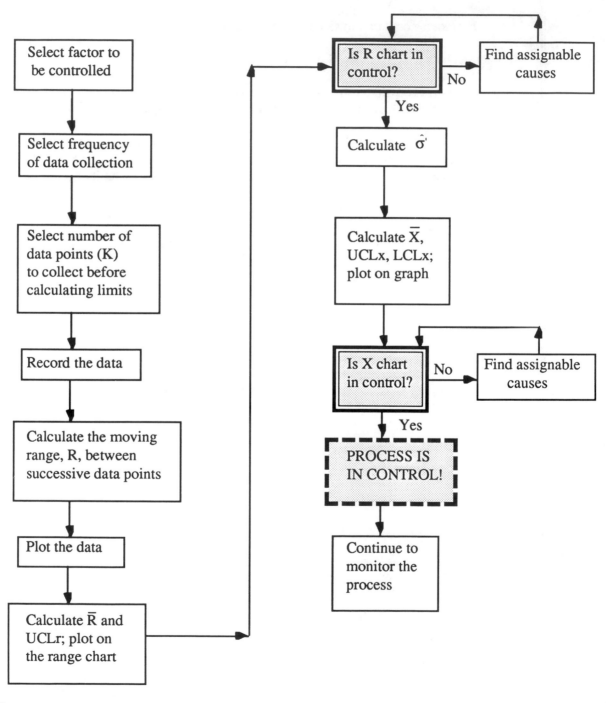

INDIVIDUALS CONTROL CHART EXAMPLE

Product Hardness

An operating unit produces plastic pellets. One parameter of interest to a customer is the hardness of the plastic. The customer has asked the operating unit to begin to monitor the hardness of the plastic pellets. The unit produces the plastic pellets in a process that begins with mixing the proper ingredients in a large blender. A sample is then taken from the blender and tested for hardness. This is considered to be a check to see if the proper amounts of ingredients were added to the blender. If the hardness is within specifications, the blender is then processed into pellets. It takes about six hours to process a blender. In addition, the operating unit only makes about 20 blenders of this product for the customer once a month. Since data are available on a limited basis only, the unit decides to use an individuals control chart (assume that the individual measurements are normally distributed). Data from the last 25 blenders are given below.

Sample Number	Hardness	Moving Range
1	45	--
2	40	5
3	46	6
4	35	11
5	38	3
6	35	3
7	37	2
8	41	4
9	50	9
10	45	5
11	43	2
12	42	1
13	46	4
14	39	7
15	33	6
16	40	
17	41	
18	42	
19	44	
20	42	
21	38	
22	42	
23	45	
24	41	
25	47	

Use these data to construct an individuals control chart for product hardness. The moving ranges have been calculated for the first 15 samples. The first 15 points have been plotted (p. 243).

After constructing the individuals chart, answer the following questions.

1. What variation is being measured on the R chart?

2. Is the R chart in or out of statistical control? What does this mean?

3. If the R chart is in statistical control, what is the estimate of the process standard deviation, $\hat{\sigma}'$? What is $\hat{\sigma}'$ measuring?

4. What variation is being examined on the X chart?

5. Is the X chart in or out of statistical control? What does this mean?

6. What do you think the customer will say about these results?

7. Is the process of checking each blender for hardness (prior to processing) to see if it is within specifications a good approach? Why or why not?

8. How would you begin to improve this process?

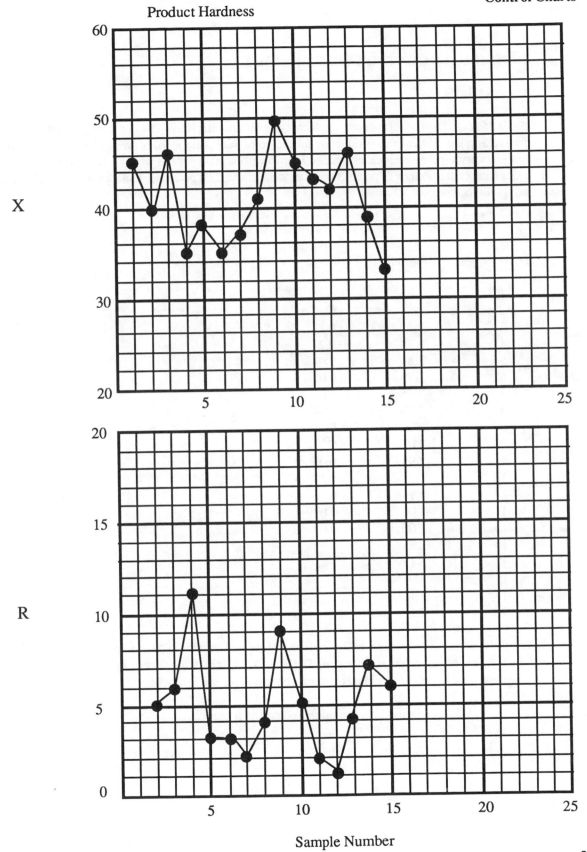

Product Hardness

X

R

Sample Number

INDIVIDUALS CONTROL CHART EXAMPLE

Product Sales

Sales of a certain product appear to be increasing on a regular basis. A sales manager is interested in tracking the sales of this product over time to be sure that the sales are really increasing and not just varying in some predictable fashion. The sales manager has weekly sales figures. These sales figures are pounds of product sold per week (in million pounds). Data for the last 20 weeks are given below.

Week Number	Sales (million lbs)	Moving Range
1	2.76	
2	2.34	0.42
3	2.45	0.11
4	2.74	0.29
5	2.81	0.07
6	2.79	0.02
7	2.85	0.06
8	2.89	0.04
9	2.78	0.11
10	3.01	0.23
11	2.95	0.06
12	3.05	0.10
13	3.11	0.06
14	3.02	0.09
15	3.00	0.02
16	2.95	
17	3.17	
18	3.04	
19	3.12	
20	3.13	

Use these data to construct an individuals chart for product sales. The moving ranges for the first 15 weeks have been calculated for you. The first 15 points have also been plotted for you (p. 246). After constructing the individuals control chart, answer the following questions.

1. What variation is being measured on the R chart?

2. Is the R chart in or out of statistical control? What does this mean?

3. If the R chart is in statistical control, what is the estimate of the process standard deviation, $\hat{\sigma}'$? What is $\hat{\sigma}'$ measuring?

4. What variation is being examined on the X chart?

5. Is the X chart in or out of statistical control? What does this mean?

6. What are some possible reasons for the way these charts look?

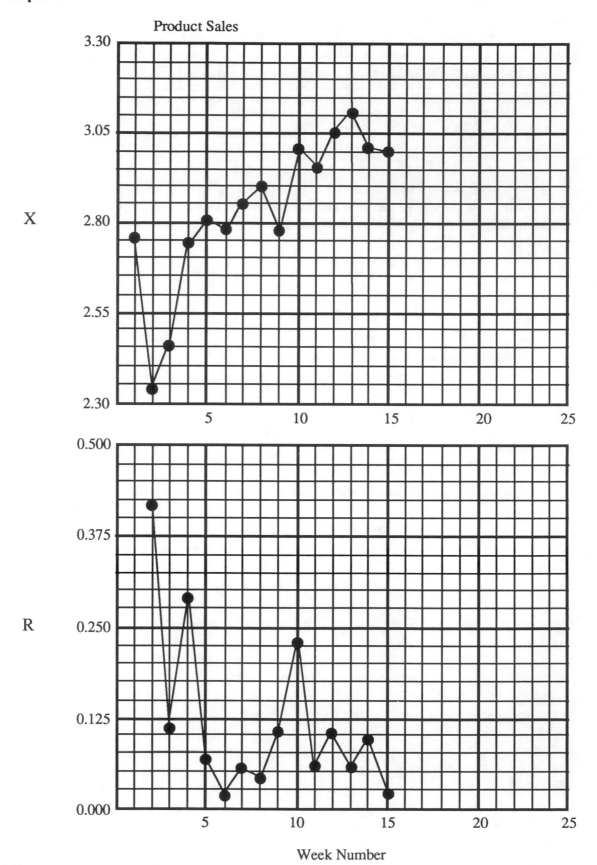

Product Sales

X

R

Week Number

INDIVIDUALS CONTROL CHART EXAMPLE

Reaction Run Times

A team has been organized to study the effect reaction run time has on a certain product parameter. The team has discovered that reaction run time has a major impact on this parameter. If reaction run time can be controlled to within certain limits, the product parameter will remain within specification limits. To determine what the normal variation is for reaction run time, the team examines data for the last 20 batches. The data, in minutes, are given below.

Run Number	Time (Min)	Moving Range
1	136	
2	125	11
3	133	8
4	137	4
5	126	11
6	131	5
7	134	3
8	133	1
9	135	2
10	139	4
11	134	5
12	129	5
13	132	3
14	134	2
15	133	1
16	135	
17	129	
18	132	
19	135	
20	134	

Use these data to construct an individuals chart for reaction run time. The moving ranges for the first 15 samples have been calculated for you. The first 15 points have also been plotted (p.248). After constructing the individuals chart, answer the following questions.

1. What variation is being measured on the R chart?

2. Is the R chart in or out of statistical control? What does this mean?

3. If the R chart is in statistical control, what is the estimate of the process standard deviation, $\hat{\sigma}'$? What is $\hat{\sigma}'$ measuring?

4. What variation is being examined on the X chart?

5. Is the X chart in or out of statistical control? What does this mean?

6. What should be done next to improve the process?

Reaction Run Times

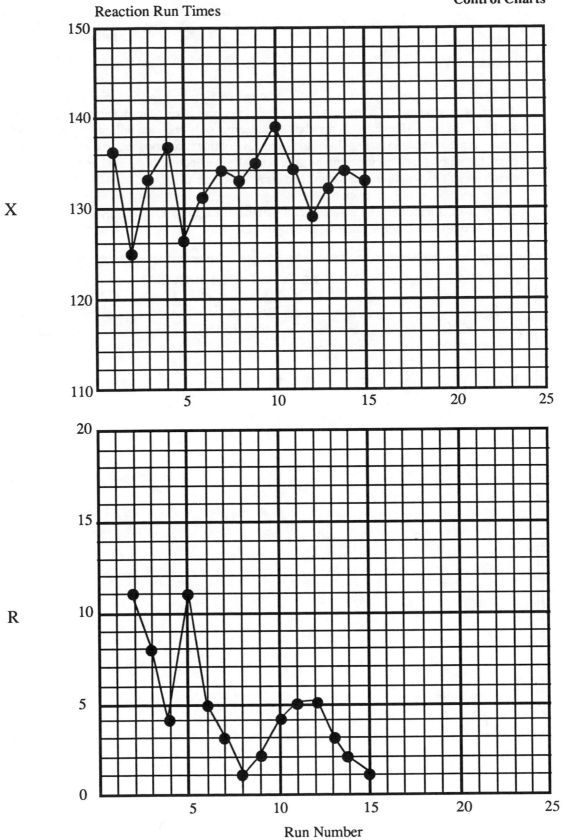

X

R

Run Number

Chapter 17

SUMMARY

The individuals control chart is a useful problem solving tool to use when you are trying to determine how a process is behaving or to monitor a process over time. It can be used when there are limited data available. This may be due to having only one data point to represent a given situation at a time, or to testing that is too expensive or that takes too long. To use an individuals chart, the individual measurements must be normally distributed. In general, individuals control charts are not as sensitive to process changes as $\overline{X}$-R charts.

The X chart is a measure of the long term variation in a process. The moving range chart is a measure of the short-term variation in the process.

APPLICATIONS OF INDIVIDUALS CONTROL CHARTS

Think about your own work or work area. What are three possible applications of individuals control charts? Why are these better than $\overline{X}$-R charts for these applications? Write down your responses below.

1.

2.

3.

18

MOVING AVERAGE AND MOVING RANGE CHARTS

The $\overline{X}$-R chart is used when you have data frequently available, when you can rationally subgroup the data, and when you want to detect differences in subgroups over time. The individuals chart is used when you have only one data point to represent a situation at a given time (infrequent data) and when the individual measurements are normally distributed. There will be times when you have infrequent data and the individual measurements are not normally distributed. For example, a process may be running near a minimum or a maximum. This often leads to skewed distributions. The $\overline{X}$-R chart can't be used since data are not available frequently. The individuals chart can't be used since the individual measurements are not normally distributed. What kind of chart can you use in these situations? These situations are handled by the moving average and moving range chart. This type of chart is introduced in this chapter.

OBJECTIVES

In this chapter you will learn:

1. What the moving average and moving range chart is
2. When to use the moving average and moving range chart
3. How to construct the moving average and moving range chart

The moving average and moving range chart is a type of control chart used with variables data. It is actually two charts. One chart is for the moving average. The other chart is for the moving range.

INTRODUCTION TO THE MOVING AVERAGE AND MOVING RANGE CHART

The moving average and moving range control chart is a method of looking at sources of process variation. One source is the variation in the subgroup averages. The other source is the variation in the subgroup ranges.

The moving average and moving range control chart is similar to the $\overline{X}$-R chart. One major difference is the method used for subgrouping the data. In the moving average and moving range chart, the same sample result is used in several different subgroups. To understand how this works, consider the example below.

Figure 1 is an example of a moving average and moving range chart. In this example, the purity of a stream on the top of a distillation column is being monitored. This purity normally runs close to 98%. Samples are taken once every 12 hours from this stream. This gives data too infrequently to use an $\overline{X}$-R chart. A histogram of individual measurements indicates that the distribution is skewed so an individuals control chart can't be used. The moving average and moving range chart was selected. A subgroup size of four was selected. The results from the first four samples are used to form the first subgroup as shown below.

Figure 1

PURITY: MOVING AVERAGE CHART (n = 4)

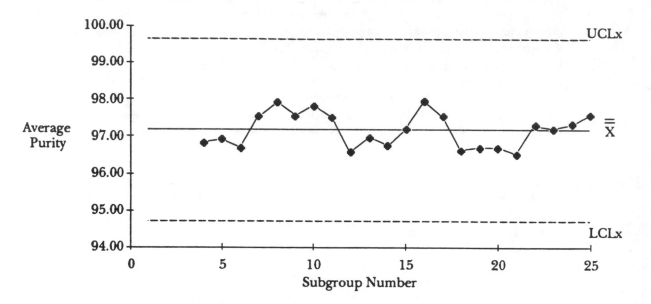

MOVING RANGE CHART (n = 4)

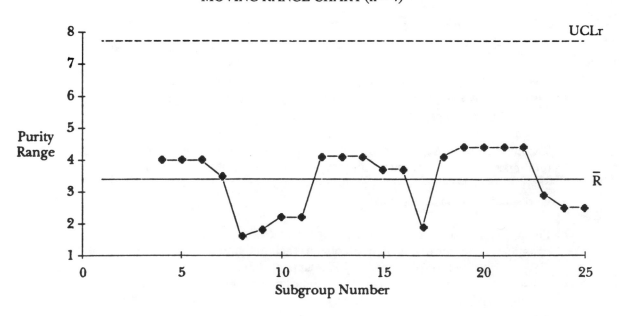

Subgroup Number	X_1	X_2	X_3	X_4	$\overline{X}$	R
1	97.6	98.5	94.5	96.7	96.83	4.0

The average $(\overline{X})$ and range (R) of these four samples are calculated and shown above. Suppose the next sample pulled has a purity of 98.0. To form the next subgroup, the first sample in the subgroup above is removed and replaced with the new sample result.

Subgroup Number	X_2	X_3	X_4	X_5	$\overline{X}$	R
2	98.5	94.5	96.7	98.0	96.93	4.0

The average and range of these four samples for subgroup two are then calculated. This procedure is repeated for the next sample pulled. The manner in which subgroups are formed for a subgroup size of 4 is shown below.

Subgroup Number	Samples in Subgroup			
1	X_1	X_2	X_3	X_4
2	X_2	X_3	X_4	X_5
3	X_3	X_4	X_5	X_6
4	X_4	X_5	X_6	X_7
5	X_5	X_6	X_7	X_8

Note that each sample result is reused in subsequent subgroups up to n times (the subgroup size). The averages and ranges calculated for each subgroup are called moving averages and ranges since data are being reused.

The top part of Figure 1 is the moving average chart for this example. The subgroup averages $(\overline{X})$ are plotted on this chart. The overall process average $(\overline{\overline{X}})$ has been calculated and plotted as a solid line. Control limits have also been calculated and plotted as dashed lines. The bottom part of Figure 1 is the moving range chart. The subgroup ranges, the average range $(\overline{R})$ and the control limits have been plotted.

The next step is to determine if the process is in control. The subgroups used in a moving average and moving range chart are not independent since data are being reused. Because of this, the only test that can be used to determine if a process is in statistical control is to see if there are points beyond the control limits. As can be seen in Figure 1, there are no points beyond the control limits. The process is in statistical control. This means that the subgroup averages are consistent over time, i.e., they are not significantly different from the overall process average $(\overline{\overline{X}})$. The subgroup ranges are also consistent over time, i.e., they are not significantly different from the average range $(\overline{R})$.

The charts in Figure 1 appear to cycle. This pattern is very common with the moving average and moving range chart. This type of chart is not as sensitive as $\overline{X}$-R or individuals control charts.

WHEN TO USE THE MOVING AVERAGE AND MOVING RANGE CHART

Moving average and moving range charts can be used in the second step of the problem solving model. This step involves analyzing how the process is behaving.

Moving average and moving range charts should be used if there is only one data point at a time to represent a given situation and if the individual measurements are not normally distributed. This permits you to form a subgroup for each sample taken and to determine the stability of the process.

Since data are reused and the subgroups are not independent, the moving average and moving range chart is not as sensitive to changes as the $\overline{X}$-R and individuals chart. For this reason, the moving average and moving range chart should be used only if the other charts cannot be used.

STEPS IN CONSTRUCTION OF THE MOVING AVERAGE AND MOVING RANGE CHART

The steps in constructing the moving average and moving range chart are given below. A process flow diagram of these steps is shown in Figure 2. The steps are the same as the steps in constructing the $\overline{X}$-R chart. The only differences are the manner in which the subgroups are formed and the tests used to interpret the charts for statistical control.

1. Gather the data.

    a. Select the subgroup size (n). Typical subgroup sizes are 4 to 5. The concept of rational subgrouping should be considered. The objective is to minimize the amount of variation within a subgroup. This helps us "see" the variation in the averages chart more easily.

    b. Select the frequency with which the data will be collected. Data should be collected in the order in which it is generated.

    c. Select the number of subgroups (k) to be collected before control limits are calculated (at least 20).

---

d. For each new sample result, remove the oldest sample result from the previous subgroup and add the new sample result to form the new subgroup. This is shown below for a subgroup size of 4.

Subgroup Number	Samples in Subgroup			
1	X_1	X_2	X_3	X_4
2	X_2	X_3	X_4	X_5
3	X_3	X_4	X_5	X_6
4	X_4	X_5	X_6	X_7
5	X_5	X_6	X_7	X_8

e. For each subgroup, calculate the subgroup moving average:

$$\overline{X} = \Sigma X_i/n = (X_1 + X_2 + ... + X_n)/n$$

where X_1, X_2, etc. are the individual sample results and n is the subgroup size.

f. For each subgroup, calculate the subgroup moving range:

$$R = X_{max} - X_{min}$$

where X_{max} is the maximum individual sample result in the subgroup and X_{min} is the minimum individual sample result in the subgroup.

2. Plot the data.

   a. Select the scales for the x and y axes for both the moving average and the moving range charts.

   b. Plot the subgroup ranges on the moving range chart and connect consecutive points with a straight line.

   c. Plot the subgroup averages on the moving average chart and connect consecutive points with a straight line.

3. Calculate the overall process averages and control limits.

   a. Calculate the average moving range ($\overline{R}$):

   $$\overline{R} = \Sigma R_i/k = (R_1 + R_2 + ... + R_k)/k$$

   where R1, R2, etc. are the ranges for subgroups 1, 2, etc. and k is the number of subgroups.

where R_1, R_2, etc. are the ranges for subgroups 1, 2, etc. and k is the number of subgroups.

b. Plot $\bar{R}$ on the moving range chart as a solid line and label.

c. Calculate the overall process average ($\bar{\bar{X}}$):

$$\bar{\bar{X}} = \Sigma\bar{X}_i/k = (\bar{X}_1 + \bar{X}_2 + ... + \bar{X}_k)/k$$

where X_1, X_2, etc. are the subgroup averages for subgroups 1, 2, etc.

d. Plot $\bar{\bar{X}}$ on the moving average chart as a solid line and label.

e. Calculate the control limits for the moving range chart. The upper control limits is given by UCLr. The lower control limit is given by LCLr.

$$UCLr = D_4\bar{R}$$

$$LCLr = D_3\bar{R}$$

where D_4 and D_3 are control chart constants that depend on subgroup size.

f. Plot the control limits on the moving range chart as dashed lines and label.

g. Calculate the control limits for the moving average chart. The upper control limit is given by UCLx. The lower control limit is given by LCLx.

$$UCLx = \bar{\bar{X}} + A_2\bar{R}$$

$$LCLx = \bar{\bar{X}} - A_2\bar{R}$$

where A_2 is a control chart constant that depends on subgroup size.

h. Plot the control limits on the moving average chart as dashed lines and label.

4. Interpret both charts for statistical control.

    a. Always consider variation first. If the moving range chart is out of control, the control limits on the moving average chart are not valid since you do not have a good estimate of $\bar{R}$.

    b. Points beyond the control limits is the only test that applies (both charts).

5. Calculate the process standard deviation, if appropriate.

    a. If the moving chart is in statistical control, the process standard deviation, s', can be calculated as:

$$\hat{\sigma}' = \bar{R}/d_2$$

    where d_2 is a control chart constant that depends on subgroup size.

To calculate control limits and to estimate the process standard deviation, you must use the control chart constants D_4, D_3, A_2, and d_2. These control chart constants depend on the subgroup size (n) and are the same as those used for the $\bar{X}$-R charts.

Chapter 18

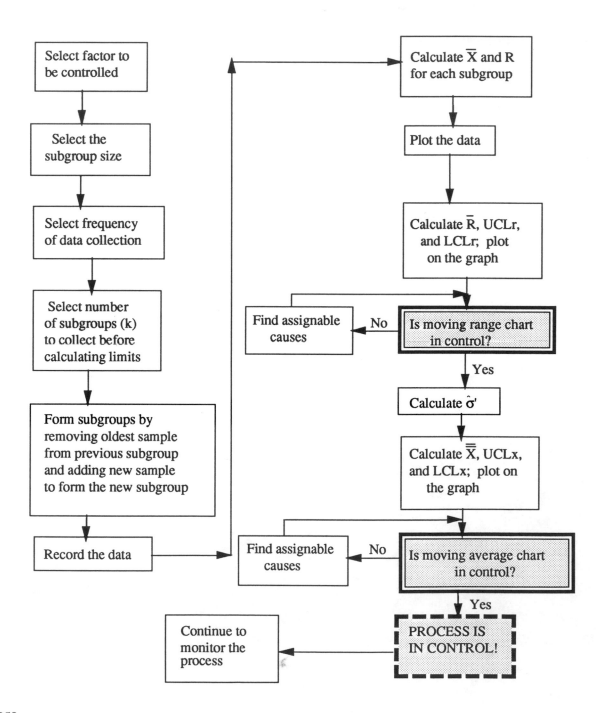

Figure 2

STEPS IN CONSTRUCTION OF THE MOVING AVERAGE AND MOVING RANGE
CHART

MOVING AVERAGE AND MOVING RANGE EXAMPLE

Paraffin Content of Kerosene

The paraffin content of incoming kerosene has a major impact on the operating parameters in an operating unit in a plant. A team in the unit decided to track this paraffin content. New lots of kerosene are received every four to seven days and tested for percent paraffin. Past data indicated that the distribution of percent paraffin measurements is skewed. The team decided to use a moving average and moving range chart. A subgroup size of five was selected. Historical data from the last 25 lots of kerosene are given below.

Lot Number	Percent Paraffin	Moving Average	Moving Range
1	21.0		
2	20.2		
3	22.7		
4	19.7		
5	22.4	21.20	3.0
6	21.9	21.38	3.0
7	20.7	21.48	3.0
8	19.8	20.90	2.7
9	21.0	21.16	2.6
10	18.5	20.38	3.4
11	20.0	20.00	2.5
12	23.5	20.56	5.0
13	18.7	20.34	5.0
14	23.5	20.84	5.0
15	19.0	20.94	4.8
16	18.9	20.72	4.8
17	19.7	19.96	4.8
18	18.5	19.92	5.0
19	18.0	18.82	1.7
20	19.5	18.92	1.7
21	18.6	18.86	1.7
22	19.5	18.82	1.5
23	20.7		
24	22.0		
25	21.5		

Use these data to construct a moving average and moving range chart. Calculations through Lot Number 22 have been completed and plotted for you (p. 261). After constructing the chart, answer the following questions.

1. Is the moving average and moving range chart the correct chart to use in this situation? Why?

2. What variation is being examined on the moving range chart?

3. Is the moving range chart in or out of statistical control? What does this mean?

4. If the moving range chart is in control, what is the estimate of the process standard deviation, $\hat{\sigma}'$? What is $\hat{\sigma}'$ measuring?

5. What variation is being examined on the moving average chart?

6. Is the moving average chart in or out of statistical control? What does this mean?

7. What should be done next?

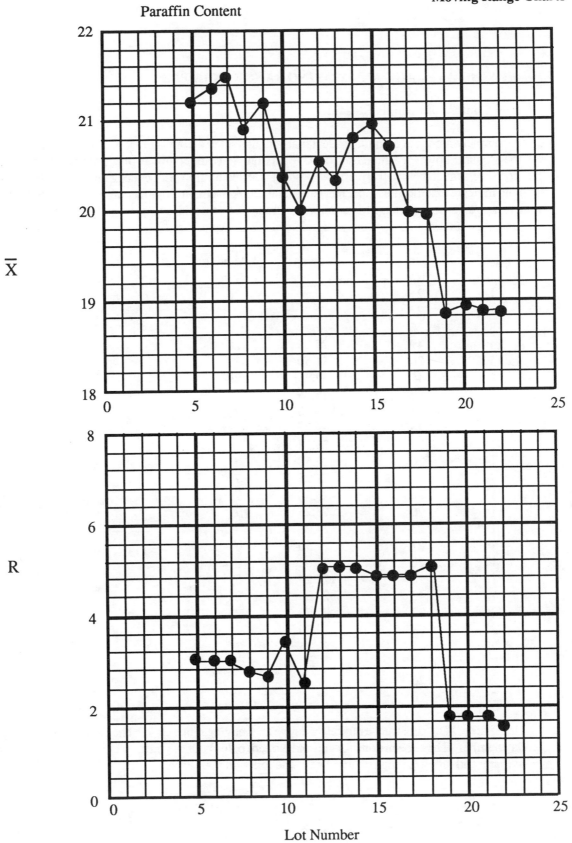

Paraffin Content

$\overline{X}$

R

Lot Number

MOVING AVERAGE AND MOVING RANGE EXAMPLE

Iron Content of Product Stream

A company produces a liquid product that contains small amounts of iron. The specification on the product allows for a maximum of 100 ppm iron. Since the company produces the product with iron levels much below 100 ppm, the product is tested for iron only once per day. Past historical data indicate that the individual measurements are skewed. An engineer decides to use a moving average and moving range to examine the process. She decides to use a subgroup size of four. Data from the past 25 days are available and are shown below.

Day Number	ppm Iron	Moving Average	Moving Range
1	21		
2	22		
3	25		
4	49	29.25	28
5	30	31.50	27
6	45	37.25	24
7	23	36.75	26
8	24	30.50	22
9	25	29.25	22
10	21	23.25	4
11	20	22.50	5
12	27	23.25	7
13	26	23.50	7
14	20	23.25	7
15	23	24.00	7
16	21	22.50	6
17	20	21.00	3
18	25	22.25	5
19	43	27.25	23
20	22	27.50	23
21	28	29.50	21
22	24	29.25	21
23	25		
24	26		
25	20		

Using these data, construct a moving average and moving range chart. The calculations through day number 22 have been done for you and plotted (p.264). Once you have completed the moving average and moving range chart, answer the following questions.

1. Is the moving average and moving range chart the correct chart to use in this situation? Why?

2. What variation is being examined on the moving range chart?

3. Is the moving range chart in or out of statistical control? What does this mean?

4. If the moving range chart is in control, what is the estimate of the process standard deviation, $\hat{\sigma}'$? What is $\hat{\sigma}'$ measuring?

5. What variation is being examined on the moving average chart?

6. Is the moving average chart in or out of statistical control? What does this mean?

7. What should be done next?

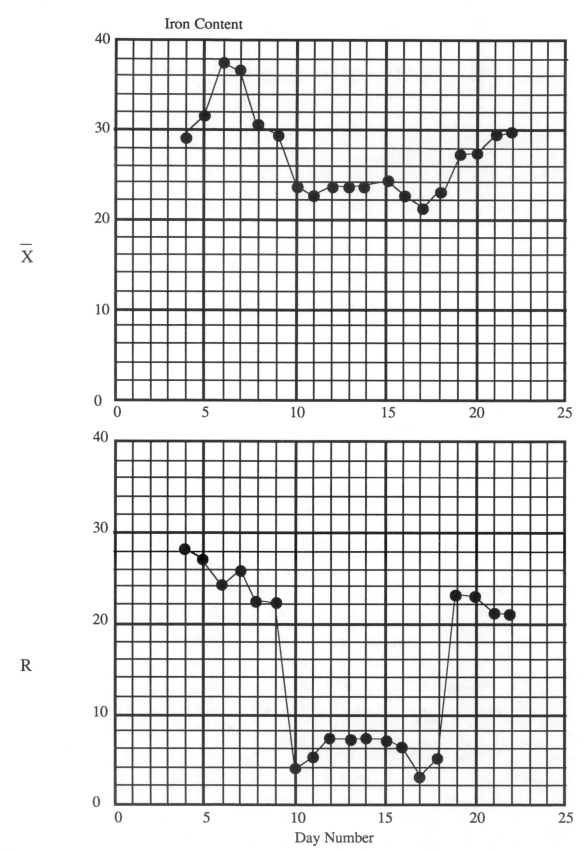

Iron Content

$\overline{X}$

R

Day Number

SUMMARY

The moving average and moving range chart is a type of control chart that can be used to determine if a process is in statistical control. It should be used whenever there is only one data point at a time to represent a given situation and the individual measurements are not normally distributed. The moving average and moving range chart reuses sample results in subsequent subgroups. Thus, the subgroups are not independent and the only test for out-of-control situations that applies is to look for points beyond the control limits. The moving average and moving range chart is not as sensitive to changes in the process as $\overline{X}$-R and individuals charts. Moving average and moving range charts should be used only if the other charts cannot be used.

APPLICATIONS OF MOVING AVERAGE AND MOVING RANGE CHARTS

Think about the processes at work. What are three possible applications of the moving average and moving range chart and why is it the appropriate chart? Record your responses below.

1.

2.

3

19
$\overline{X}$-s CHARTS

For variables data, the range is most frequently used to measure the variation within a subgroup. As the size of the subgroup increases, the range becomes a poorer measure of variation. In this case, the standard deviation (s) of the individual measurements in the subgroup should be used as the measure of subgroup variation. The $\overline{X}$-s chart is appropriate in this type of situation. The $\overline{X}$-s chart is introduced in this chapter.

OBJECTIVES

In this chapter you will learn:

1. What the $\overline{X}$-s chart is
2. When to use the $\overline{X}$-s chart
3. How to construct the $\overline{X}$-s chart

The $\overline{X}$-s chart is a type of control chart that is used with variables data. Like most other variable control charts, it is actually two charts. One chart is for the subgroup averages ($\overline{X}$). The other chart is for the subgroup standard deviations (s).

INTRODUCTION TO THE $\overline{X}$-s CHART

The $\overline{X}$-s chart is a method of looking at sources of variation. One chart looks at variation in the subgroup averages. The other chart examines variation in the subgroup standard deviations.

The $\overline{X}$-s chart is very similar to the $\overline{X}$-R chart. The major difference is that the subgroup standard deviation is plotted when using the $\overline{X}$-s chart, while the subgroup range is plotted when using the $\overline{X}$-R chart. The constants used to calculate the control limits and to estimate the process standard deviation are different for the $\overline{X}$-s chart than for the $\overline{X}$-R chart. As with the $\overline{X}$-R chart, frequent data and a method of rationally subgrouping the data are required to use the $\overline{X}$-s chart.

Figure 1 is an example of a $\overline{X}$-s chart. In this example, reaction yield from a batch reactor is being monitored. Five batches are made each day. This provides frequent data and a method of rationally subgrouping the data. A subgroup size of 5 is used. The first subgroup is composed of the five batches made on day 1. The percent yield is measured for each individual batch. The subgroup average ($\overline{X}$) is the average of the five individual batch yields. The subgroup standard deviation (s) is the standard deviation of the five individual batch yields. The top part of Figure 1 is the $\overline{X}$ chart. The subgroup averages are plotted on this chart. In addition, the overall process average ($\overline{\overline{X}}$) and the control limits have been calculated and plotted. The bottom part of Figure 1 is the s chart. The subgroup standard deviations are plotted on this chart. The average standard deviation ($\overline{s}$) and the control limits have been calculated and plotted.

Figure 1

$\overline{\text{X}}$ CHART (n = 5)

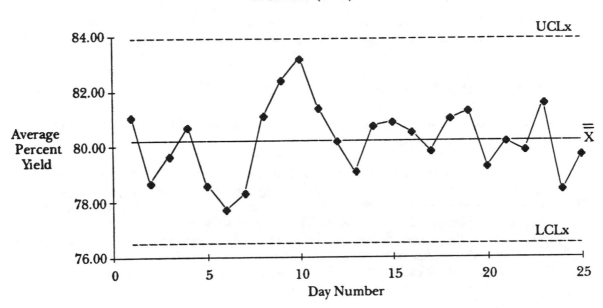

s CHART (n = 5)

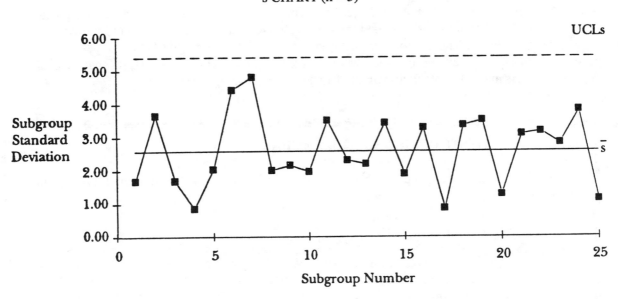

The process represented by Figure 1 is in statistical control. Since the s chart is in control, this means that the variation between individual batch yields (the within day variation) is consistent over time. We don't know what the next s value will be, but we do know that it will be between 0 and 5.4 with a long-term average of 2.8. Since the $\overline{X}$ chart is in control, the variation in subgroup averages (the day-to-day variation) is consistent over time. The next result will be between 76.5 and 83.9 with a long-term average of 80.2. The process is predictable in the near future.

WHEN TO USE THE $\overline{X}$-s CHART

$\overline{X}$-s charts are used in the second step of the problem solving model. This step involves analyzing how the process is behaving.

$\overline{X}$-s charts can be used whenever you can use $\overline{X}$-R charts. This includes situations when data are frequently available and when you can rationally subgroup the data. If the individual values are not normally distributed, the $\overline{X}$-s chart can be used since subgrouping the data tends to make the subgroup averages normally distributed. Like the R chart, the s chart is a measure of the short-term variation in the process. Subgroups should be formed to minimize the amount of variation in the s chart. This causes the $\overline{X}$ chart to do the work in detecting process changes. s charts give patterns similar to the R charts. The conclusions reached will generally be the same from either the s or the R chart. Control limits on the $\overline{X}$ chart will generally be the same, regardless if the limits are based on s or R.

The $\overline{X}$-s chart should be used instead of the $\overline{X}$-R chart if the subgroup size is larger than 10. In these cases, the standard deviation is better than the range as a measure of the variation between individual measurements in a subgroup.

Deciding whether to use s or R is largely based on convenience. Ranges are easier to calculate (by hand) and easier to understand for most people. Ranges are also easier to plot. However, ranges are affected more by outliers in the data. Standard deviations give better estimates of the process standard deviation, particularly for larger subgroup values. Calculators and personal computers make the calculation of standard deviations relatively easy.

STEPS IN CONSTRUCTION OF THE $\overline{X}$-s CHART

The steps in constructing the $\overline{X}$-s chart are given below. A process flow diagram of the steps is shown in Figure 2.

1. Gather the data.

   a. Select the subgroup size (n). Typical subgroup sizes are 4 to 5. The concept of rational subgrouping should be considered. The objective is to minimize the amount of variation within a subgroup. This helps us "see" the variation in the averages chart more easily.

---

b. Select the frequency with which the data will be collected. Data should be collected in the order in which they are generated (in most cases).

c. Select the number of subgroups (k) to be collected before control limits are calculated (at least 20).

d. For each subgroup, record the individual, independent sample results.

e. For each subgroup, calculate the subgroup average:

$$\overline{X} = \Sigma X_i/n = (X_1 + X_2 + ... + X_n)/n$$

where X_1, X_2, etc. are the individual sample results and n is the subgroup size.

f. For each subgroup, calculate the subgroup standard deviation:

$$s = \sqrt{\Sigma(X_i - \overline{X})^2/(n - 1)}$$

2. Plot the data.

a. Select the scales for the x and y axes for both the $\overline{X}$ and the s charts.

b. Plot the subgroup standard deviations on the s chart and connect consecutive points with a straight line.

c. Plot the subgroup averages on the $\overline{X}$ chart and connect consecutive points with a straight line.

3. Calculate the overall process averages and control limits.

a. Calculate the average standard deviation ($\bar{s}$):

$$\bar{s} = \Sigma s_i/k = (s_1 + s_2 + ... + s_k)/k$$

where s_1, s_2, etc. are the standard deviations for subgroups 1, 2, etc. and k is the number of subgroups.

b. Plot $\bar{s}$ on the s chart as a solid line and label.

c. Calculate the overall process average ($\overline{\overline{X}}$):

$$\overline{\overline{X}} = \Sigma \overline{X}_i/k = (\overline{X}_1 + \overline{X}_2 + ... + \overline{X}_k)/k$$

where X_1, X_2, etc. are the subgroup averages for subgroups 1, 2, etc.

d. Plot $\overline{\overline{X}}$ on the $\overline{X}$ chart as a solid line and label.

e. Calculate the control limits for the s chart. The upper control limit is given by UCLs. The lower control limit is given by LCLs.

$$UCLs = B_4\overline{s}$$

$$LCLs = B_3\overline{s}$$

where B_4 and B_3 are control chart constants that depend on subgroup size.

f. Plot the control limits on the s chart as dashed lines and label.

g. Calculate the control limits for the $\overline{X}$ chart. The upper control limit is given by UCLx. The lower control limit is given by LCLx.

$$UCLx = \overline{\overline{X}} + A_3\overline{s}$$

$$LCLx = \overline{\overline{X}} - A_3\overline{s}$$

where A_3 is a control chart constant that depends on subgroup size.

h. Plot the control limits on the X chart as dashed lines and label.

4. Interpret both charts for statistical control.

a. Always consider variation first. If the s chart is out of control, the control limits on the $\overline{X}$ chart are not valid since you do not have a good estimate of $\overline{s}$. All tests for statistical control apply to the $\overline{X}$ chart. Points beyond the control limits, number of runs and length of runs apply to the s chart.

5. Calculate the process standard deviation, if appropriate.

a. If the s chart is in statistical control, the process standard deviation, $\hat{\sigma}'$, can be calculated as:

$$\hat{\sigma}' = \overline{s}/c_4$$

where c_4 is a control chart constant that depends on subgroup size.

To calculate control limits and to estimate the process standard deviation, you must use the control chart constants B_4, B_3, A_3, and c_4. These control chart constants depend on the subgroup size (n). These control chart constants are summarized in Table 1. For example, if your subgroup is 4, then $B_4 = 2.266$, $A_3 = 1.628$, and $c_4 = 0.9213$. There is no value for B_3. This simply means that the s chart has no lower control limit when the subgroup size is 4.

Figure 2

STEPS IN CONSTRUCTION OF THE $\overline{\text{X}}$-s CHART

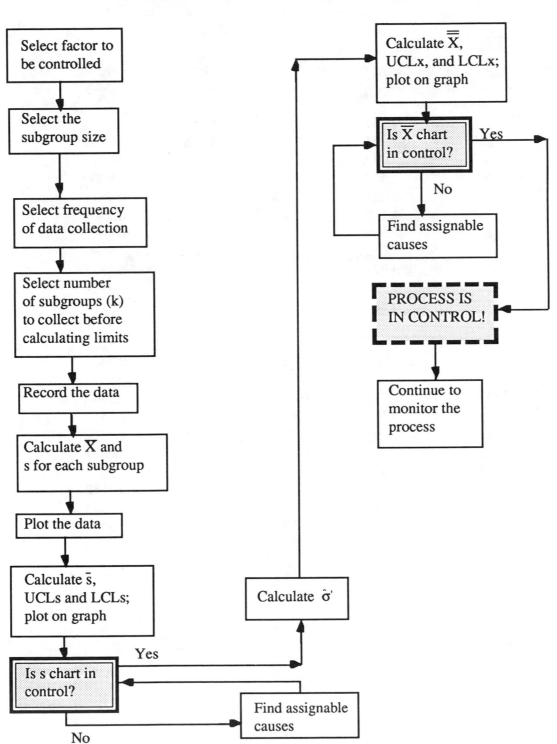

Table 1

FACTORS FOR USE WITH $\overline{X}$-s CHARTS

Subgroup Size (n)	A_3	B_3	B_4	c_4
2	2.659		3.267	0.7979
3	1.954		2.568	0.8862
4	1.628		2.266	0.9213
5	1.427		2.089	0.9400
6	1.287	0.030	1.970	0.9515
7	1.182	0.118	1.882	0.9594
8	1.099	0.185	1.815	0.9650
9	1.032	0.239	1.761	0.9693
10	0.975	0.284	1.716	0.9727
11	0.927	0.321	1.679	0.9754
12	0.886	0.354	1.646	0.9776
13	0.850	0.382	1.618	0.9794
14	0.817	0.406	1.594	0.9810
15	0.789	0.428	1.572	0.9823
16	0.763	0.448	1.552	0.9835
17	0.739	0.466	1.534	0.9845
18	0.718	0.482	1.518	0.9854
19	0.698	0.497	1.503	0.9862
20	0.680	0.510	1.490	0.9869
21	0.663	0.523	1.477	0.9876
22	0.647	0.534	1.466	0.9882
23	0.633	0.545	1.455	0.9887
24	0.619	0.555	1.445	0.9892
25	0.606	0.565	1.435	0.9896

$\overline{X}$-s CHART EXAMPLE

Multiple Suppliers

A plant has multiple suppliers for one "commodity" chemical. This chemical is delivered in railcars by all the suppliers. The railcars are weighed before and after unloading to determine the weight of the chemical received. The suppliers bill the plant based on weights obtained from their loading cells. There have been some discrepancies between the weights obtained at the plant and the weights measured by the suppliers. There also appear to be some differences between suppliers. Over the last month, seven shipments have been received from each supplier. It was decided to use the $\overline{X}$-s chart to look for differences between suppliers. Each subgroup is composed of the shipments received from one supplier. The differences in weights measured at the plant and the weights reported by the suppliers are given below.

Subgroup (Supplier) Number	Plant - Supplier Weights (1,000 lbs)							$\overline{X}$	s
	1	2	3	4	5	6	7		
1	1.7	-1.6	0.8	-0.3	0.2	1.2	0.7	0.39	1.09
2	0.3	0.6	0.1	-0.5	0.6	-1.2	0.3	0.03	0.66
3	-0.9	-1.5	-1.7	-0.4	-0.8	-0.1	-0.5	-0.84	0.58
4	0.3	0.5	-0.2	-0.4	0.2	0.3	0.1	0.11	0.31
5	-0.2	0.1	-1.0	0.5	0.2	0.9	-0.7	-0.03	0.66
6	0.1	0.3	-0.2	0.3	0.5	0.1	-0.2		
7	-0.4	0.8	0.3	-0.9	-0.5	0.4	0.2		

Use these data to construct the $\overline{X}$-s chart to compare suppliers. The first 5 subgroup averages and standard deviations have been calculated and plotted for you (p. 275). After constructing the $\overline{X}$-s chart, answer the following questions.

1. What variation is being examined on the s chart?

2. Is the s chart in or out of statistical control? What does this mean?

3. If the s chart is in statistical control, what is the estimate of the process standard deviation, $\hat{\sigma}'$? What is $\hat{\sigma}'$ measuring?

4. What variation is being examined on the $\overline{X}$ chart?

5. Is the $\overline{X}$ chart in or out of control? What does this mean?

6. What should be done next?

7. What do you think about the method of subgrouping used in this example? Are there other methods of subgrouping the data that will give you as much or more information?

8. Does this method of subgrouping affect the tests you can apply for statistical control?

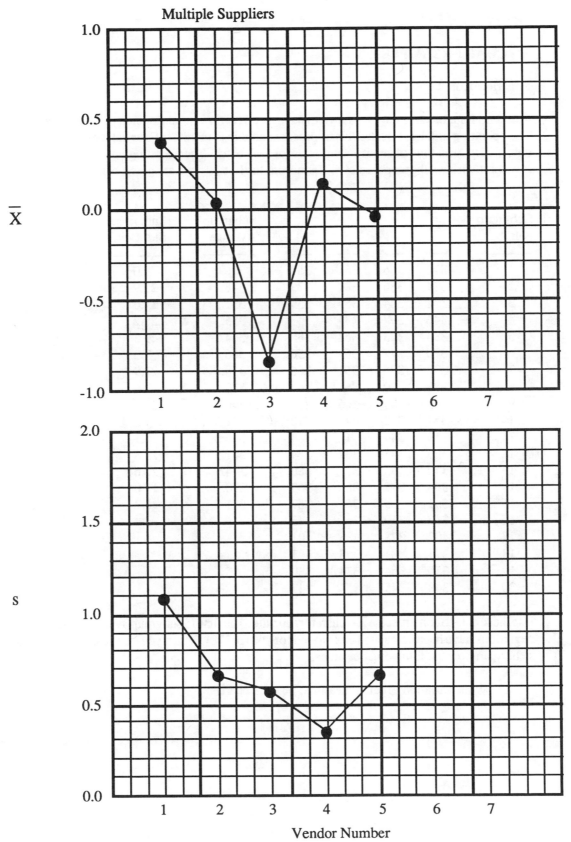

$\overline{X}$-s CHART EXAMPLE

Reaction Run Time

Run time for a batch reactor is known to have an effect on several finished product parameters. The operating unit has selected this in-process variable to monitor. Four batches are made each day, giving a run time of about 360 minutes for each batch. The operating unit decides to use the $\overline{X}$-s chart with a subgroup size of 4. Data from the last 20 days are given below.

Subgroup Number	1	2	3	4	Subgroup Average	Subgroup Std. Dev.
1	316	366	361	381	356.0	28.0
2	334	336	384	339	348.3	23.9
3	334	364	369	339	351.5	17.6
4	346	397	335	343	355.3	28.2
5	384	323	363	388	364.5	29.8
6	370	372	369	356	366.8	7.3
7	332	386	351	366	358.8	22.9
8	383	380	366	366	373.8	9.0
9	399	405	366	416	396.5	21.5
10	371	372	373	370	371.5	1.3
11	359	344	317	364	346.0	21.1
12	395	345	362	344	361.5	23.8
13	342	393	377	386	374.5	22.6
14	364	372	385	360	370.3	11.0
15	353	365	361	367	361.5	6.2
16	362	369	366	345		
17	402	378	388	401		
18	371	381	353	347		
19	350	364	374	330		
20	348	345	351	347		

Use these data to construct the $\overline{X}$-s chart for run time. The first 15 subgroup averages and standard deviations have been calculated and plotted for you (p. 278). When you have constructed the $\overline{X}$-s chart, answer the following questions.

1. What variation is being measured on the s chart?

2. Is the s chart in or out of statistical control? What does this mean?

3. If the s chart is in statistical control, what is the estimate of the process standard deviation, $\hat{\sigma}'$? What is $\hat{\sigma}'$ measuring?

4. What variation is being examined on the $\overline{X}$ chart?

5. Is the $\overline{X}$ chart in or out of statistical control? What does this mean?

6. What should be done next to improve the process?

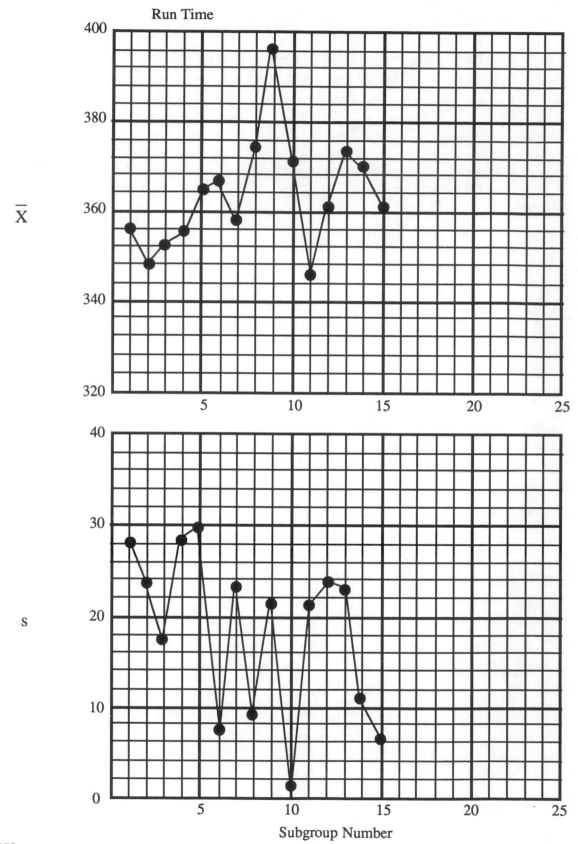

Run Time

SUMMARY

The $\overline{X}$-s chart is a control chart that may be used in place of the $\overline{X}$-R chart if the subgroup size is larger than 10. Other conditions for use of the $\overline{X}$-s chart include data that are frequently available and a method of rationally subgrouping the data. The $\overline{X}$ chart is a measure of the long-term variation in the process. The s chart is a measure of the short-term variation in the process.

APPLICATIONS OF $\overline{X}$-s CHARTS

Thing about your own work or processes. What are some possible applications of $\overline{X}$-s charts? Record you responses below.

1.

2.

3.

20
CONTROL CHARTS WITH TREND LINES

In the classical sense, if a process is in control, there is only one average and one measure of the variation about that average. However, some processes produce data that trend upwards or downwards. Examples of such processes include fixed costs (these will increase due to inflation) or catalyst yields (these may decrease over time). How do you handle these situations? Control charts with trend lines can be applied in these situations. This type of control chart is introduced in this chapter.

OBJECTIVES

In this chapter you will learn:

1. What a control chart with a trend line is
2. When to use a control chart with a trend line
3. How to construct a control chart with a trend line

The trend control chart is a type of control chart that can be used with variables data. It consists of two charts. One chart monitors the variation in the individual response or subgroup average. The other chart monitors the variation in the range between individual responses or within the subgroup.

INTRODUCTION TO THE TREND CONTROL CHART

The classical Shewhart charts consist of two charts. One chart monitors variation in a subgroup range, or the moving range, between successive samples. The objective of this chart is to determine if the short-term variation about the average is consistent over time. The other chart monitors variation in the subgroup average or the individual sample result. The objective of this chart is to determine if the process average is consistent over time. If the process is in statistical control, the average and variation about the average do not change over time, as shown below in Figure 1.

Figure 1

VARIATION OF A STABLE PROCESS

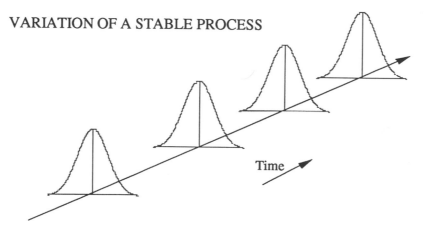

Time

There are many naturally occurring processes, however, where the average value is not consistent over time but changes in a predictable fashion. The average either increases or decreases with time. Examples of these types of processes include catalyst activity, financial items affected by inflation, machine wear, etc. If parameters from these processes are charted using classical Shewhart charts, the process would appear to be out of control with respect to the average. However, this would not be true since the process naturally changes its average over time. The short-term variation measured by the range chart does not change. This situation is depicted in Figure 2. This figure demonstrates how the average value changes over time while the variation about that average remains consistent over time.

Figure 2

VARIATION IN A PROCESS WITH A TREND

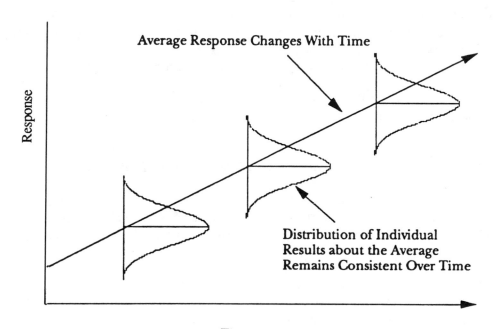

The trend control chart provides a method of handling this type of situation. A trend control chart is two charts. One chart is for either subgroup averages or individual sample results. The other chart is for the subgroup ranges or ranges between successive individual sample results.

Figure 3 is an example of a trend control chart. In this example, a department manager is monitoring his fixed cost expenditures over time. He expects these costs to increase due to inflation but wants a method of determining if his costs are rising faster than inflation or if any assignable causes are present. An individuals control chart can't be used since the process would appear to be out of control due to the upward trend caused by inflation. The department manager selected a trend control chart. Data on fixed cost expenditures are available on a monthly basis. He has data for the past 20 months.

The top part of Figure 3 is a plot of the monthly fixed cost expenditures. As can be seen, the data points are trending upward. To determine the center line for a trend control chart, the equation of the best-fit, least- squares line must be determined. Once this has been determined, this line is plotted on the chart. As can be seen in Figure 3, this line slopes upward. The control limits can then be calculated and plotted. As shown in Figure 3, these control limits have the same slope as the center line. Procedures for determining the center line and control limits are given below.

The bottom part of Figure 3 is the moving range chart. It is constructed exactly the same as for the individuals control chart.

The process represented by Figure 3 is in statistical control. What does this mean for a trend control chart? The interpretation of the range chart is the same as before. Since the range chart is in statistical control, the short-term variation is consistent over time. The ranges between successive samples are not significantly different from the average range. The trend chart itself is in statistical control. This means that the monthly fixed cost expenditures are not significantly different from the average predicted by the best-fit equation for that point in time. The department manager can predict what his fixed costs will be in the future.

DETERMINING THE BEST-FIT EQUATION FOR THE CENTER LINE

The method of determining the equation for the center line is given below. This method assumes that individual sample results are being plotted. The same method holds if subgroup averages are being plotted except that the subgroup average $(\overline{X})$ is used in place of the individual sample result (X).

The equation for the center line is determined using the method of least squares. The equation for a line using simple regression between two variables is usually represented as $y = mx + b$ where y is the dependent variable, x is the independent variable, m is the slope of the line and b is the y intercept. Since X is used as the symbol for the parameter being monitored by control charts, the equation of the line will be written as:

$$X_t = mt + b$$

where X_t = parameter being plotted (dependent variable)
          t = time (independent variable)
          m = slope of the line
          b = y intercept

Figure 3

TREND CONTROL CHART EXAMPLE: COST EXPENDITURES

TREND CHART: MONTHLY EXPENDITURES

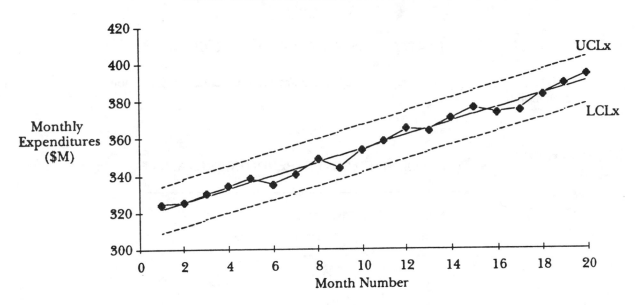

RANGE CHART (n = 2)

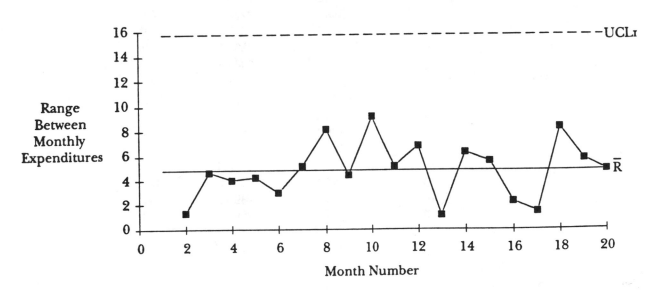

The slope of the line, m, represents the change in the average value of the parameter between samples. The value of m is obtained from the equation below.

$$m = [\Sigma tX - ((\Sigma t \Sigma X)/k)]/[\Sigma t^2 - ((\Sigma t)^2/k)]$$

where k is the number of subgroups. The data used to generate Figure 3 is shown in Table 1. Using these data, the value of m can be calculated as:

$$m = [77242.5-((210)(7125.3)/20)]/[2870-(210^2/20)] = 3.65$$

This slope implies that the average fixed cost expenditures will increase by $3.65 thousand dollars each month.

Table 1

DATA FOR FIXED COST EXPENDITURES EXAMPLE

Month Number	Fixed Costs $ thousand	tX	t^2
1	324.3	324.3	1
2	325.6	651.2	4
3	330.2	990.6	9
4	334.2	1336.8	16
5	338.4	1692.0	25
6	335.4	2012.4	36
7	340.5	2383.5	49
8	348.6	2788.8	64
9	344.2	3097.8	81
10	353.4	3534.0	100
11	358.5	3943.5	121
12	365.3	4383.6	144
13	364.2	4734.6	169
14	370.5	5187.0	196
15	376.1	5641.5	225
16	373.8	5980.8	256
17	375.2	6378.4	289
18	383.5	6903.0	324
19	389.3	7396.7	361
20	394.1	7882.0	400
Sum 210	7125.3	77242.5	2870
Average 10.50	356.27		
Std. Dev. 5.92	21.79		

The value of b is determined from the equation below.

$$b = \overline{X} - m(\Sigma t_i)/k$$

where $\overline{X}$ is the overall average. Using the data in Table 1, the value of b is calculated as:

$$b = \overline{X} - m(\Sigma t_i)/k = 356.27 - 3.65(210)/20 = 317.945$$

The equation for the center line is then given by:

$$X_t = mt + b = 3.65t + 317.945$$

The procedure above is a simple linear regression for determining the relationship between two variables. The question one must ask is "Is this relationship significant?" For example, if there was no relationship between X_t and t, one would expect the value of m to be zero. However, it will not be zero due to the variation in the process. One must ask "Is the value of m significantly different from zero?" If it is, then there is a statistically significant relationship between X_t and t. If it is not, there is no relationship between X_t and t.

To determine if there is a statistically significant relationship between two variables, the simple linear correlation coefficient (R_{yx}) can be used. R_{yx} is the same coefficient used with scatter diagrams. The value of R_{yx} is determined by:

$$R_{yx} = (\Sigma t_i X_i - nt\overline{X})/(n - 1)s_t s_x$$

where s_t is the standard deviation of the t values and s_x is the standard deviation of the X values. Using the data in Table 1, the value of R_{yx} can be calculated as:

$$R_{yx} = [(210)(7125.3) - (20)(10.5)(356.27)]/(19)(5.92)(21.79) = 0.991$$

This value of R_{yx} is then compared to the critical values shown in Table 2. If the calculated value is larger than the table value, a statistically significant correlation exists. For the data in Table 1, there are 18 degrees of freedom (df = n - 2, where n = number of paired samples). The critical value for R_{yx} from Table 2 is .444. Since 0.991 is larger than this, there is a statistically significant correlation between fixed costs expenditures and time.

Even if there is a statistically significant correlation between two variables, it may not be useful in a practical sense. It will only be useful if most of the variation in X_t is explained by the variation in t. The fraction of variation in X_t explained by the variation in t is given by R^2 which is equal to R_{yx}^2. In general, if R^2 is greater than 0.80, the correlation will be useful from a practical standpoint. However, this decision is up to the person developing the chart. For the fixed cost expenditure example, the value of R^2 is

Table 2

CRITICAL VALUES FOR THE LINEAR CORRELATION COEFFICIENT

Degrees of Freedom	Probability	Degrees of Freedom	Probability
1	.997	21	.413
2	.950	22	.404
3	.878	23	.396
4	.811	24	.388
5	.754	25	.381
6	.707	26	.374
7	.666	27	.367
8	.632	28	.361
9	.602	29	.355
10	.576	30	.349
11	.553	35	.325
12	.532	40	.304
13	.514	45	.288
14	.497	50	.273
15	.482	60	.250
16	.468	70	.232
17	.456	80	.217
18	.444	90	.205
19	.433	100	.195
20	.423		

The probabilities listed are for $\alpha = 0.05$. If a calculated value of R_{yx} is larger or equal to the table value, there is a statistically significant correlation between x and y. α is the probability that we will conclude that there is not a correlation when there actually is one.

$0.991^2 = 0.982$. Thus, 98.2% of the variation in fixed cost expenditures can be explained by the variation in time. The equation for the center line is useful from a practical viewpoint.

DETERMINING CONTROL LIMITS FOR THE TREND CHART

The procedure for determining control limits for trend control charts is given below. The range chart control limits depend on what is being plotted. If the individual sample results are being plotted, the control limits for the range chart are:

$$UCLr = 3.267\overline{R} \quad LCLr = None$$

Note that these are the same as for the moving range chart associated with the individuals control chart. If subgroup averages are being plotted, the control limits for the range chart are the same as for the $\overline{X}$-R chart and are given by:

$$UCLr = D_4\overline{R} \quad LCLr = D_3\overline{R}$$

where D_4 and D_3 are control chart constants that depend on subgroup size. (The s chart for the $\overline{X}$-s chart could also be used to determine short-term variation.)

The control limits for the trend chart itself depend upon what is being charted. For individual sample results, the control limits are:

$$UCLx = X_t + 2.66\overline{R} = mt + b + 2.66\overline{R}$$

$$LCLx = X_t - 2.66\overline{R} = mt + b - 2.66\overline{R}$$

If subgroup averages are being plotted, the control limits are:

$$UCLx = X_t + A_2\overline{R} = mt + b + A_2\overline{R}$$

$$LCLx = X_t - A_2\overline{R} = mt + b - A_2\overline{R}$$

where A_2 is a control chart constant that depends on subgroup size.

With classical Shewhart charts, an estimate of the process standard deviation can be obtained from the range chart if it is in control. The process standard deviation is given by $\hat{\sigma}' = \overline{R}/d_2$. The process standard deviation is a measure of the variation in the individual measurements. With a trend control chart, this can't be done. The value obtained for the standard deviation would be based on the short-term variation. Since the average changes over time, the true variation in the process is much larger.

WHEN TO USE THE TREND CONTROL CHART

The trend control chart can be used in the second step of the problem solving model. This step involves analyzing how the process is behaving.

The trend control chart should be used when the process has an average that increases or decreases with time in a predictable fashion. It can be used with either individual sample results or subgroups. Care should be taken when using trend charts with individual sample results. The underlying assumption is that the individual sample results at a given time are normally distributed. This will not be possible to check in most cases since the average changes with time.

STEPS IN CONSTRUCTION OF THE TREND CONTROL CHART

The steps in constructing a trend control chart are given below. The procedure assumes that individual sample results are being plotted. Similar procedures hold if subgroup averages are being plotted. A process flow diagram of the steps is given in Figure 4.

1. Gather the data.

    a. Select the frequency with which the data will be collected. Data should be collected in the order in which it is generated.

    b. Select the number of data points (k) to be collected before control limits are calculated (at least 20).

    c. Record the individual sample result.

    d. Calculate the moving range (R_{i+1}) between consecutive sample results:

    $$R_{i+1} = |X_{i+1} - X_i|$$

    where X_{i+1} is the result for sample i+1 and X_i is the result for sample i. The range value is always positive.

2. Plot the data.

    a. Select the scales for the x and y axes for both the trend and R charts.

    b. Plot the ranges on the R chart and connect consecutive points with a straight line.

    c. Plot the individual sample results on the trend chart and connect consecutive points with a straight line.

---

3. Calculate the average range and control limits for the range chart.

    a. Calculate the average range ($\bar{R}$):

$$\bar{R} = \Sigma R_i/(k - 1) = (R_2 + R_3 + ... + R_k)/(k - 1)$$

       where R_2, R_3, etc. are the ranges between samples 2 and 1, samples 3 and 2, etc. and k is the number of subgroups.

    b. Plot $\bar{R}$ on the range chart as a solid line and label.

    c. Calculate the control limits for the R chart. The upper control limit is given by UCLr. The lower control limit is given by LCLr.

$$UCLr = 3.267\bar{R}$$

$$LCLr = none$$

    d. Plot the upper control limit on the R chart as a dashed line and label.

4. Determine the equation for the center line and control limits for the trend chart.

    a. Calculate the overall process average ($\bar{X}$):

$$\bar{X} = \Sigma X_i/k = (X_1 + X_2 + ... + X_k)/k$$

       where X_1, X_2, etc. are the individual sample results 1, 2, etc.

    b. Calculate m and b:

$$m = [\Sigma tX - ((\Sigma t\Sigma X)/k)]/[St^2 - ((\Sigma t)^2/k)]$$

$$b = \bar{X} - m(\Sigma t_i)/k$$

    c. Plot $X_t = mt + b$ on the X chart as a solid line and label.

    d. Determine if the best-fit line is statistically significant by calculating R_{yx}:

$$R_{yx} = (\Sigma t_i X_i - nt\bar{X})/(n - 1)s_t s_x$$

    e. Compare R_{yx} to the critical value in the table of linear regression coefficients. If R_{yx} is greater than the critical value, there is a statistically significant correlation between X and t. If not, there is no relation between X and t and the trend chart cannot be used.

f. Determine R^2 to see if the fit is of practical importance:

$$R^2 = R_{yx}{}^2$$

If R^2 is greater than 0.80, the fit is generally useful from a practical point of view.

g. Calculate the control limits for the trend chart. The upper control limit is given by UCLx. The lower control limit is given by LCLx.

$$UCLx = X_t + 2.66\overline{R} = mt + b + 2.66\overline{R}$$

$$LCLx = X_t - 2.66\overline{R} = mt + b - 2.66\overline{R}$$

h. Plot the control limits on the trend chart as dashed lines and label.

5. Interpret both charts for statistical control.

a. Always consider variation first. If the R chart is out of control, the control limits on the trend chart are not valid since you do not have a good estimate of $\overline{R}$.

b. All tests for statistical control apply to the trend chart. However, the data on the range chart are not independent. Each data point is used twice. The only test that is valid for the range chart is points beyond the control limits.

Figure 4

STEPS IN CONSTRUCTION OF THE TREND CONTROL CHART

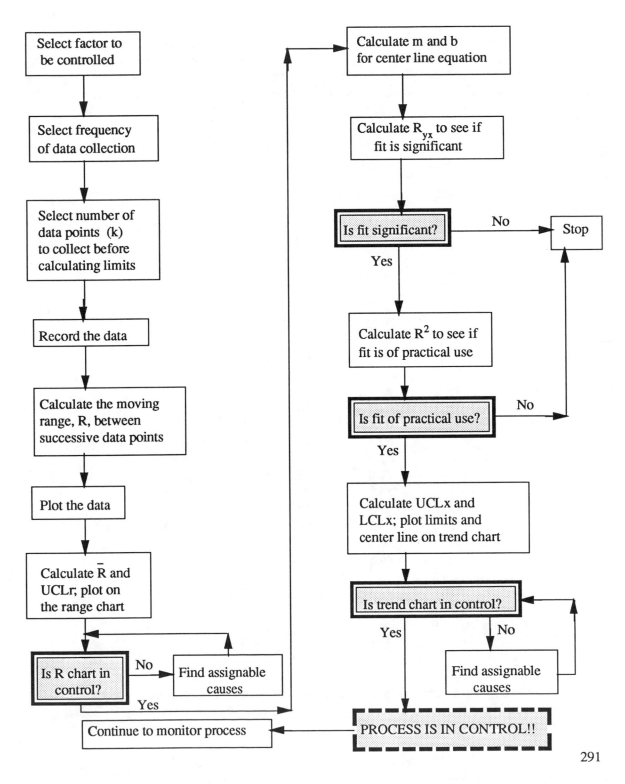

Chapter 20

TREND CONTROL CHART EXAMPLE

Reaction Yields

A reactor in a continuous process contains a catalyst that has declining yields over time. The decrease in yield can be offset for some time by increasing reaction temperature. This adds cost, however, to the production of the product. The lowest acceptable yield from the reactor is 85%. New catalyst produces yields of around 89%. This yield drops off with time. In the past, the temperature of the reactor was increased anytime the reaction yield dropped below 88%. In an effort to save utility costs, an operating unit decided to let the reaction yield drop closer to 85% before changing the reaction temperature. Personnel in the unit decided to use a trend chart to monitor reaction yield. Data from the past 20 samples are given below.

Sample Number	% Yield	tX	t^2	Moving Range
1	89.3	89.3	1	
2	89.5	179.0	4	0.20
3	89.2	267.6	9	0.30
4	89.4	357.6	16	0.20
5	88.7	443.5	25	0.70
6	88.9	533.4	36	0.20
7	88.6	620.2	49	0.30
8	88.3	706.4	64	0.30
9	88.5	796.5	81	0.20
10	87.9	879.0	100	0.60
11	87.6	963.6	121	0.30
12	88.2	1058.4	144	0.60
13	87.8	1141.4	169	0.40
14	87.3	1222.2	196	0.50
15	87.0	1305.0	225	0.30
16	86.8	1388.8	256	0.20
17	87.5	1487.5	289	0.70
18	86.4	1555.2	324	1.10
19	86.2	1637.8	361	0.20
20	86.8	1736.0	400	0.60

Sum	210	1759.9	18368.4	2870
Average	10.50	88.00		0.42
Std. Dev.	5.92	1.03		

Use these data to construct a trend control chart. Most of the calculations have been completed for you above. The points have been plotted for you (p. 294). When you finish constructing the chart, answer the following questions.

1. What variation is being measured on the R chart?

2. Is the R chart in statistical control? What does this mean?

3. What variation is being measured on the trend chart?

4. What is the equation for the center line? Is it statistically significant? If yes, is it of any practical use?

5. Is the trend chart in statistical control? What does this mean?

6. If the yield can never get below 85%, at what value of the yield should the reaction temperature be adjusted?

7. What, if anything, should be done next to improve this process?

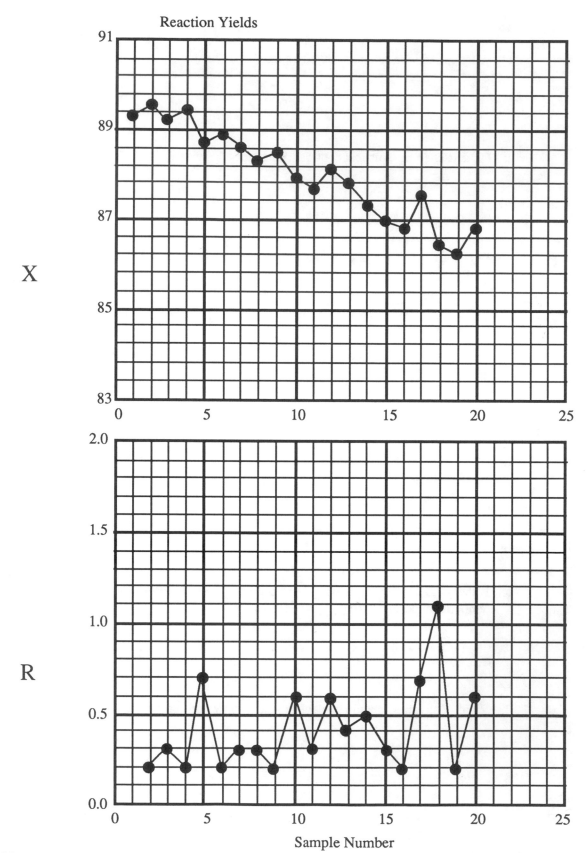

Reaction Yields

X

R

Sample Number

TREND CONTROL CHART EXAMPLE

Product Sales

A product manager has been monitoring sales of a certain product. Sales have been increasing steadily due to improving economic conditions. The manager would like to monitor sales to determine if any assignable causes are present in the process. She decides to use a trend control chart to help account for the improving economic conditions. She has data for the last 20 months. The data are shown below.

Month Number	Sales (MM lb)	tX	t^2	Moving Range
1	4.5	4.5	1	
2	4.6	9.2	4	0.10
3	5.7	17.1	9	1.10
4	4.5	18.0	16	1.20
5	7.3	36.5	25	2.80
6	5.6	33.6	36	1.70
7	8.3	58.1	49	2.70
8	6.5	52.0	64	1.80
9	5.7	51.3	81	0.80
10	6.8	68.0	100	1.10
11	7.9	86.9	121	1.10
12	8.3	99.6	144	0.40
13	9.2	119.6	169	0.90
14	8.6	120.4	196	0.60
15	8.1	121.5	225	0.50
16	8.5	136.0	256	0.40
17	9.1	154.7	289	0.60
18	9.5	171.0	324	0.40
19	8.7	165.3	361	0.80
20	9.0	180.0	400	0.30
Sum	210	146.4	1703.3	2870
Average	10.50	7.32		1.02
Std. Dev.	5.92	1.69		

Use these data to construct a trend control chart. Most of the calculations have been done for you. The points have also been plotted (p. 297). When you have completed the chart, answer the following questions.

1. What variation is being measured on the R chart?

2. Is the R chart in statistical control? What does this mean?

3. What variation is being measured on the trend chart?

4. What is the equation for the center line? Is it statistically significant? If yes, is it of any practical use?

5. Is the trend chart in statistical control? What does this mean?

6. What, if anything, should be done next to improve this process?

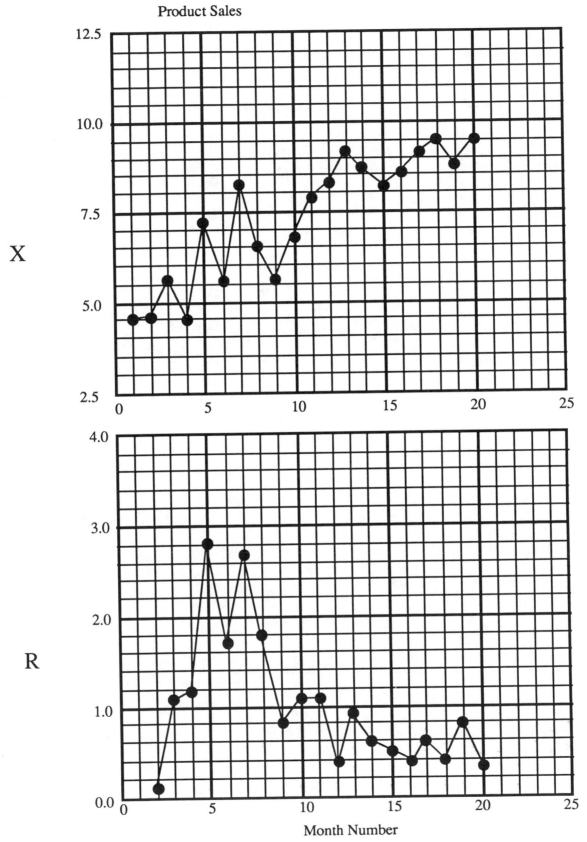

Product Sales

X

R

Month Number

SUMMARY

Trend control charts were introduced in this chapter. Trend control charts are useful for processes with averages that change over time in a predictable fashion. Trend control charts can be used to determine how a process is behaving. Either individual sample results or subgroup averages can be used with trend control charts. A best-fit equation is determined for the center line of the trend chart. This equation is used to predict what the average will be for the next points in time.

APPLICATIONS OF TREND CONTROL CHARTS

Think about your processes at work. What are three possible applications of trend control charts.

1.

2.

3.

21

CUMULATIVE SUM CONTROL CHARTS

In the classical sense, if a process is in statistical control, it means that there is only one average, one measure of the variation, and one shape of the distribution. Control charts and histograms can be used to estimate these population parameters (average, standard deviation, and distribution shape). In the process industries, there are many variations (such as raw material variations and swings in ambient temperature) that can tend to move a process off the average without changing the variation about that average. The classical Shewhart charts will often not detect these shifts quickly, particularly if the zones tests and runs tests are not applied. The cumulative sum (CUSUM) control chart is one method of detecting when the process average has shifted from the target. The CUSUM control chart is introduced in this module.

OBJECTIVES

In this chapter you will learn:

1. What the CUSUM control chart is
2. When to use the CUSUM control chart
3. How to construct the CUSUM control chart

The CUSUM control chart provides a method of detecting when the process average has shifted from the target. It is a type of control chart that is used with variables data. Unlike most other variables control charts, the CUSUM chart is only one chart. In fact, a chart is not often used. The CUSUM chart discussed in this chapter is the basic CUSUM chart. Other control schemes exist for CUSUM charts.

INTRODUCTION TO THE CUSUM CONTROL CHART

The CUSUM control chart is a method of looking at variation. A CUSUM control chart looks at deviations of individual sample results or subgroup averages from the target value. The cumulative sum is simply the sum of these deviations over successive samples. The first step is to determine what shift from the target you want to detect quickly. This shift is usually one standard deviation of the parameter being plotted. Sample results near the target value are considered to be an indication that the process is operating on target. These result in subtractions from the cumulative sum unless it is already zero. Sample results that are beyond a certain value (which depends on the shift desired to be detected) are considered evidence that the process is not operating on target, and these are added to the cumulative sum. If the cumulative sum equals or exceeds the control limit (or action limit), an out-of-control situation is indicated (DuPont, 1987).

Figure 1 is an example of a CUSUM chart. It looks considerably different from other types of control charts. The center line on the CUSUM chart is usually zero. Three different points are usually plotted on the chart. One point is the deviation of the sample from the target. This is determined by subtracting the target value from the sample result. Another point is the cumulative sum on the high side (SH). The other point is the cumulative sum on the low side (SL). Usually -SL is plotted. There are four additional lines plotted on the chart. The lines marked "K" represent the area where the process is assumed to be operating on target. Anytime a deviation from target lies in this area, it results in subtractions from the cumulative sum of the deviations. Anytime a deviation falls between the lines marked "K" and "H," it is an indication that the process may be operating off target, and this results in additions to the cumulative sum of the deviations. The lines marked "H" are the action limits. These action limits are like control limits. Anytime the cumulative sum is greater than this value, the process is out of control.

Figure 1

EXAMPLE OF CUSUM CONTROL CHART

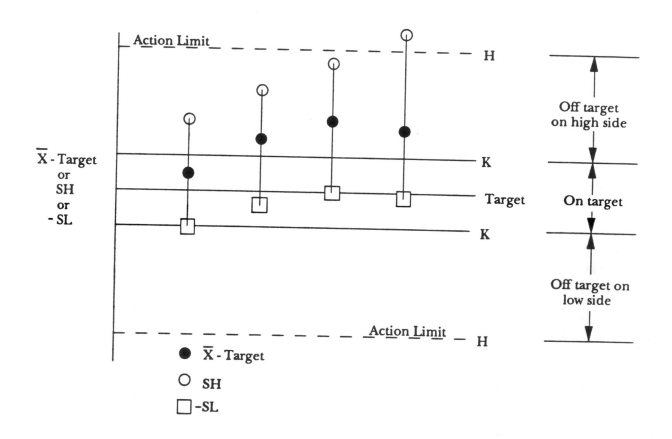

To understand how the CUSUM control chart works and the terminology associated with the CUSUM control chart, consider the following example. Suppose the target value for a product response is 100. Samples are taken four times per day. An $\overline{X}$-R chart is being used to monitor the variation in the product response. The data for the past 25 days are given in Table 1 (these data were randomly generated using a process with an average of 100 and a standard deviation of 10). The control limit calculations are also given in the table. The process is in statistical control.

Table 1

DATA USED TO CONSTRUCT THE $\overline{X}$-R CHART ON PRODUCT RESPONSE

Subgroup Number	1	2	3	4	$\overline{X}$	Range
1	102.2	114.1	113.8	108.5	109.6	11.9
2	100.0	99.2	89.2	118.0	101.6	28.8
3	100.1	97.4	106.4	109.4	103.3	11.9
4	100.3	113.5	91.4	123.3	107.1	31.9
5	79.4	101.9	90.6	103.5	93.8	24.1
6	112.2	107.4	107.1	102.0	107.2	10.1
7	115.8	89.2	105.9	94.8	101.4	26.6
8	103.4	79.6	111.3	94.6	97.2	31.6
9	108.1	103.1	95.0	99.5	101.4	13.1
10	104.3	88.0	83.0	100.3	93.9	21.3
11	98.8	120.1	88.0	106.2	103.3	32.1
12	100.7	103.0	89.4	104.3	99.4	14.9
13	96.4	96.1	110.3	99.2	100.5	14.2
14	115.1	92.6	88.1	93.7	97.4	27.0
15	77.0	84.8	103.5	103.9	92.3	26.9
16	91.2	115.0	82.4	99.6	97.1	32.6
17	98.7	94.2	80.0	88.9	90.5	18.7
18	92.2	104.8	91.1	103.3	97.9	13.7
19	99.3	101.1	108.8	109.4	104.7	10.2
20	89.4	109.2	107.2	103.4	102.3	19.8
21	110.4	106.7	86.8	94.4	99.6	23.6
22	107.6	119.0	114.2	95.5	109.1	23.5
23	94.5	83.7	110.9	110.2	99.8	27.1
24	106.3	103.9	104.3	97.7	103.0	8.6
25	92.3	106.6	96.9	104.2	100.0	14.3

$\overline{\overline{X}} = 100.54$

$\overline{R} = 20.75$

$UCLx = \overline{\overline{X}} + A_2\overline{R} = 115.66$

$LCLx = \overline{\overline{X}} - A_2\overline{R} = 85.41$

$UCLr = D_4\overline{R} = 47.34$

$\hat{\sigma}' = \overline{R}/d_2 = 10.07$

Suppose the process average shifts to 95 instead of 100. Samples continue to be taken from the process. Subgroup averages and ranges are calculated and plotted on the $\overline{X}$-R chart. The data for the next 10 subgroups are given below in Table 2.

Table 2

DATA WITH PROCESS AVERAGE EQUAL TO 95

Subgroup Number	1	2	3	4	$\overline{X}$	Range
26	113.2	96.5	109.4	89.6	102.2	23.5
27	91.5	104.5	83.8	110.2	97.5	26.4
28	111.1	89.4	95.0	107.5	100.7	21.8
29	89.9	100.1	91.5	103.0	96.1	13.1
30	79.5	103.1	100.5	123.3	101.6	43.8
31	104.4	85.0	95.7	73.3	89.6	31.2
32	90.6	90.0	109.6	94.6	96.2	19.6
33	79.5	82.5	111.1	100.4	93.4	31.5
34	87.6	97.3	83.6	89.4	89.5	13.7
35	93.3	93.8	91.6	108.6	96.8	17.0

The $\overline{X}$-R chart for the 35 subgroups is shown in Figure 2. Control limits were based on the first 25 subgroups. The chart indicates that the process is in statistical control. The shift of the process average from 100 to 95 has not yet been "seen" by the control chart. The CUSUM chart provides a method of detecting these types of shifts.

The CUSUM chart can be used to detect small shifts in the process average. Before constructing a CUSUM chart for this example, we introduce the following terminology associated with the CUSUM.

- Aim = Target value
- $\overline{X}_i$ = average of subgroup i
- DELTA = shift from aim that we would like to detect quickly (usually one standard deviation of the variable being monitored)
- $\hat{\sigma}'_{\overline{X}}$ = routine random variability in the subgroup averages
- H = action limits = $4.5\hat{\sigma}'_{\overline{X}}$
- K = allowable slack or deadband = DELTA/2
- SH = cumulative sum of the high side difference; used to detect a deviation from target on the high side; if the value is negative, SH is set to zero; if SH is greater than H, an out-of-control situation is indicated; SH for the ith subgroup is given by:

$$SH(i) = Max[0,(\overline{X}_i - AIM) - K + SH(i - 1)]$$

Figure 2

$\overline{X}$-R CHART FOR PRODUCT RESPONSE

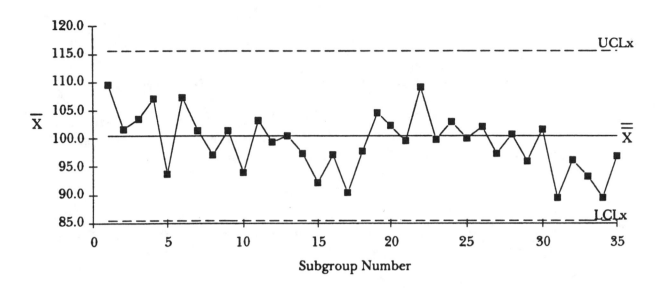

RANGE CHART FOR PRODUCT RESPONSE (n = 4) WITH 35 SUBGROUPS

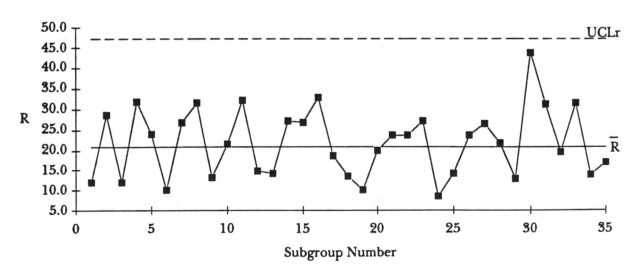

- SL = cumulative sum of the low side difference; used to detect a deviation from target on the low side; if the value is negative, SL is set to zero; if SL is greater than H, an out-of-control situation is indicated; SL for the ith subgroup is given by:

$$SL(i) = Max[0, -(\overline{X}_i - AIM) - K + SL(i - 1)]$$

- NH = number of consecutive subgroups for which SH is greater than zero
- NL = number of consecutive subgroups for which SL is greater than zero

For this example, AIM = 100. DELTA will be one standard deviation ($\hat{\sigma}'_{\overline{X}}$) of the subgroup averages. There are a number of ways to determine this value. The method used here has been suggested by James Lucas (DuPont, 1987). It uses the mean square successive difference to estimate $\hat{\sigma}'_{\overline{X}}$:

$$\hat{\sigma}'_{\overline{X}} = \sqrt{\Sigma(\overline{X}_{i+1} - \overline{X}_i)^2 / 2(k - 1)}$$

where $\overline{X}_{i+1}$ and $\overline{X}_i$ are the subgroup averages for subgroups i+1 and i, respectively, and k is the number of subgroups. For this example:

$$DELTA = \hat{\sigma}'_{\overline{X}} = \sqrt{\Sigma(\overline{X}_{i+1} - \overline{X}_i)^2 / 2(k - 1)} = \sqrt{1077.17/48} = 4.74$$

The action limit (or control limit) is given by 4.5 $\hat{\sigma}'_{\overline{X}}$. Thus:

$$H = 4.5\hat{\sigma}'_{\overline{X}} = 4.5(4.74) = 21.33$$

The allowable slack or deadband is K:

$$K = DELTA/2 = 4.74/2 = 2.37$$

The method for calculating SH and SL is given below. The initial values of SH and SL are zero, i.e., SH(0) = SL(0) = 0. The average of subgroup 1 is 109.6. SH and SL for this subgroup are:

$$SH(1) = Max[0,(\overline{X}_1 - AIM) - K + SH(0)] = Max[0,(109.6 - 100) - 2.37 + 0]$$
$$= Max[0,7.23]$$

$$SH(1) = 7.23$$

$$SL(1) = Max[0,-(\overline{X}_1 - AIM) - K + SL(0)] = Max[0,-(109.6 - 100) - 2.37 + 0]$$
$$= Max[0,-11.43]$$

$$SL(1) = 0$$

The values of SH(1) and SL(1) are compared to H. If either one is larger than H, there is an out-of-control situation. Since both are less than H = 21.33, the process is assumed to be operating on target. The average of subgroup 2 is 101.6. SH and SL are calculated as:

$$SH(2) = Max[0,(\overline{X}_2 - AIM) - K + SH(1)] = Max[0,(101.6 - 100) - 2.37 + 7.23]$$

$$SH(2) = Max[0, 6.46] = 6.46$$

$$SL(2) = Max[0, -(\overline{X}_2 - AIM) - K + SL(1)] = Max[0, -(101.6 - 100) - 2.37 + 0]$$

$$SL(2) = Max[0, -3.97] = 0$$

This process continues for all the subgroups. The results for all 35 subgroups are shown in Table 3.

For subgroup number 34, the value of SL exceeds H. This means that there is an out-of-control point and that the process is no longer operating on target. The reason for the out-of-control situation should be identified and corrected. If the reason for the assignable cause cannot be found, the process must be adjusted to bring it back on target. An estimate of the current process can be obtained using the following:

$$Current\ process\ average = AIM - K - (SL/NL) = 100 - 2.37 - (21.85/4) = 92.2$$

Whenever SH is greater than H, the current process average can be calculated from:

$$Current\ process\ average = AIM + K + (SH/NH)$$

Figure 3 is the CUSUM chart for this example. Plotted on the chart are the values of SH, SL and $\overline{X}_i$ - AIM. Also plotted on the chart are H and K. Note that it is not necessary to chart the data. The information can be tracked in tabular form. However, charting the data does help visualize what is occurring in the process.

Once the process has been adjusted or the assignable cause removed, the process should be operating on target. The values of SH and SL can be set back to zero. However, there is a technique called the Fast Initial Response (FIR) in which the values of SH and SL are not set back to zero. Instead, the values are set to an initial value SU(0). Usually, SU(0) = H/2. Using this technique provides a method of permitting a faster response to off-target operation after a process change or the removal of the assignable cause.

Table 3

CUSUM RESULTS

Sample Number	$\overline{X}$	SH	NH	SL	NL
1	109.6	7.27	1	0.00	0
2	101.6	6.50	2	0.00	0
3	103.3	7.47	3	0.00	0
4	107.1	12.21	4	0.00	0
5	93.8	3.69	5	3.78	1
6	107.2	8.50	6	0.00	0
7	101.4	7.56	7	0.00	0
8	97.2	2.42	8	0.40	1
9	101.4	1.49	9	0.00	0
10	93.9	0.00	0	3.72	1
11	103.3	0.90	1	0.00	0
12	99.4	0.00	0	0.00	0
13	100.5	0.00	0	0.00	0
14	97.4	0.00	0	0.26	1
15	92.3	0.00	0	5.60	2
16	97.1	0.00	0	6.16	3
17	90.5	0.00	0	13.33	4
18	97.9	0.00	0	13.11	5
19	104.7	2.30	1	6.07	6
20	102.3	2.24	2	1.39	7
21	99.6	0.00	0	0.00	0
22	109.1	6.72	1	0.00	0
23	99.8	4.19	2	0.00	0
24	103.0	4.85	3	0.00	0
25	100.0	2.50	4	0.00	0
26	102.2	2.32	5	0.00	0
27	97.5	0.00	0	0.16	1
28	100.7	0.00	0	0.00	0
29	96.1	0.00	0	1.50	1
30	101.6	0.00	0	0.00	0
31	89.6	0.00	0	8.04	1
32	96.2	0.00	0	9.45	2
33	93.4	0.00	0	13.70	3
34	89.5	0.00	0	21.85	4
35	96.8	0.00	0	22.65	5

Figure 3

CUSUM CHART FOR PRODUCT RESPONSE

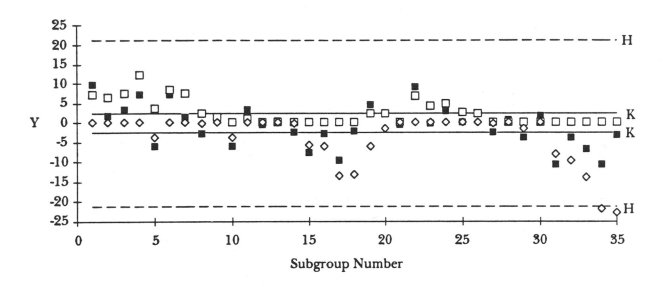

$Y = \overline{X}_i$ - Target or SH or -SL. The solid squares are $\overline{X}_i$ - Target. The empty squares are SH. The empty diamonds are -SL.

WHEN TO USE A CUSUM CONTROL CHART

The CUSUM control chart can be used in the second step of the problem solving model. This step involves analyzing how the process is behaving.

CUSUM charts can be used with subgroup averages or with individual sample results. They should be used when you want to detect small shifts in the process average (usually 0.5 to 2 standard deviations of the parameter being plotted) quickly. If the zones tests and runs tests are applied to other variable control (such as the $\overline{X}$-R chart), the ability of these charts to detect small process shifts approaches the ability of the CUSUM chart.

Range charts are not normally used in conjunction with CUSUM charts. Thus, there is not a method of monitoring short-term variation when using the CUSUM chart.

There are other control schemes for the CUSUM chart. These control schemes are based on the average run length (ARL). The average run length is the average number of samples taken until an out-of-control point occurs. To minimize the number of false out-of-control points, this number should be large when the process is operating on target. To provide quick signals of process average shifts, this number should be small if the average has shifted by more than one standard deviation. Different values of H and K result in different ARL for when the process is on target and when the process average has shifted.

307

STEPS IN CONSTRUCTION OF A CUSUM CONTROL CHART

The steps in constructing a CUSUM chart are given below. A process flow diagram of the steps is given in Figure 4. The steps below assume that subgroup averages are being monitored. Similar steps apply if individual sample results are being used in place of subgroup averages. These steps use the basic design techniques of a CUSUM control chart.

1. Determine the process target (AIM).

2. Use historical data to calculate the routine process variation. At least 35 subgroups should be used. It is better to have 60 - 100. The process variation is estimated using the following equation:

$$\hat{\sigma}'_{\overline{X}} = \sqrt{\Sigma(\overline{X}_{i+1} - \overline{X}_i)^2 / 2(k - 1)}$$

where $\overline{X}_{i+1}$ and $\overline{X}_i$ are the subgroup averages for subgroups i+1 and i, respectively, and k is the number of subgroups.

3. Determine the value of DELTA, the shift from aim that you want to detect quickly. Usually:

$$DELTA = \hat{\sigma}'_{\overline{X}}$$

4. Determine the values of K, the allowable slack or deadband, and H, the action limits. Usually:

$$K = DELTA/2$$

$$H = 4.5\hat{\sigma}'_{\overline{X}}$$

5. Select the scales for the chart and plot K and H on the CUSUM chart.

6. For each subgroup, calculate the following:

    a. The subgroup average:

    $$\overline{X} = \Sigma X_i / n = (X_1 + X_2 + ... X_n)/n$$

    where X_1, X_2, etc. are the individual sample results and n is the subgroup size.

    b. The deviation from target for $\overline{X}_i$ = the ith subgroup average:

    $$\text{Deviation from target} = \overline{X}_i - AIM$$

c. The cumulative sum on the high side (SH):

$$SH(i) = Max[0,(\overline{X}_i - AIM) - K + SH(i - 1)]$$

d. The cumulative sum on the low side (SL)

$$SL(i) = Max[0,-(\overline{X}_i - AIM) - K + SL(i - 1)]$$

Note: $SH(0) = SL(0) = 0$

7. Plot the deviation from target, $SH(i)$ and $SL(i)$ on the CUSUM chart.

8. Interpret the chart for statistical control. If $SH(i)$ or $SL(i)$ is greater than H, an out-of-control situation exists.

9. If an out-of-control situation exists, estimate the current process average.

If $SH > H$,

$$\text{Current process average} = AIM + K + (SH/NH)$$

where NH equals the number of consecutive subgroups for which SH is greater than zero.

If $SL > H$,

$$\text{Current process average} = AIM - K - (SL/NL)$$

where NL equals the number of consecutive subgroups for which SL is greater than zero.

10. Once the process has been adjusted or the assignable cause removed, reset the values of SH and SL to:

$$SH = SL = H/2$$

Figure 4

STEPS IN CONSTRUCTION OF A CUSUM CHART

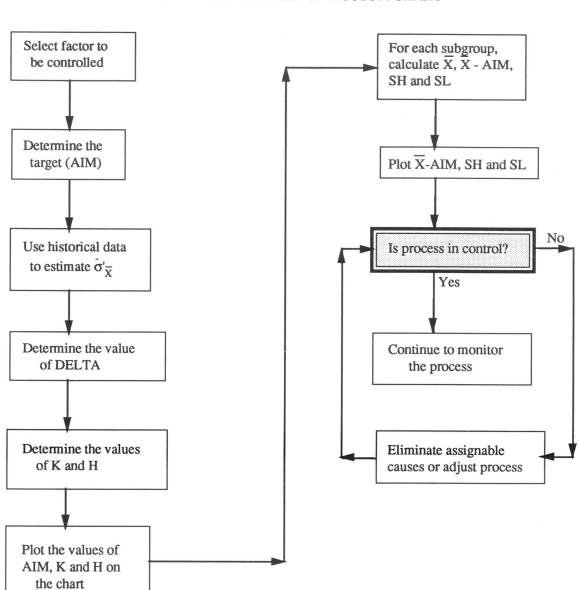

CUSUM CHART EXAMPLE

Reaction Yield

A team in an operating unit is trying to keep the yield from a given reaction as close to 80% as possible. Deviations from this target value can cause processing problems downstream. The team had been using a $\overline{X}$-R chart to monitor the process. It was wondering if the CUSUM chart could help them detect deviations faster. Five batches of product are made each day. A subgroup size of 5 was used. Data from the last 25 days are given below. The $\overline{X}$-R chart for those 25 days is given on page 312. It appears to be in statistical control.

Subgroup Number	n = 5					Subgroup Average	Subgroup Range
	1	2	3	4	5		
1	81.3	80.4	78.6	83.1	81.8	81.04	4.5
2	74.3	76.4	82.4	77.8	82.5	78.68	8.2
3	78.7	77.4	79.4	81.6	81.0	79.62	4.2
4	80.4	81.7	81.4	79.7	80.2	80.68	2.0
5	79.4	75.6	80.3	80.2	77.4	78.58	4.7
6	85.0	75.4	73.8	75.8	78.6	77.72	11.2
7	78.5	86.2	77.1	73.3	76.4	78.30	12.9
8	81.7	84.0	80.2	78.6	80.9	81.08	5.4
9	84.5	82.4	78.8	83.2	83.0	82.38	5.7
10	82.7	80.5	85.9	82.7	84.0	83.16	5.4
11	78.4	83.1	80.1	78.5	86.6	81.34	8.2
12	82.9	82.4	78.9	78.2	78.4	80.16	4.7
13	75.6	80.1	81.1	78.3	80.4	79.10	5.5
14	78.2	76.4	82.3	81.7	85.1	80.74	8.7
15	81.8	80.6	79.1	79.3	83.6	80.88	4.5
16	75.2	82.2	79.6	83.6	81.9	80.50	8.4
17	78.6	80.1	80.6	79.3	80.4	79.80	2.0
18	82.3	80.8	79.7	76.5	85.6	80.98	9.1
19	83.0	83.6	75.2	83.3	81.3	81.28	8.4
20	77.6	79.1	78.7	80.8	80.2	79.28	3.2
21	75.0	81.0	82.9	80.0	81.9	80.16	7.9
22	82.7	78.8	81.2	74.8	81.7	79.84	7.9
23	76.9	82.5	82.5	81.4	84.4	81.54	7.5
24	78.1	82.9	73.7	81.5	75.9	78.42	9.2
25	79.9	78.7	81.3	80.0	78.5	79.68	2.8

Use these data to develop a CUSUM control table. Is the process in statistical control? If it is out of control, use the fast initial response to restart the CUSUM procedure.

REACTION YIELD SOLUTION

$\overline{X}$ CHART (n=5)

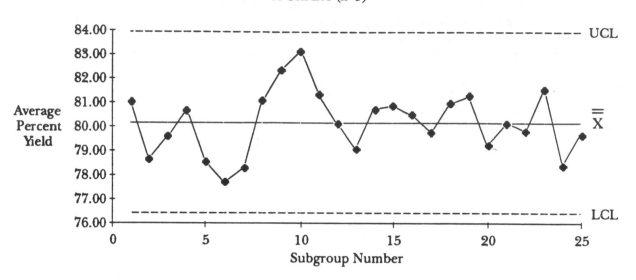

RANGE CHART (n = 5)

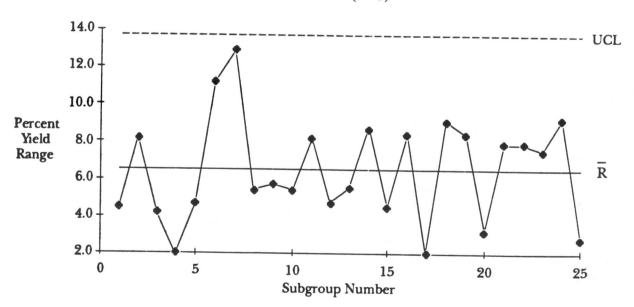

CUSUM CHART EXAMPLE

Moisture Measurement

A critical processing parameter in the production of a solid material is the moisture content of the solid. The moisture is measured using an infrared device. The laboratory personnel wanted to determine how stable the measurement device was. They set aside a large sample of solid to use as a control. The moisture content of this control was measured once per shift. The results from running the control for the last 25 shifts are given below.

Sample Number	$\%H_2O$	Range
1	0.10	
2	0.14	0.04
3	0.10	0.04
4	0.12	0.02
5	0.12	0.00
6	0.12	0.00
7	0.11	0.01
8	0.12	0.01
9	0.10	0.02
10	0.12	0.02
11	0.12	0.00
12	0.10	0.02
13	0.11	0.01
14	0.12	0.01
15	0.11	0.01
16	0.11	0.00
17	0.12	0.01
18	0.11	0.01
19	0.13	0.02
20	0.12	0.01
21	0.12	0.00
22	0.13	0.01
23	0.14	0.01
24	0.12	0.02
25	0.13	0.01

Use these data to construct a CUSUM control scheme. Assume that the target value is 0.11 Is the measurement system in statistical control?

SUMMARY

This chapter has introduced the CUSUM control chart. The procedure introduced was the basic design for the CUSUM chart. This type of chart looks at deviations of individual sample results or subgroups averages from the process target value. It provides a method of quickly detecting small shifts from the process target.

APPLICATIONS OF CUSUM CHARTS

Think about your processes at work. What are three possible applications of CUSUM charts. Record your responses below.

22

EXPONENTIALLY WEIGHTED MOVING AVERAGE CONTROL CHARTS

The classical Shewhart charts ($\overline{X}$-R, $\overline{X}$-s, etc.) were developed for use with the manufacturing of discrete parts. While these charts are often useful in the process industries, there are times when they are not applicable. Being in control means, in the classical sense, that there are only one average and one measure of the variation about that average. In the process industries, there are many variations (such as variations in raw materials) that continuously tend to move the process off target or aim. The classical Shewhart chart will not detect many of these shifts, particularly if the zones tests or runs tests are not applied. The exponentially weighted moving average (EWMA) control chart is one method of detecting these small shifts. The EWMA is different from other control charts in that it predicts what the next point will be. The EWMA control chart is introduced in this chapter.

OBJECTIVES

In this chapter you will learn:

1. What the EWMA statistic is
2. When to use the EWMA statistic as a control chart
3. How to construct the EWMA control chart

The EWMA control chart provides a method of detecting shifts from the process target or aim. It does this by providing different weighting to past data points. The EWMA also can be used as a forecasting tool since it predicts what the next point on the chart will be. This makes it possible to use the EWMA chart as a dynamic process control tool.

INTRODUCTION TO THE EWMA CHART

The EWMA control chart is a method of looking at variation. Unlike most variable control charts, there is only one chart. This chart examines the variation in the process average. The objective of using the EWMA is to discover small shifts in the process average. Normally, the "process average" used with the EWMA is the desired value of the process average, i.e., the target or the aim (Hunter, 1986). There is normally not a range chart associated with the EWMA chart.

Calculating and Plotting the EWMA

Figure 1 shows a comparison of an individuals control chart with the EWMA control chart. In this example, a sample from a process stream is pulled once a day and tested for a certain product characteristic. It is very important that the process produce a product with a value of 30 for this characteristic. It is desired that one be able to detect small shifts from this target value of 30.

Figure 1

EXPONENTIALLY WEIGHTED MOVING AVERAGE CHART

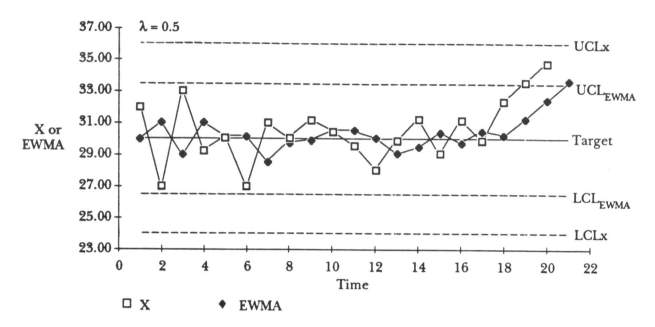

Table 1

DATA FOR FIGURE 1

Time	X	EWMA	ε_t	ε_t^2	Time	X	EWMA	ε_t	ε_t^2
1	32.0	30.00	2.00	4.000	11	29.6	30.54	-0.94	0.875
2	27.0	31.00	-4.00	16.000	12	28.1	30.07	-1.97	3.871
3	33.0	29.00	4.00	16.000	13	29.9	29.08	0.82	0.666
4	29.3	31.00	-1.70	2.890	14	31.3	29.49	1.81	3.269
5	30.1	30.15	-0.05	0.002	15	29.1	30.40	-1.30	1.679
6	27.0	30.13	-3.13	9.766	16	31.2	29.75	1.45	2.108
7	31.0	28.56	2.44	5.941	17	29.9	30.47	-0.57	0.329
8	30.1	29.78	0.32	0.102	18	32.4	30.19	2.21	4.897
9	31.2	29.94	1.26	1.586	19	33.6	31.29	2.31	5.320
10	30.5	30.57	-0.07	0.005	20	34.8	32.45	2.35	5.538
					21		33.62		

$\hat{\sigma}' = 2.02002475$ $\Sigma \varepsilon_t^2 = 84.846$

$UCL_{EWMA} = 33.46$

$LCL_{EWMA} = 26.54$

With the individuals chart, the actual sample results are plotted each day. With the EWMA chart, the EWMA is charted. The EWMA is a statistic that gives less weight to older data points. The EWMA is defined as the present predicted value plus λ times the present observed error:

$$\text{EWMA} = \hat{x}_{t+1} = \hat{x}_t + \lambda\varepsilon_t = \hat{x}_t + \lambda(x_t - \hat{x}_t)$$

where $\hat{x}_{t+1}$ = predicted value at time t+1

x_t = observed value at time t

$\hat{x}_t$ = predicted value at time t

$\varepsilon_t = x_t - \hat{x}_t$ = observed error at time t

1 = weighting constant $(0 < 1 \le 1)$

The EWMA is the predicted value. The data used to construct Figure 1 are shown in Table 1. This table will be used to help understand how the EWMA is calculated and plotted. Assume that, prior to taking the first sample, the process is in statistical control. In addition, assume that λ is known to be 0.5. The method for determining λ will be covered later. To begin the chart, the EWMA is set at the target value. Thus, for time 1 (or sample number 1), EWMA = 30. This value is plotted in Figure 1 as a "•" at time 1. Sample number 1 is then pulled from the process, the product characteristic determined, and the result (X = 32) plotted as a 'Ⴇ' at time 1. The error for time 1 is given by $\varepsilon_1 = x_1 - \hat{x}_1 = 32 - 30 = 2$. The next predicted value of X is given by:

$$\text{EWMA} = \hat{x}_2 = x_1 + \lambda\varepsilon_1 = \hat{x}_1 + \lambda(x_1 - \hat{x}_1) = 30 + 0.5(32 - 30) = 31$$

This value of EWMA is then plotted for time 2. Once the actual value for sample number 2 is obtained, it is also plotted on the chart. The actual value was 27. The next predicted value can now be calculated:

$$\text{EWMA} = \hat{x}_3 = \hat{x}_2 + \lambda\varepsilon_2 = \hat{x}_2 + 1(x_2 - \hat{x}_2) = 31 + 0.5(27 - 31) = 29$$

This process continues for each sample. A summary of the calculations for the first six times is given below.

Time	Actual Value	EWMA
t = 1	32.0	30.00 (Target)
t = 2	27.0	$\hat{x}_1 + \lambda(x_1 - \hat{x}_1) = 30 + 0.5(32 - 30) = 31$
t = 3	33.0	$\hat{x}_2 + \lambda(x_2 - \hat{x}_2) = 31 + 0.5(27 - 31) = 29$
t = 4	29.3	$\hat{x}_3 + \lambda(x_3 - \hat{x}_3) = 29 + 0.5(33 - 29) = 31$
t = 5	30.1	$\hat{x}_4 + \lambda(x_4 - \hat{x}_4) = 31 + 0.5(29.3 - 31) = 30.15$
t = 6	?	$\hat{x}_5 + \lambda(x_5 - \hat{x}_5) = 30.15 + 0.5(30.1 - 30.15) = 30.125$

Chapter 22

As can be seen from the above procedure, the EWMA for time t + 1 is calculated and plotted after the sample result for time t has been obtained but prior to obtaining the sample result for time t + 1. Thus, the EWMA is predicting what the next sample result will be. The result for t = 6 has been predicted to be 30.125.

Control limits can also be calculated for the EWMA. The method for doing this is described below. The control limits for this example have been calculated and plotted in Figure 1. The control limits for the individuals chart have also been added to Figure 1. The process appears to be in statistical control until the last EWMA (time = 21) is plotted. This EWMA is above the upper control limit for the EWMA. This means that something has happened to shift the process off target. The front-line personnel should discover what caused this to occur or make adjustments to bring the process back on target.

Note that none of the tests for stability have been violated if the individuals control chart is considered. Thus, the EWMA gives a faster indication of a small process average shift than the individuals control chart in this case. This is one advantage of using the EWMA. In general, the EWMA will detect smaller shifts (less than 2s') faster than the individuals control chart or the $\overline{X}$-R chart. However, if all the tests for stability (zones tests, etc.) are applied, the ability of the individuals and $\overline{X}$-R charts to detect small shifts approaches that of the EWMA chart.

Calculation of the Weighting Constant, λ

The weighting constant, λ, determines how much of the historical data the EWMA "remembers." This weighting is one of the major differences between the classical Shewhart charts, the CUSUM chart, and the EWMA chart. Before discussing how to calculate λ for the EWMA, these differences will be explored.

The differences in weighting for the various charts are shown in Figure 2. When using the classical Shewhart charts, an out-of-control situation occurs if a point falls beyond the control limits. This signal that the process is out of control is due entirely to the last point plotted. None of the earlier points are included (assuming that runs and zones tests aren't applied). In this case, all the weighting is on the last point. If w_i is the weighting, $w_t = 1$ for the last point plotted and $w_i = 0$ for all previous values (i ≤ t - 1). The CUSUM chart, on the other hand, depends upon a sum to determine if there is an out-of-control point. Equal weight is given to all historical data (i.e., $w_t = 1/n$ for all points). The moving average chart is an attempt to use more historical data. For the moving average chart, equal weight is given to all samples in the subgroup. The EWMA provides a different weighting than the other control charts. The EWMA gives less and less weight to older data depending on the value of λ. If λ is close to 0, the EWMA has a long memory and approaches a CUSUM chart in its ability to detect shifts in the process. If λ is close to 1, the EWMA has a short memory and approaches a classical Shewhart control chart in its ability to detect process shifts.

318

Figure 2

WEIGHTING OF VARIOUS CONTROL CHARTS

WEIGHTING FOR SHEWHART CONTROL CHARTS

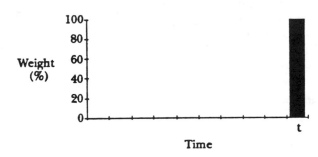

WEIGHTING FOR CUSUM CHARTS

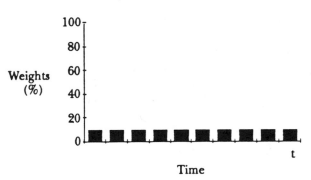

WEIGHTING FOR MOVING AVERAGE/MOVING RANGE CHARTS

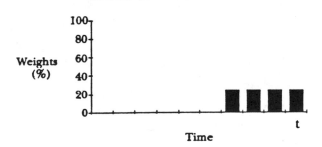

WEIGHTING FOR EWMA CHARTS

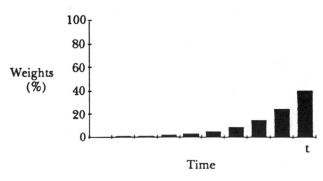

319

There are several schemes to select the value of λ. One of the easiest methods is to select the value of λ so that the sum of squares of the errors (ε_t^2) is a minimum. As shown in Table 1 for the data plotted in Figure 1, the sum of squares of the errors is 84.846 for $\lambda = 0.5$. This is not the minimum. The minimum actually occurs at $\lambda = .24$. The EWMA control chart for $\lambda = 0.24$ is shown in Figure 3. Note that since this λ is smaller than the λ used in Figure 1, more weighting is given to older data. This tends to flatten out the curve of the EWMA. Note that the point at $t = 21$ is out of control as it is in Figure 1. Usually at least 50 data points should be used to determine the value of λ.

Figure 3

EWMA CONTROL CHART

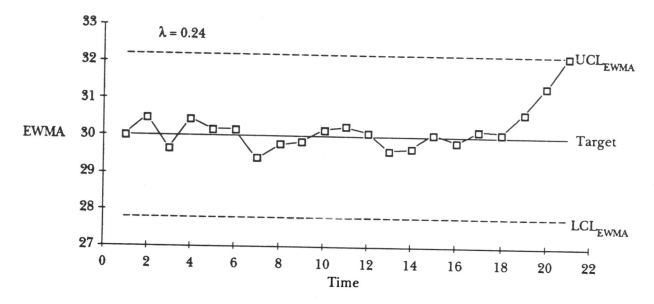

Figure 4 shows the effect of λ on the EWMA more clearly. Three different values of λ are used. As λ becomes smaller, the EWMA curve tends to flatten out more since more historical data are being used. This is similar to what happens when you increase the subgroup size with a moving average chart. Note that as λ approaches 1, the EWMA chart is very similar to the individuals chart.

The weighting given each individual point can be determined by the following equation:

$$w_i = \lambda(1 - \lambda)^{t-i}$$

For example, if $\lambda = 0.2$, the weighting given to the last data point ($t = i$) is 0.2. The weighting given to the point $t - i = 5$ is 0.0336

Figure 4

EFFECT OF WEIGHTING CONSTANT ON EWMA

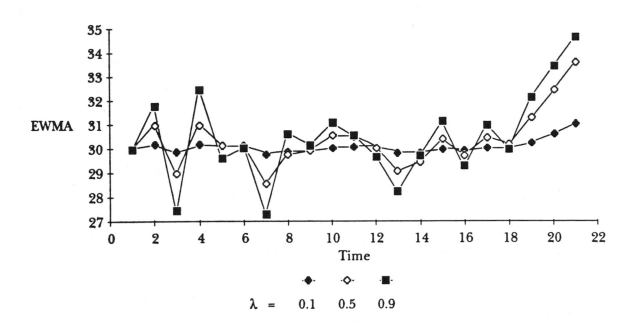

Control Limits for the EWMA

The control limits for the EWMA can be determined from the standard deviation of the individual measurements ($\hat{\sigma}'$). The relationship between the standard deviation of the EWMA and s' is given below:

$$\hat{\sigma}'_{EWMA} = \hat{\sigma}'\sqrt{1/(2 - \lambda)}$$

The three-sigma limits for the EWMA are given by:

$$UCL_{EWMA} = Target + 3\hat{\sigma}'_{EWMA}$$

$$LCL_{EWMA} = Target - 3\hat{\sigma}'_{EWMA}$$

As can be seen from the above equations, the control limits are affected by the value selected for the weighting constant. For the three weighting constants used in Figure 4, the control limits are:

λ	UCL_{EWMA}	LCL_{EWMA}
0.1	31.38	28.60
0.5	33.46	26.54
0.9	35.43	24.57

As the weighting constant becomes larger (i.e., less weight on older data), the control limits become wider.

The standard deviation of the EWMA is determined from the standard deviation of the individual measurements. The question of stability about this variation should be raised. If a moving range chart was constructed for the individual measurements, would it be in statistical control? If there is any question about this, a range chart should be constructed. The standard deviation of the individual measurements can then be calculated as $\overline{R}/d_2$.

WHEN TO USE THE EWMA CONTROL CHART

EWMA control charts can be used in the second step of the problem solving model. This step involves determining how the process is behaving.

The EWMA chart will normally be used with individuals data, not subgroup averages. However, there is nothing to prevent using the EWMA with subgroup averages. The EWMA can be used as a dynamic process control tool since it predicts what the next point will be. This can provide earlier detection of potential problems.

The EWMA can be used to detect small process shifts. The choice of l will affect how quickly the process shift is detected.

The EWMA is also often used as a forecasting tool since it has the ability to predict what the next point will be. For forecasting, the value of λ is often set at 0.2 ± 0.1.

STEPS IN CONSTRUCTION OF THE EWMA CONTROL CHART

The steps in constructing the EWMA chart for individuals data are given below. A process flow diagram of these steps is shown in Figure 5.

1. Use historical data to determine the weighting constant, λ.

   a. Collect at least 50 data points.

   b. Set the first EWMA to the target.

   c. For various values of λ, determine for each data point:

      i. The present error, $\varepsilon_t = x_t - \hat{x}_t$, where x_t is the actual value and $\hat{x}_t$ is the predicted value for time t.

      ii. The square of the error, ε_t^2.

---

    iii. The next EWMA $= \hat{x}_{t+1} = x_t + \lambda\varepsilon_t = \hat{x}_t + \lambda(x_t - \hat{x}_t)$, where $\hat{x}_{t+1}$ is the predicted value at time $t + 1$.

  d. Determine the sum of squares of the errors: $\Sigma\varepsilon_t^2$.

  e. Select λ so that the sum of squares for the error is a minimum.

2. Plot the data.

  a. Select the scales for the x and y axes for the EWMA chart.

  b. Plot the EWMA values and connect consecutive points with a straight line.

3. Calculate the control limits.

  a. Determine the standard deviation, $\hat{\sigma}'$, of the individual measurements.

  b. Determine the standard deviation of the EWMA:

$$\hat{\sigma}'_{EWMA} = \hat{\sigma}'\sqrt{\lambda/(2-1)}$$

  c. Determine the control limits for the EWMA chart. The upper control limit is given by UCL_{EWMA}. The lower control limit is given by LCL_{EWMA}.

$$UCL_{EWMA} = \text{Target} + 3\hat{\sigma}'_{EWMA}$$

$$LCL_{EWMA} = \text{Target} - 3\hat{\sigma}'_{EWMA}$$

  d. Plot the control limits on the EWMA chart as dashed lines and label.

  e. Plot the target on the EWMA chart as a solid line and label.

4. Interpret the chart for statistical control.

  a. The only test that can be applied to EWMA control charts is points beyond the control limits. This is due to the fact that the data are not independent; weighting causes data to be reused in successive values of the EWMA.

Figure 5

STEPS IN THE CONSTRUCTION OF THE EWMA CONTROL CHART

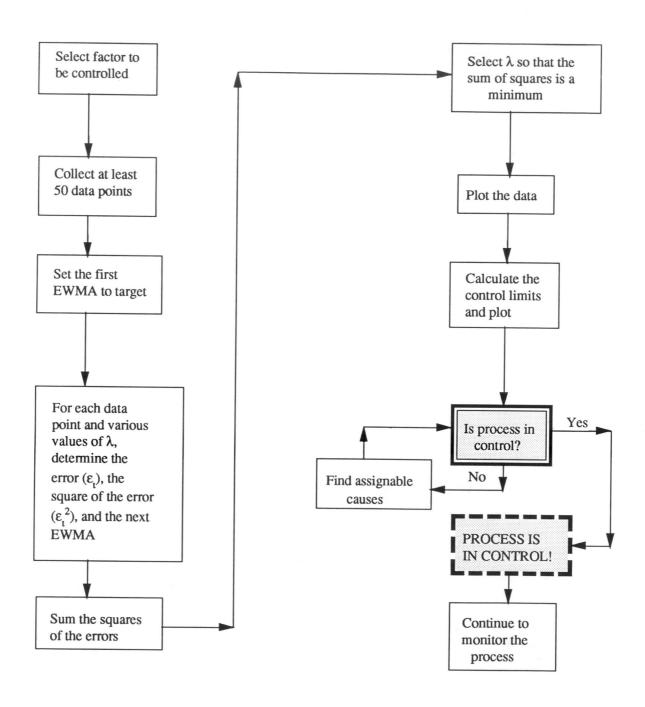

EWMA CONTROL CHART EXAMPLE

Product Sales

A district sales manager has been interested in improving the process used to predict the next month's sales for certain products. Currently, each sales person calls his/her customers to estimate the amount of product the customers will need the following month. This process takes a considerable amount of time. The sales manager decided to see if the EWMA could predict sales. She collected data for the past 20 months. Using these data, she determined that the weighting constant, λ, was 0.54. The data are given below.

Month	Sales (million lbs)	EWMA	ε_t
1	4.5	4.50	0.00
2	4.6	4.50	0.10
3	5.7	4.55	1.15
4	4.5	5.17	-0.67
5	7.3	4.81	2.49
6	5.6	6.15	-0.55
7	8.3	5.86	2.44
8	6.5	7.18	-0.68
9	5.7	6.81	-1.11
10	6.8	6.21	0.59
11	7.9	6.53	1.37
12	8.3	7.27	1.03
13	9.2	7.83	1.37
14	8.6	8.57	0.03
15	8.1	8.59	-0.49
16	8.5		
17	9.1		
18	9.5		
19	8.7		
20	9.0		
21			

Use the above data to construct the EWMA chart. Calculations have been completed for the first 15 months. The monthly sales data have been plotted along with the first 15 EWMA values (p. 327). The average of the observed monthly sales is 7.32 million lbs. The standard deviation of the individual monthly sales is 1.69. When you have completed the EWMA chart, answer the following questions.

1. What variation is being monitored on the EWMA chart?

2. What are the control limits? Note that there is no target in this case. Assume that the target is the average monthly sales.

3. Is the process in statistical control?

4. Can the EWMA chart be used to predict next month's sales?

5. What would have to be done to determine if the EWMA could predict next month's sales as well as the current process of calling each customer?

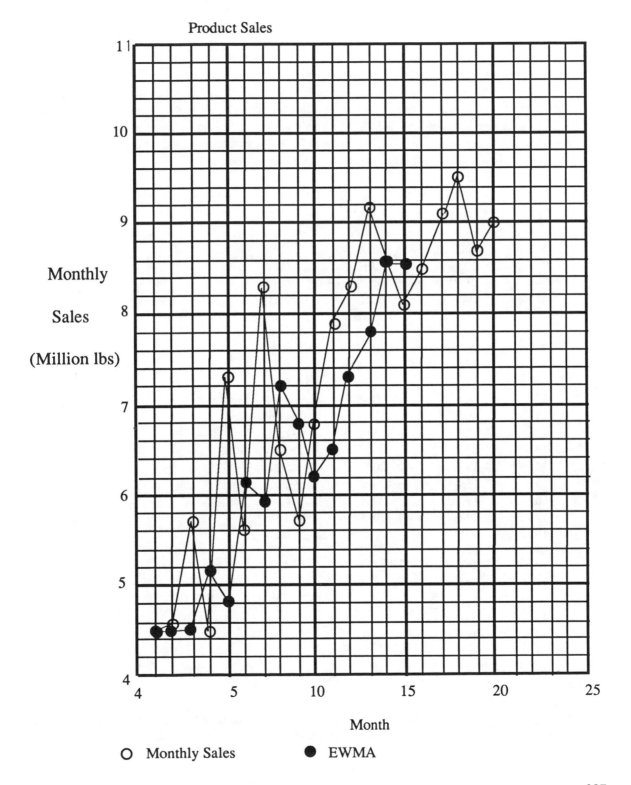

Product Sales

Month

O Monthly Sales   ● EWMA

Chapter 22

EWMA CONTROL CHART EXAMPLE

Paraffin Content

An operating unit uses molecular sieves to extract paraffins from kerosene. The sieve chamber operation (feed bed, timing, etc.) is based on the paraffin content of the incoming kerosene. If the chamber is set for the wrong paraffin content, paraffin will be lost. The paraffin content in the past had been measured once a day. Any adjustments needed were then made. A team in the unit decided to see if using the EWMA chart would help them determine when an adjustment was needed. The target value is a paraffin content of 20%. Data were collected for 20 days. The data are given below. The weighting constant has been determined to be 0.06.

Day	Paraffin Content	EWMA	ε_t
1	22.7	20.00	2.67
2	20.7	20.16	0.59
3	21.2	20.20	0.97
4	19.7	20.25	-0.51
5	18.7	20.22	-1.53
6	24.2	20.13	4.07
7	26.8	20.38	6.38
8	18.9	20.76	-1.83
9	24.5	20.65	3.84
10	24.9	20.88	4.05
11	19.2	21.12	-1.96
12	16.8	21.00	-4.25
13	23.0	20.75	2.21
14	19.8	20.88	-1.11
15	18.8	20.81	-2.01
16	19.1	20.69	-1.64
17	22.6	20.60	1.97
18	20.9	20.71	0.18
19	17.4	20.72	-3.35
20	25.6	20.52	5.10
21	22.0		
22	21.8		
23	23.2		
24	23.5		
25	26.0		
26	20.9		
27			

Use these data to construct the EWMA chart. The calculations have been completed and plotted for the first 20 samples (p. 330). When you have completed the chart, answer the following questions.

1. What variation is being monitored on the EWMA chart?

2. What are the control limits?

3. Is the process in statistical control?

4. What would you consider doing next?

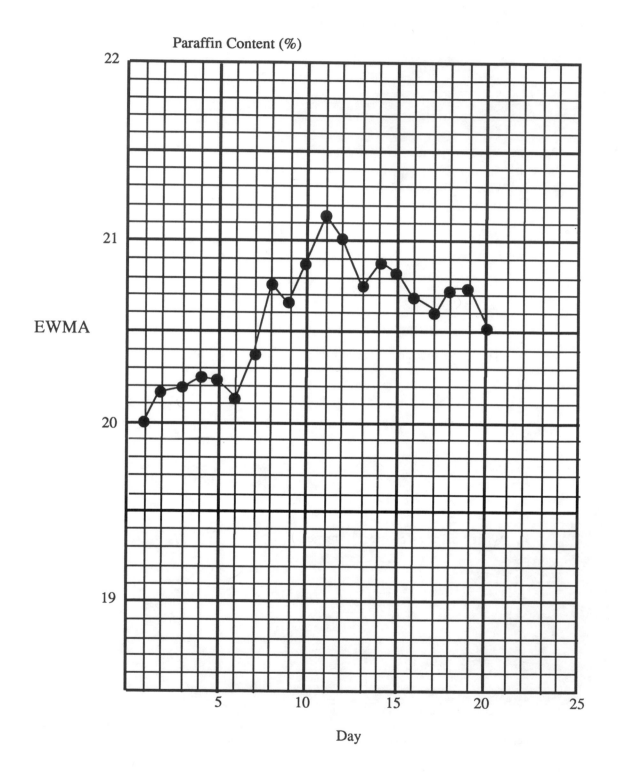

SUMMARY

This chapter introduced the exponentially weighted moving average (EWMA) chart. This chart is useful when it is necessary to detect small shifts from the process target. The chart makes use of the EWMA statistic. The EWMA predicts what the next value will be. The prediction is a combination of the last predicted value, the error and a weighting constant, λ. The weighting constant determines how much of the historical data the EWMA remembers. A range chart is usually not included with the EWMA chart. Control limits can be calculated and plotted on the chart. The only test for stability that applies is points beyond the control limits. Since the EWMA predicts the next point, it can be used as a forecasting tool and a dynamic process control tool.

APPLICATIONS OF EWMA CHARTS

Think about your processes at work. What are three possible applications of EWMA charts? Record your responses below.

1.

2.

3.

23

SPECIAL TOPICS IN VARIABLES CONTROL CHARTS

Numerous control charts using variables data have been covered thus far. Selecting the proper control chart to use for a given situation is important. A method for selecting the proper chart is given in this chapter. Rational subgrouping is very important in the use of control charts. You want to subgroup the data to explore the variation of interest to you. Rational subgrouping is introduced in this chapter. Often there are numerous products made using the same equipment. Product runs may be very short. How can control charts be applied in these situations? A method of handling these short product runs is covered in this chapter. Data from processes are often correlated with one another. This affects the calculation of the control limits. How to handle this situation is also introduced in this chapter. Control limits are usually based on three standard deviations of the variable being charted. Occasionally, you will want to use limits based on other probabilities, such as two standard deviations. This chapter introduces how to accomplish this.

OBJECTIVES

In this chapter you will learn:

1. How to select the correct variables control chart
2. What rational subgrouping is
3. How to use control charts for short product runs
4. How to calculate control limits when the data are correlated
5. How to calculate control limits other than three-sigma limits

The material in this chapter covers some special topics in variables control charts. This information will assist you in handling special situations that arise from time to time.

SELECTION OF VARIABLES CONTROL CHARTS

Numerous variables control charts have been introduced in the preceding chapters. Selection of the correct control chart is important. Figure 1 provides a guide for selecting the correct variables control chart.

If you have only measurement at a time to represent a given situation (and the distribution is normal), the individuals control chart should be used. The CUSUM and EWMA charts can also be used for individual sample results. If the distribution is not normal or if there is more than one point at a time to represent a given situation, the $\overline{X}$-R, $\overline{X}$-s, moving average/moving range, CUSUM or EWMA charts can be used. If the average is changing over time in a predictable fashion, the trend control chart can be used.

Figure 1

SELECTION OF VARIABLE CONTROL CHARTS

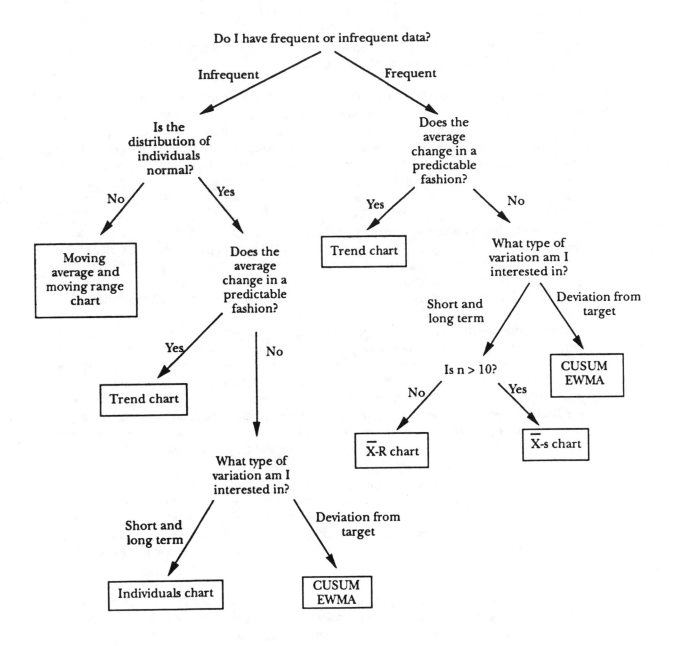

RATIONAL SUBGROUPING

Rational subgrouping is a primary topic in the study of sampling procedures. You want to take data in a way that will yield the most information possible. A major objective of sampling is to estimate population parameters (the average, standard deviation, and shape of the distribution). Samples represent the population. Thoughtful sampling is important to ensure that the population being studied is the one you actually have in mind.

Samples should be selected in a way that explores the variation of interest to you. The variation of interest might be shift to shift, day to day, hour to hour, batch to batch, etc. One possible way of looking at control charts is that they are a statistical test to determine if the variation from subgroup to subgroup is consistent with the average variation within the subgroups. A basic principle is to choose the subgroups in a way that will give the maximum chance for the measurements in each subgroup to be alike and the maximum chance for the subgroups to be different.

The most obvious rational subgrouping basis is the order of production. Even then, you have a choice of how to take samples. For example, you could take a group of five items, wait a while, then take five more, etc. Or you could take every third item until you had five, wait a while, then continue taking every third until you had five, etc. The first method is more likely to miss a process change that occurred and got corrected; the second might miss small process cycles.

For an example of the importance of subgrouping, suppose that you are making a chemical product in a batch reactor. There is always variation from batch to batch. If the product is well mixed in the reactor, the within-batch variation may be small. If subgroups were set up using four samples from one batch, the $\overline{X}$ chart would probably be out of control. In this instance, it is not reasonable to expect the batch-to-batch variability to be consistent or the same as the within-batch variability. It would make more sense in this case to use an individuals chart. The batch averages would be plotted on the X (individuals) chart and the moving range between successive batch averages would be plotted on the range chart (Wheeler and Chambers, 1986).

The basic idea behind rational subgrouping is to let the $\overline{X}$ chart do the work. The $\overline{X}$ chart can be effective as long as the subgroups are chosen in such a way that there is minimum chance for change to occur within a subgroup and maximum chance for a change to occur between subgroups. All sources of variation contributing to the range chart should also be sources of variation in the $\overline{X}$ chart. This relationship will be true as long as production run order is maintained.

It is not necessary to have the subgroups always in production run order. For example, an $\overline{X}$-s chart could be used to compare different suppliers, with a subgroup being material sent by one supplier. Even if production run order is not maintained, the test for points beyond the control limits can still measure stability.

RATIONAL SUBGROUPING EXAMPLE

Comparing Subgrouping Plans

Suppose there is a manufacturing plant with four different continuous reactors. The reactors make the same product. As shown in the figure below, the product from the four reactors is blended downstream.

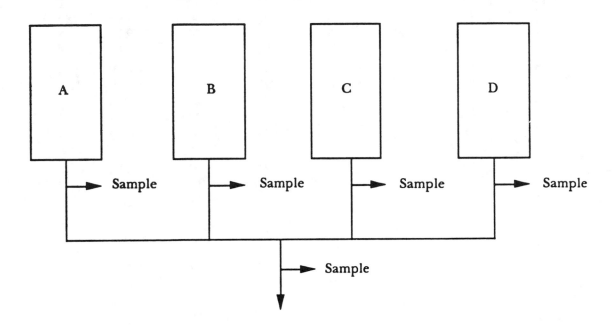

A certain quality characteristic, X, is of interest to a customer. There are several potential methods of subgrouping the data to be used in control charts. Three different subgrouping plans are presented below. Which one do you think is best?

Plan A: Select the first subgroup to consist of four samples from reactor A, the second subgroup to consist of four samples from reactor B, the third subgroup to consist of four samples from reactor C and the fourth subgroup to consist of four samples from reactor D. Repeat this sequence over time.

Plan B: Select one sample from each reactor to form the subgroup of four each time.

Plan C: Select each subgroup to consist of four samples from the blended stream.

Results from each sampling plan are given below (University of Tennessee, 1986).

Plan A: Table 1 shows the results obtained using subgrouping plan A. The control limit calculations are also included in the table. Figure 2 is the $\overline{X}$-R chart for these data.

Table 1

DATA FOR SUBGROUPING PLAN A

Subgroup Number	Reactor	1	2	3	4	$\overline{X}$	R
1	A	17	18	19	20	18.5	3
2	B	24	20	19	20	20.8	5
3	C	25	27	27	22	25.3	5
4	D	26	23	24	23	24.0	3
5	A	21	19	20	20	20.0	2
6	B	19	18	24	22	20.8	6
7	C	25	25	26	23	24.8	3
8	D	23	26	27	24	25.0	4
9	A	20	21	15	17	18.3	6
10	B	20	19	21	20	20.0	2
11	C	23	20	25	19	21.8	6
12	D	26	23	29	23	25.3	6
13	A	19	19	19	21	19.5	2
14	B	16	19	18	22	18.8	6
15	C	26	27	24	21	24.5	6
16	D	24	27	25	27	25.8	3
17	A	18	18	23	19	19.5	5
18	B	17	20	19	15	17.8	5
19	C	24	26	28	26	26.0	4
20	D	24	24	22	25	23.8	3
21	A	23	19	21	19	20.5	4
22	B	22	20	19	22	20.8	3
23	C	25	25	23	26	24.8	3
24	D	24	25	28	26	25.8	4
25	A	17	22	21	18	19.5	5
26	B	19	21	20	24	21.0	5
27	C	22	24	27	26	24.8	5
28	D	25	24	23	27	24.8	4
29	A	23	18	18	20	19.8	5
30	B	19	19	19	18	18.8	1
31	C	22	27	27	25	25.3	5
32	D	25	27	23	25	25.0	4

$\overline{\overline{X}} = 22.20$

$\overline{R} = 4.16$

$\text{UCLx} = \overline{\overline{X}} + A_2\overline{R} = 22.20 + (0.729)(4.16) = 25.23$

$\text{LCLx} = \overline{\overline{X}} - A_2\overline{R} = 22.20 - (0.729)(4.16) = 19.17$

$\text{UCLr} = D_4\overline{R} = 2.282(4.16) = 9.49$

Figure 2

$\overline{X}$ CHART FOR SUBGROUP PLAN A (n = 4)

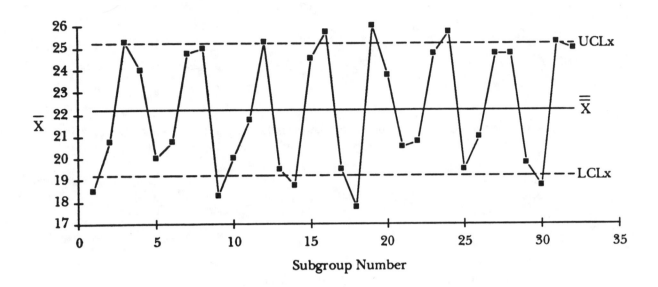

RANGE CHART FOR SUBGROUP PLAN A (n = 4)

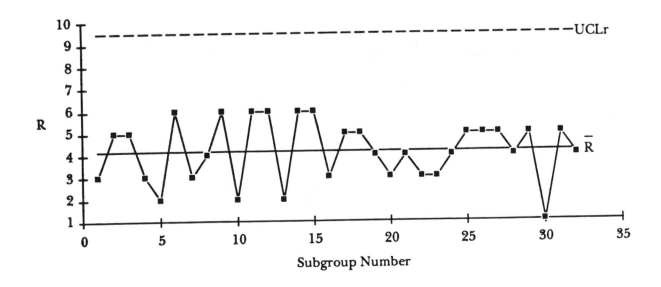

Chapter 23

Using the data and the chart, answer the following questions on subgrouping plan A.

1. Is the range chart in statistical control? What does this mean?

2. If the range chart is in statistical control, what is the estimate of $\hat{\sigma}'$? What is $\hat{\sigma}'$ measuring?

3. Is the $\overline{X}$ chart in statistical control? What does this mean?

4. What explanations do you have for the way the $\overline{X}$ chart looks?

Plan B: Table 2 lists the data obtained from subgrouping plan B. The control limit calculations are also included in the table. Figure 3 is the $\overline{X}$-R chart based on these data.

Table 2

DATA FOR SUBGROUPING PLAN B

Subgroup Number	1	2	3	4	$\overline{X}$	R
1	19	19	25	27	22.5	8
2	22	22	20	25	22.3	5
3	18	19	26	26	22.3	8
4	16	18	25	27	21.5	11
5	21	18	25	23	21.8	7
6	20	21	24	24	22.3	4
7	22	20	26	24	23.0	6
8	20	19	22	26	21.8	7
9	21	20	24	29	23.5	9
10	21	22	25	26	23.5	5
11	21	21	25	24	22.8	4
12	17	16	25	23	20.3	9
13	18	21	22	26	21.8	8
14	23	23	26	23	23.8	3
15	19	18	24	22	20.8	6
16	21	20	26	24	22.8	6
17	19	20	25	27	22.8	8
18	17	20	26	28	22.8	11
19	16	24	23	27	22.5	11
20	20	19	26	26	22.8	7

$\overline{\overline{X}} = 21.48$

$\overline{R} = 7.02$

$\text{UCLx} = \overline{\overline{X}} + A_2\overline{R} = 21.48 + (0.729)(7.02) = 26.60$

$\text{LCLx} = \overline{\overline{X}} - A_2\overline{R} = 21.48 - (0.729)(7.02) = 16.36$

$\text{UCLr} = D_4\overline{R} = 2.282(7.02) = 16.03$

Figure 3

$\overline{X}$ CHART FOR SUBGROUP PLAN B (n = 4)

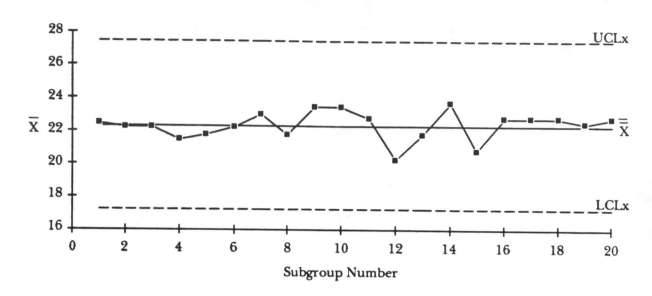

RANGE CHART FOR SUBGROUP PLAN B (n = 4)

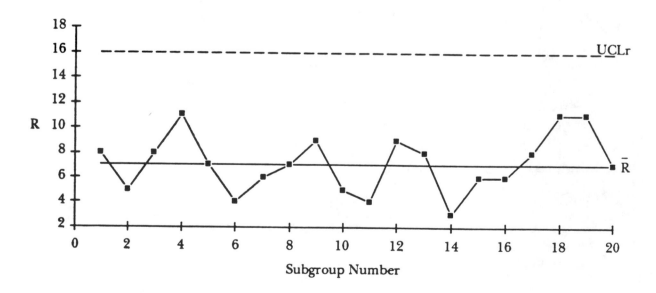

Using the data and the chart, answer the following questions on subgrouping plan B.

1. Is the range chart in statistical control? What does this mean?

2. If the range chart is in statistical control, what is the estimate of $\hat{\sigma}'$? What is $\hat{\sigma}'$ measuring?

3. Is the $\overline{X}$ chart in statistical control? What does this mean?

4. What explanations do you have for the way the $\overline{X}$ chart looks?

5. How can you explain the difference in $\overline{R}$ for subgrouping plan A and subgrouping plan B?

Plan C: Table 3 lists the data obtained from subgrouping plan C. The control limit calculations are also included in the table. Figure 4 is the $\overline{X}$-R chart based on these data.

Table 3

DATA FOR SUBGROUPING PLAN C

Subgroup Number	1	2	3	4	$\overline{X}$	R
1	27	23	19	25	23.5	8
2	18	24	26	20	22.0	8
3	28	19	21	25	23.3	9
4	17	21	22	21	20.3	5
5	26	19	18	27	22.5	9
6	25	22	17	18	20.5	8
7	20	21	18	21	20.0	3
8	26	26	19	20	22.8	7
9	18	26	24	21	22.3	8
10	18	19	23	22	20.5	5
11	22	21	22	23	22.0	2
12	19	24	23	24	22.5	5
13	19	26	23	28	24.0	9
14	27	26	23	24	25.0	4
15	25	23	22	20	22.5	5
16	19	25	20	21	21.3	6
17	21	19	24	19	20.8	5
18	20	23	23	20	21.5	3
19	17	21	21	23	20.5	6
20	24	17	20	19	20.0	7

$\overline{\overline{X}} = 21.02$

$\overline{R} = 6.10$

$UCLx = \overline{\overline{X}} + A_2\overline{R} = 21.02 + (0.729)(6.10) = 25.47$

$LCLx = \overline{\overline{X}} - A_2\overline{R} = 21.02 - (0.729)(6.10) = 16.58$

$UCLr = D_4\overline{R} = 2.282(6.10) = 13.92$

Figure 4

$\overline{X}$ CHART FOR SUBGROUP PLAN C (n = 4)

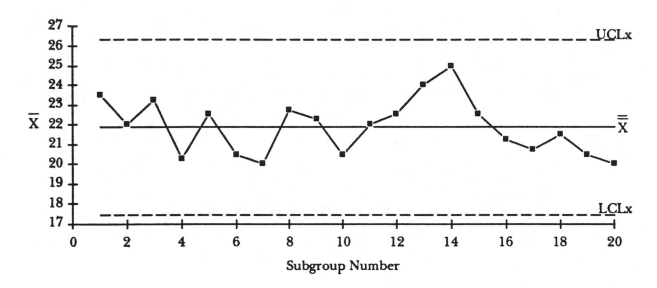

RANGE CHART FOR SUBGROUP PLAN C (n = 4)

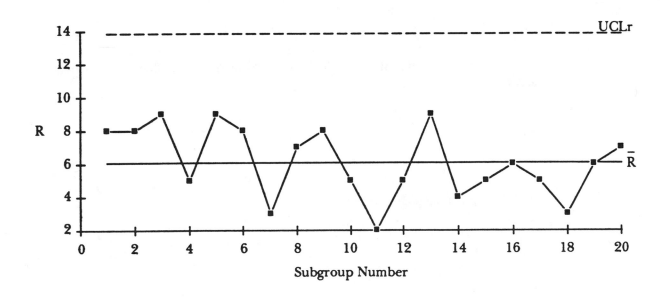

Chapter 23

Using the data and the chart, answer the following questions on subgrouping plan C.

1. Is the range chart in statistical control? What does this mean?

2. If the range chart is in statistical control, what is the estimate of $\hat{\sigma}'$? What is $\hat{\sigma}'$ measuring?

3. Is the $\overline{X}$ chart in statistical control? What does this mean?

4. What explanations do you have for the way the $\overline{X}$ chart looks?

5. How can you explain the differences in $\overline{R}$ for subgrouping plan A, subgrouping plan B, and subgrouping plan C?

6. Which subgrouping plan is best?

SHORT-RUN CONTROL CHARTS

The same process is often used to make different products by changing operating conditions, raw materials, etc. The time spent making any one product may vary from several hours to a week or longer. Specifications are often different for each product. How can control charts be applied in these situations? Is it necessary to have a control chart for each product? If this is so, it may take a long time to get enough data on any one product to construct a control chart. For these situations, an individuals control chart can be used. The result monitored on the chart is the deviation of the product or process parameter from the target value. The same chart is used for all products.

SHORT-RUN CONTROL CHART EXAMPLE

Compounding Operation

In the production of plastic compounds, the same compounding line is used to produce many products. Usually only three or four blendings of any one product are made at a time. One parameter of interest is the hardness of the resulting compound. Data from the last 20 blendings are given below. Five different products were made during this time.

Blending Number	Product Code	Hardness Target	Hardness Result	Deviation	Moving Range
1	A100	40	41	1	
2	A100	40	43	3	2
3	A100	40	38	-2	5
4	A100	40	40	0	2
5	A014	34	35	1	1
6	A014	34	32	-2	3
7	A014	34	36	2	4
8	B203	47	51	4	2
9	B203	47	47	0	4
10	B203	47	45	-2	2
11	B203	47	48	1	3
12	B203	47	47	0	1
13	B203	47	46	-1	1
14	B090	38	40	2	3
15	B090	38	39	1	1
16	B090	38	36		
17	B090	38	38		
18	B102	42	41		
19	B102	42	42		
20	B102	42	45		

Use these data to construct an individuals chart based on deviations from the target value. The calculations for the first 15 points have been completed and plotted for you (p.347). When you have completed the chart, answer the following questions.

1. Is the range chart in statistical control? What does this mean?

2. If the range chart is in statistical control, what is the estimate of $\hat{\sigma}'$? What is $\hat{\sigma}'$ measuring?

3. Is the $\overline{X}$ chart in statistical control? What does this mean?

4. If the centerline on the $\overline{X}$ chart is not 0, what does this mean?

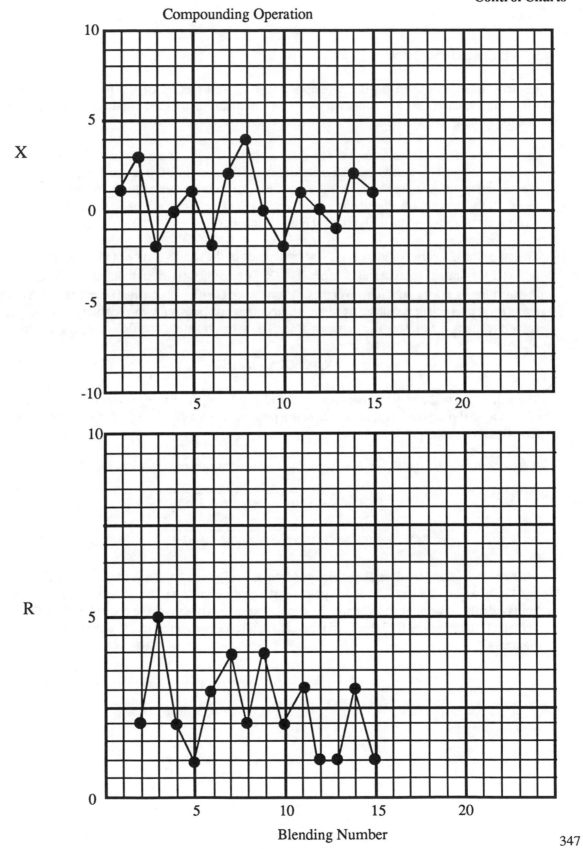

Compounding Operation

Blending Number

AUTOCORRELATION

Control limits are based on the assumption that the distribution of the plotted parameter (individual sample results or subgroup averages) is normally distributed. Another assumption is that the samples are independent. This assumption is often overlooked.

In the process industries, samples are often correlated with one another. This is called autocorrelation. This means that if a measurement is made at time t, it is usually related to the measurement made at time t - 1. For example, a high result is often followed by another high result. If the data are correlated, the control limits calculated using the basic formulas will not be correct. An adjustment to the control limits is required.

To determine if a correlation exists, a scatter diagram of the results can be made. The result at time t (X_t) is plotted versus the result at time t - 1 (X_{t-1}). Note that it is the individual sample results that are plotted, not the subgroup averages. A correlation coefficient can then be calculated. The correlation coefficient is a measure of the linear correlation between two variables, x and y. In autocorrelation, x and y are the same variable. The autocorrelation coefficient, ρ, is given by:

$$\rho = \Sigma(X_t - \overline{\overline{X}})(X_{t-1} - \overline{\overline{X}})/\Sigma(X_t - \overline{\overline{X}})^2$$

The value of ρ will always be between -1 and 1. If ρ is negative, it means there is a negative correlation. If ρ is positive, it means there is a positive correlation. The closer ρ is to zero, the less the correlation.

If the correlation is significant, the control limit equations become (Hunter, 1988):

$$UCL = \overline{\overline{X}} + 3\sqrt{(\hat{\sigma}'^2/n)}\sqrt{1+[2(n - 1)\rho/n]}$$

$$LCL = \overline{\overline{X}} - 3\sqrt{(\hat{\sigma}'^2/n)}\sqrt{1+[2(n - 1)\rho/n]}$$

where $\hat{\sigma}'$ is the standard deviation estimated from the range chart and n is the subgroup size. ρ must be between -0.5 and 0.5. If ρ is positive, the control limits will be wider than the control limits calculated without considering the correlation. If ρ is negative, the control limits will be tighter. Determining whether a correlation is significant is covered in the chapter on scatter diagrams.

AUTOCORRELATED DATA EXAMPLE

Product Response

An operating unit has been keeping an $\overline{X}$-R chart on a certain product response, X. Data from the last 25 subgroups are given below.

Subgroup Number	1	2	3	4	$\overline{X}$	R
1	98.0	103.1	105.1	97.2	100.84	7.90
2	111.8	94.0	81.8	99.0	96.66	30.01
3	91.3	109.7	111.8	106.3	104.78	20.53
4	95.7	89.9	100.3	116.8	100.64	26.89
5	119.7	83.7	73.3	96.1	93.20	46.41
6	97.9	97.3	95.3	102.2	98.17	6.90
7	116.2	113.8	122.3	100.0	113.07	22.33
8	99.2	89.2	107.2	100.1	98.94	17.98
9	97.6	106.4	115.8	100.3	105.02	18.23
10	113.8	91.4	114.7	79.4	99.80	35.30
11	81.3	90.6	94.1	112.2	94.55	30.86
12	119.6	107.1	109.1	115.8	112.91	12.52
13	105.0	105.9	100.7	103.4	103.73	5.17
14	83.0	111.3	105.9	108.1	102.07	28.26
15	111.2	95.0	94.6	104.3	101.27	16.63
16	92.3	83.0	83.3	106.0	91.16	23.03
17	104.8	120.1	108.1	106.2	109.80	15.29
18	106.9	103.0	92.4	104.3	101.65	14.58
19	100.8	96.1	106.4	99.2	100.64	10.33
20	114.3	92.6	80.7	93.7	95.33	33.67
21	70.7	84.8	88.2	103.9	86.92	33.19
22	95.1	115.0	97.5	99.6	101.81	19.94
23	98.4	94.2	74.2	88.9	88.92	24.17
24	81.1	104.8	96.0	103.3	96.28	23.72
25	102.5	101.1	109.9	109.4	105.77	8.80

The averages and control limits based on these data are given below.

$\overline{\overline{X}} = 100.16$

$\overline{R} = 21.31$

$UCLx = \overline{\overline{X}} + A_2\overline{R} = 115.69$

$LCLx = \overline{\overline{X}} - A_2\overline{R} = 84.63$

$UCLr = D_4\overline{R} = 48.63$

$\hat{\sigma}' = \overline{R}/d_2 = 10.35$

The question about whether or not the data are correlated has been raised. To determine this, the product response at time t was plotted versus the product response at time t - 1. The resulting plot is shown below.

PLOT OF TIME t RESULTS VERSUS TIME t - 1 RESULT

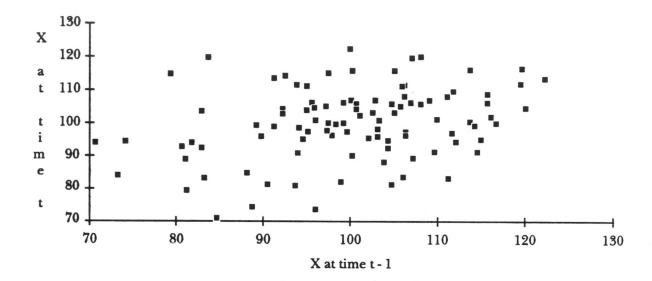

As can be seen in the figure there does appear to a positive correlation. A high product response is often followed by another high product response. A low product response is often followed by another low product response. This correlation is statistically significant.

The value of ρ can be calculated from the data as:

$$\rho = \Sigma(X_t - \overline{\overline{X}})(X_{t-1} - \overline{\overline{X}})/\Sigma(X_t - \overline{\overline{X}})^2 = 5059.8/13367.13 = 0.379$$

Using this value of ρ, what are the modified control limits?

The $\overline{X}$-R chart for this example is shown in Figure 5.

Figure 5

$\overline{X}$ CHART WITH MODIFIED CONTROL LIMITS

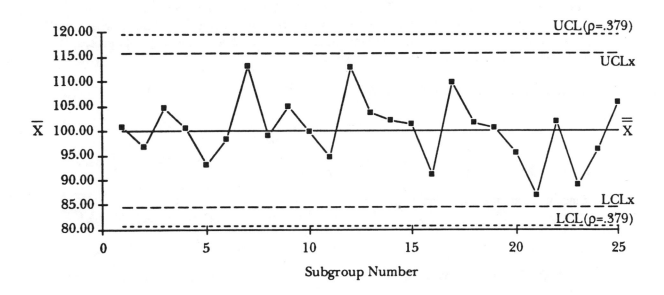

RANGE CHART (n = 4)

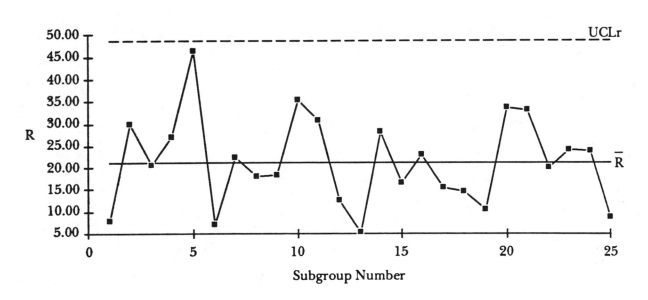

DETERMINING LIMITS OTHER THAN THREE-SIGMA LIMITS

The three-sigma limits on control charts really represent a probability. The limits are set so that the probability of getting a point beyond the control limits is less than or equal to 3 out of a 1,000 if a process is in statistical control. These limits represent a tradeoff between the cost of looking for an assignable cause when one is not present and the cost of not looking for one when it is present. In some cases, you may be willing to take more of a risk in concluding that a process is unstable because an out-of-control point exists.

The control limits for the $\overline{X}$-R chart are given by:

$$UCLx = \overline{\overline{X}} + A_2\overline{R}$$

$$LCLx = \overline{\overline{X}} - A_2\overline{R}$$

Since these are the three-sigma limits,

$$A_2\overline{R} = 3\hat{\sigma}'_{\overline{X}}$$

To obtain control limits based "a" sigma, use $(a/3)A_2\overline{R}$. Thus, two-sigma control limits for the $\overline{X}$-R chart are given by:

$$UCLx = \overline{\overline{X}} + (2/3)A_2\overline{R}$$

$$LCLx = \overline{\overline{X}} - (2/3)A_2\overline{R}$$

For two-sigma limits, the probability of getting an out-of-control point for a stable process is 5 out of 100.

SUMMARY

How to select the proper variables control chart was introduced in this chapter. The selection includes considering the amount of data available at any one given time and whether or not the distribution of individual measurements is normal. The concept of rational subgrouping was also introduced. The major idea is to subgroup the data in a manner so that the opportunity for variation within a subgroup is minimized and the opportunity for variation between subgroups is maximized. It was also shown that an individuals control chart could be used to monitor deviations from target values for short product runs. Using this approach allows you to keep a chart on an ongoing basis for different types of products. Autocorrelated data were also covered. If the data are autocorrelated, the control limits must be modified slightly to account for this. The method of determining limits other than three-sigma limits was also covered.

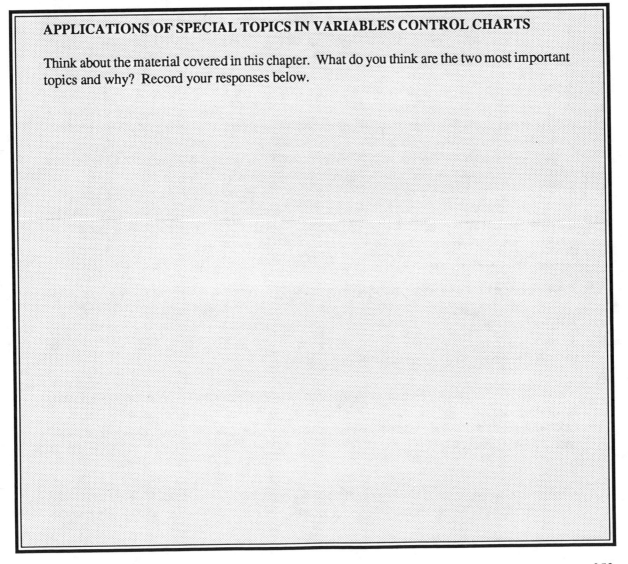

APPLICATIONS OF SPECIAL TOPICS IN VARIABLES CONTROL CHARTS

Think about the material covered in this chapter. What do you think are the two most important topics and why? Record your responses below.

24

CONTROL STRATEGIES

The point you just plotted on your control chart is above the upper control limit. Your chart is telling you that there is an assignable cause present in your process. You are supposed to find out what caused this point to be above the upper control limit. Where do you start looking?. There are thousands of things which could have caused this point to be above the upper control limit. A control strategy provides a method of helping you look for causes of out-of-control points. This module introduces the concept of control strategies.

OBJECTIVES

In this chapter you will learn:

1. What a control strategy is
2. Who should develop the control strategy
3. How the control strategy is updated
4. The benefits of using a control strategy

Control charts tell us two basic things: when to take action and when to leave our process alone. We should take action when our process shows signs of assignable causes of variation. When our control charts show us that there is an out-of-control situation, it is the responsibility of those people closest to the process to find the cause for the out-of-control situation and eliminate the cause from the process. This prevents the cause from occurring again. Sounds fairly easy, doesn't it? However, bringing a process back into control is usually not that simple. For example, if you think about all the possible reasons why a process may go out of control, you will probably develop a fairly long list. Time constraints and costs probably wouldn't allow you to look into all the possible causes. In most cases, you do know that some causes are more likely to occur than others. Control strategies help you look into the more likely causes.

CONTROL STRATEGIES: DEFINITION AND USAGE

Control strategies are specific action plans for bringing a process back into control. The strategies usually consist of five to ten steps that help you find reasons for assignable causes and, most importantly, help you do something about the causes.

A non-manufacturing example of a control strategy is shown in Figure 1. This control strategy is for operating supplies variance. Operating supplies is often a category used in budgeting. Operating supplies variance is the difference between what was budgeted for operating supplies and what was actually spent.

The first column of the control strategy is a statement of the out-of-control situation. In this example, the out-of-control situation is that the operating supplies variance is below the lower control limit. Notice that the out-of-control situation talks only about the variance below the

Figure 1

NON-MANUFACTURING CONTROL STRATEGY: OPERATING SUPPLIES VARIANCE

Out-of-Control Situation	Possible Causes	Action To Be Taken	Results of Investigation	Investigated by
Operating Supplies Variance is above upper control limit	1. Miscoding on vouchers for expenditures	1. Check coding	1. Coding is correct ___ Yes ___ No	___
	2. Expenditure occurred during month different than budgeted	2. Compare month expenditure was budgeted in to month it actually occurred	2. Budget differs from expenditure by ___ months	___
	3. Excessive overnight mail	3. Compare overnight mail actual to budget	3. Overnight budget minus actual = ___	___
	4. Unbudgeted mass mailings occurred	4. Check to see if mass mailings occurred	4. Mass mailing ___ Yes ___ No	___
	5. Item budgeted but estimate inaccurate	5. Compare bugeted item amount to actual spent	5. Budgeted item cost minus actual cost = ___	___

lower control limit (or seven points in a row below the average). The control strategy for points above the upper control limits (or seven points in a row above the average) might be different. You might also have different control strategies depending on which type of chart (range or average) the out-of-control situation occurred.

The second column lists the five to ten most probable causes for the out-of-control situation described. The third column indicates what action needs to be taken to investigate each cause. The results or information from the action taken is recorded in the fourth column. The final column is for the initials or the name of the person who took action.

A manufacturing example of a control strategy is given in Figure 2. The out-of-control situation in this case is a point below the lower control limit. The setup of the strategy is essentially the same as for the non-manufacturing case. The first column is the out-of-control situation. The second column lists the eight most probable causes. The third column tells you what action to take. The results from the action taken are written in the next column. The last column is for the initials or name of the person taking the action.

Control strategies should be developed and used by the local work force. Those people closest to the process will know best what are the most likely causes of an out-of-control situation and what to do about it. In addition, out-of-control situations should be addressed immediately. For this reason, the responsibility for using control strategies also lies with the local work force.

Control strategies should be updated on a regular basis. If one of the causes currently on the control strategy has never surfaced as a reason for an out-of-control point, then that cause should be deleted from the strategy and replaced by another possible cause (and action to be taken for pursuing the new cause).

BENEFITS OF CONTROL STRATEGIES

There are various benefits to control strategies. Control strategies provide a method of systematically looking for assignable causes. This helps front line personnel to do their jobs better. Also, control strategies help you increase the understanding of your processes. For example, if you use and update control strategies over time, you will establish a database that will evaluate the causes of out-of-control points over the long term. This increased knowledge about your processes will be useful for training new hires. They will have a standardized approach for controlling their processes instead of working by trial and error. Control strategies also provide a permanent record of what happened after each out-of-control situation.

SUMMARY

Control strategies are specific plans for what to do when your process shows the presence of assignable causes. This plan describes the out-of-control situation, possible causes, how to check each cause and the result of your check. All control charts in use should have a control strategy. It is the responsibility of those people closest to the process to develop and use the control strategies.

Figure 2

MANUFACTURING CONTROL STRATEGY: PERCENT PURITY

Out-of-Control Situation	Possible Causes	Action To Be Taken	Results of Investigation	Investigated by
Percent purity beyond the control limits	1. Incorrect Calculations	1. Recheck all Calculations	1. Caculations correct ____ yes ____ no	____
	2. Too much catalyst added	2. Check batch records; check scales	2. Amount of catalyst added ____ lbs.; scales accurate ____ yes ____ no	____
	3. Too little catalyst added	3. Check batch records, check scales, check for leaks in line	3. Amount of catalyst added ____ lbs.; scales accurate ____ yes ____ no; Leaks, ____ yes ____ no	____
	4. Purity of catalyst too low	4. Check purity records	4. Purity ____	____
	5. New batch of catalyst	5. Check records	5. New batch ____ yes ____ no	____
	6. Incorrect temperature	6. Check batch records	6. Temperature ____	____
	7. Incorrect pressure	7. Check batch records	7. Pressure ____	____
	8. Lab error	8. Ask lab about test	8. Test method in control ____ yes ____ no	____

APPLICATIONS OF CONTROL STRATEGIES

Think about some parameter you might construct a control chart on to monitor over time. Now suppose you get an out of control situation. Begin the development of a control strategy for one of these charts by determining two possible causes for an out-of-control situation and the action you would take for these causes.

Control Chart Parameter:

Out-of-Control Situation:

      Cause   Action To Take

1.

2.

25

USE OF CONTROL CHARTS

How will I ever find time to do all this charting? What possible use can these control charts be to me? How will they help me do my job? What variables should I chart? These types of questions and concerns are quite common. The purpose of this chapter is to provide answers to these and other similar questions.

OBJECTIVES

In this chapter, you will learn:

1. What statistical management is
2. How to start a control chart
3. How to maintain a control chart
4. What variables should be charted
5. When to be concerned about the stability of a process

The purpose of this chapter is to provide a general understanding of the use of control charts and why they are valuable tools. They help us manage our business in a more effective way.

STATISTICAL MANAGEMENT

A major barrier to the use of control charts is the failure of people to understand the information contained in variation (for a more detailed discussion of this section, see Deming, 1982). If you understand this information, you will realize that the type of action required to reduce assignable cause variation is totally different from the type of action required to reduce common cause variation. You will also understand why costs decrease as quality improves. You will also understand why management that blames all its problems on the employees is totally wrong.

To understand variation, it is essential to understand the difference between a process that is in statistical control (stable) and a process that is out of control (unstable). To determine the difference, we must know how to plot points and then decide by some rational method if the process is in control. There are many things we can plot, including weekly sales figures, customer complaints, product responses, inventory, absenteeism, accidents, accounts receivable, turnover, etc.

A key to understanding variation is to plot the points in the order they were produced. For example, you might plot accounts receivable on a weekly basis. After plotting a sufficient number of points (at least 20), the next step is to question whether the process that produced the data is in statistical control. Why do we want to consider statistical control? The major reason is that the type of action required to improve a process is different for a process that is in control than for a process that is out of control.

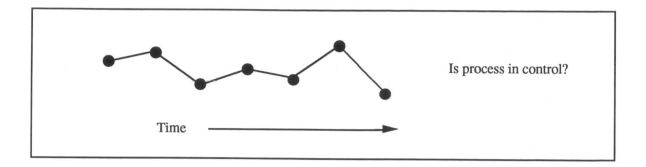

A major fault, which exists in many places today, is the assumption that every event (defect, mistake, accident) is attributable to someone or is related to one event. For example, suppose this month's expenditures versus budget (the variance) was slightly higher than last month's variance. Why did this occur? Is it attributable to one person or one event? We often spend a lot of time looking for these types of answers. The fact is, however, that most troubles with service and production lie in the system and are not attributable to one person or one event.

What do we mean by "the system"? The system includes many things. It includes people, machines, methods, materials, measurement and the environment (Shewhart's concept of a process). It also includes customers, suppliers, management style, shareholders, and the government.

There are two types of variation. Common cause variation is variation due to the system. To reduce common cause variation is the responsibility of management. Dr. Deming has estimated that 85% of problems are due to common cause variation, i.e., due to faults in the system. Assignable cause variation is variation due to sporadic, unnatural events. Assignable cause variation is the responsibility of the front-line personnel to correct. Dr. Deming has estimated that 15% of problems are due to assignable cause variation.

The problem becomes how to determine if results are due to common cause variation or assignable cause variation. For example, suppose this month's sales are less than last month's sales. Is this due to common cause variation or assignable cause variation? Confusion between common and assignable causes leads to frustration, greater variability and higher costs.

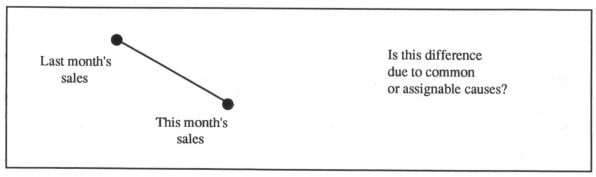

Good management requires the knowledge of how to separate the two kinds of causes. The only good way to do this is through the use of control charts.

There are two types of mistakes we can make when looking at data. One mistake is to assume that a data point is due to an assignable cause when in fact it is due to common cause (mistake 1). For example, we could make this mistake if we looked at each month's sales figures and tried to find one reason why sales were up or down. The second type of mistake is to assume that a data point is due to common cause when in fact it is due to assignable causes (mistake 2). For example, we could make this mistake by never looking at reasons for changes in sales figures, assuming that the changes are always due to the normal variation in the sales process.

There is no hope of avoiding both these mistakes all the time. In other words, we will make the first mistake sometimes and the second mistake sometimes. What we need are some rules to help us minimize the loss due to these mistakes. These rules are the control limits on control charts.

A control chart sends us a statistical signal. This signal tells us if there is an assignable cause present or if the observed variation is attributable to the system. For example, if a point is beyond the control limits, this is a signal that an assignable cause is present. The front-line personnel should find out the reason for the assignable cause and permanently remove it from the system so it won't occur again. Is it possible that this out-of-control point is not due to an assignable cause and is really due to common causes (mistake 1)? The answer to this is yes, but it is highly improbable. Assume that a process is in statistical control (only common cause variation present). If you plot 1000 points from this process, only three will be beyond the control limits. Therefore, anytime a point is beyond the control limits, there is only a 3 in 1000 chance that it is due to common causes. Since this probability is low, it is most likely due to an assignable cause. Thus, the control limits help minimize how often we make mistake 1.

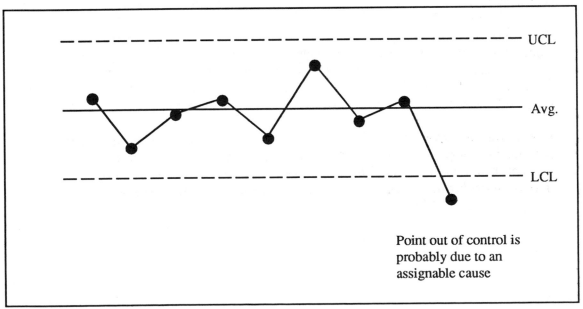

Point out of control is
probably due to an
assignable cause

We will make mistake 2 more often than mistake 1. Points that lie between the control limits are assumed to be due to common cause variation. It is possible that an assignable cause is present from time to time but the point will still lie between the control limits. We are willing to make this mistake more often because it does not require any action from us. If a point is out of control, it takes time and energy to find out what happened. We want to be sure that there is an assignable cause present before we start looking for it. The control limits help us accomplish this.

Control charts help us determine the type of variation present. This is important to know because the strategies for improving a process that is in statistical control are different from the strategies for improving a process that is out of statistical control.

If a process is out of control, the discovery and removal of assignable causes is usually the responsibility of the front-line personnel (those closest to the process). However, some can be removed only by management. It is important to remember that removing assignable causes from a process is not improving the process. You are merely restoring the process to where it should be.

A process is in statistical control if there are no assignable causes present. Such a process is random but is predictable in the near future. For example, we don't know what the exact value of the next data point will be, but we do know that it will fall between the control limits. We also know what the long-term average of the process will be. No action is taken on the "ups" and "downs" in the process. Being in statistical control is not a natural state. It is obtained by eliminating assignable causes one by one. In addition, costs are predictable and at a minimum for a process that is in control. Productivity is at a maximum.

The responsibility for improving a process in statistical control is management's. Front-line personnel may have excellent suggestions on how to do this. Improving a process that is in control may mean changing the average or reducing variation. It is a never-ending process. The system must be changed to improve the process. Examples of system changes include new machines, new methods, and new materials.

What happens if you use the wrong strategy for improving a process? For example, suppose your process is in control. One point comes in below the average. You decide to adjust the process to compensate for this and you continue to adjust the process anytime a point is below average. What have you done to the process? In reality, you have increased the variation in the process. It is possible that you have created a process that is now out of control.

Suppose your process is out of control and you try to improve the process by changing the system. For example, suppose you want to decrease the average turnaround time for memos. To accomplish this, you buy a new software package. The next point you get is below the average. Have you improved the process? You don't know, because the original process was not in control.

In summary, control charts are valuable tools because they provide us with a basis for action. They tell us when to leave the process alone and when to look for problems. In addition, the strategies for improving a process are different, depending on the state of control of a process.

STARTING A CONTROL CHART

The amount of time required to start a control chart will vary, depending on the problem. However, the amount of time required to keep the chart up to date is minimal in most cases. For example, if you are examining weekly sales figures, you will only plot one point per week once the chart has been established. If you are looking at the fraction of invoices with errors on a daily basis, you will only plot one point per day. In a manufacturing plant, no more than one point per hour will probably be plotted. Upkeep of the charts is not time consuming. Looking for assignable causes can be.

Before starting a control chart, you should determine what the objective of the chart is. For example, are you trying to reduce the time it takes to pay invoices or perhaps improve the accuracy of invoices? Charts should never be started without some objective in mind. In other words, you have defined some problem and are trying to solve it. Process flow diagrams should be completed prior to starting a control chart. These flow diagrams are useful in determining where data can be collected.

Once you have established an objective, the next step is to select the type of chart required. This depends on the type of data you have. Figure 1 represents a process flow diagram of how to select the right control chart for your application. Other decisions to be made include how often the data will be collected and the size of the subgroups.

If at all possible, use constant subgroup sizes. This makes charting much easier. In many cases, you can take random samples to make the subgroup sizes the same. For example, if you are looking at errors on invoices, you could take a random sample of 25 per day. This would make the subgroup size the same each day. The conclusions you reach from the control chart will be the same as if you had inspected each invoice (assuming the subgroup you select is large enough).

Historical data will often be available. For example, sales figures are available from the past. Historical data, if available, can and should be used to begin a control chart. There will be times when no historical data are available. In this case, data collection must begin.

Once sufficient data (including historical data, minimum of 20 points) are available and plotted, the overall process average and control limits can then be calculated and added to the control chart. The chart can then be used to determine if the process is in statistical control. If there are out-of-control points, you may want to try and find out what caused them. However, in some cases, this will be difficult. For example, suppose you are charting weekly sales figures. After 20 weeks, you calculate control limits and add them to the chart. Suppose week two is out of control. It may be difficult to find out what happened 18 weeks ago. It may not be worth the time and energy it would take.

Figure 1

SELECTION OF CONTROL CHARTS

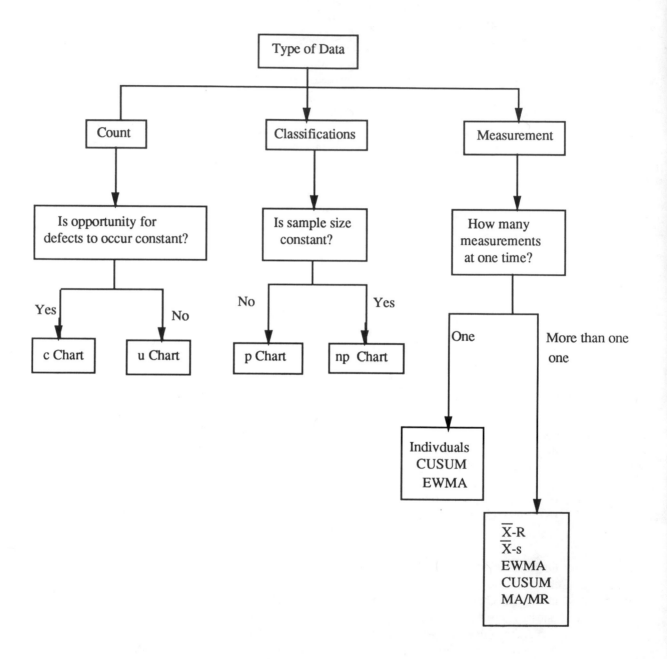

MAINTAINING THE CONTROL CHART

The control chart has now been established. Each time a new data point is available, it is plotted. Now if an out-of-control point occurs, front-line personnel should look for the cause. When an out-of-control point has just occurred, it will be easier to find the reason why.

Two possibilities exist once the chart has been established. One possibility is that the process is in control. This will normally not be the case since being in statistical control is not the natural state. However, if this is true, you can begin looking at methods of changing the system to improve the process. The second possibility is that the process is out of control. If this is the case, assignable causes should be identified and removed. Control strategies are useful for accomplishing this. Over time, the process will come into control. This requires at least 20 points in a row being in statistical control. At this point, the control limits should be recalculated using the most recent data. Then you can begin looking at methods of changing the system to improve the process.

An important point to remember is that control charts do not tell us how to improve our processes. They do tell us if any changes we make have an effect. The problem solving model provides a guide to help us determine how to improve our processes.

When is it appropriate to stop charting? There are several cases when you may want to stop charting. One case is if the chart is not helping you. There is no sense in plotting points on a control chart if you aren't being helped. Another case may be after the process has been improved sufficiently. Caution should be taken here, however. Remember that being in control is not the natural state. If you stop charting, it is possible that the process will reach a state where it is no longer in statistical control. You may want to consider keeping the chart, but taking data less frequently.

WHAT MANUFACTURING PROCESS VARIABLES SHOULD BE CHARTED?

Many process parameters can be monitored using control charts. This section deals with manufacturing plants. There are three major types of manufacturing variables. They are process variables, process responses, and product responses. Each is discussed below.

Process variables: Process variables are the process parameters over which there is direct control. They are the parameters set by the front-line personnel, i.e., the "knobs" used to control or adjust the process. In statistical terms, process variables are the independent variables. Process variables include variables such as temperature and pressure, whose levels are determined by set point controllers.

There are two types of process variables. Fixed process variables are those controlled at set conditions. For example, if a furnace is controlled at one set temperature at all times, the furnace temperature is a fixed process variable. Adjustable process variables are those whose target values are changed to achieve a different end result in the product. An example of an adjustable process variable is reactor temperature in polyvinyl chloride polymerization. The reactor temperature is adjusted to produce the desired molecular weight resin.

Process variables are not responses. They do not have the random variation that is required for control chart usage. Thus, control charts are not needed for process variables. Control is obtained through operator monitoring and log sheets. One may want to show control over the process variables by use of run charts, such as those obtained from strip chart recorders.

It is possible to use process variables data to analyze the frequency of adjustment needed by operators to maintain the process variables at set points. This will identify process variables that exhibit frequent problems. Ways to correct this type of problem include repairing the controller, installing a more accurate controller, or increasing the frequency of operator monitoring.

One question that must be addressed about process variables is "Are the process variables at their optimum setting?" Experimental design techniques should be used to answer this question once the process is stable.

Process responses: Process responses are measurements determined primarily on-line that relate to the quality of the product being produced. In statistical terms, process responses are dependent variables. They are affected by process variable settings, raw materials used, the environment, etc. Process responses can be controlled only indirectly.

In some cases, process responses correlate with important product characteristics. Correlations can be determined by use of scatter diagrams. If correlations exist between a process response and an important quality characteristic, control charts should be used to monitor the process response over time.

Product responses: Product responses are measurements made on the product for purposes of controlling the process or controlling the product to be shipped. These measurements are normally measured off-line, such as in the laboratory. Examples include purity, color, bulk density, etc. Control charts should be used to monitor important product responses.

In most cases, manufacturing units will begin a quality improvement process by monitoring product responses. The objective should be to move the monitoring upstream to process responses once correlations have been established. Monitoring the process responses and having the process variables set at the optimum settings will ensure that the product is made right the first time.

PROCESS STABILITY

How concerned should you be if the control chart you are examining has several out-of-control points? The answer depends on the intended use of the control chart. Unless you are involved in the direct control of the process, you should not be concerned about a control chart that has less than 5% of the points out of control. This subject is addressed in more detail below.

In previous chapters, methods have been introduced to determine if a process was in statistical control. In addition, it has been stated that being in control is not the natural state of a process. It is obtained only by working to permanently remove assignable causes from the process.

This is the dilemma we face. Being in statistical control in the strict sense means no points beyond the control limits and no evidence of non-random patterns on the control chart. However, since being in control is not the natural state, we expect to see out of control situations from time to time. If this is the case, will our processes ever be in control? How many out-of-control points should we expect and what should our reaction be? This depends on the intended use of the control chart.

For the front-line personnel in a manufacturing operation, a control chart is used as a basis for action. There are four possible actions that can be taken as a result of an out-of-control situation. These four are:

1. Investigative action: Search for the assignable cause of variation
2. Corrective action: Removal of a specific known assignable cause
3. Preventive action: Permanent removal of a known assignable cause to prevent recurrence (preferred action)
4. Compensating action: Process adjustments when the assignable cause is unknown to bring the process back into control

When an out-of-control situation occurs, the front-line personnel will attempt to discover why. Over time, more and more assignable causes will be removed through preventive action. As a result, the number of out-of-control situations will decrease over time. For the front-line personnel, an out-of-control situation always requires action. Control strategies are very beneficial in helping find assignable causes.

Others within a company may have uses for the charts kept by the front-line personnel. Charts could be used by marketing to show a customer or by engineering to determine if experimental design techniques can be used. If there are out-of-control points on the chart, does this mean that the process is out of control? It is important to remember that action is being taken on out-of-control situations by the front-line personnel and that, over time, the number of out-of-control situations will decrease. When an out-of-control situation is indicated by the control chart, the process is unstable. Action will be taken by the front-line personnel to bring the process back into control. As a rule of thumb, people outside the front-line personnel can assume that a process is in statistical control if less than 5% of the points are out of control.

SUMMARY

This chapter introduced the concept of statistical management. This involves using control charts as a basis for action. Control charts help you do your job better by telling when to look for assignable causes and when to leave the process alone. Strategies for improving a process that is in control and a process that is out of control were introduced. These strategies are different. In addition, information on how to start a control chart and how to maintain a control chart over time was given. Manufacturing variables that should be charted include process parameters and product responses. In general, a process is stable if less than 5% of the points are out of control.

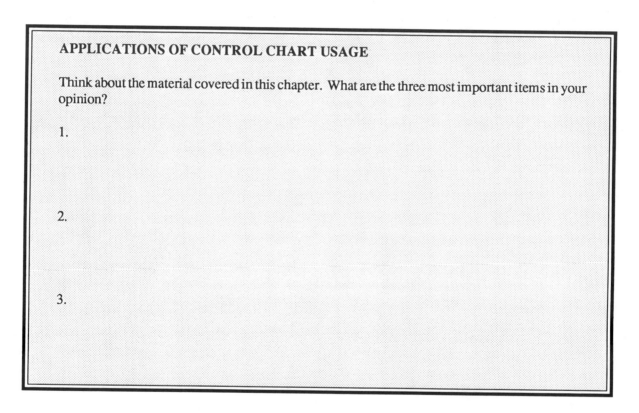

APPLICATIONS OF CONTROL CHART USAGE

Think about the material covered in this chapter. What are the three most important items in your opinion?

1.

2.

3.

26
CAUSE-AND-EFFECT DIAGRAMS

What causes variation in things we do? Why don't we arrive to work at the same time every day? Why isn't the monthly report error free each month? Why can't we produce the same product day in and day out? The cause-and-effect diagram is a good tool to use to summarize the causes of variation in our process. This module introduces the cause-and-effect diagram.

OBJECTIVES

In this chapter you will learn:

1. What a cause-and-effect diagram is
2. When to use a cause-and-effect diagram
3. How to construct a cause-and-effect diagram

Creating a cause-and-effect diagram is fun and educational. These diagrams are usually constructed as a team or group activity to get ideas from as many people as possible. As a result of everyone working on the diagram together, everyone tends to gain some new knowledge. Cause-and-effect diagrams encourage new ideas about causes of problems by helping the group think about different categories of causes. The cause-and-effect diagram also indicates how much we know about our process. If the diagram is full, we know a lot about our process. If it is sketchy, chances are we don't have a good understanding of our process. Cause-and-effect diagrams should be living documents. That is, we should actively seek causes of problems and add to the diagram as time goes on.

INTRODUCTION TO CAUSE-AND-EFFECT DIAGRAMS

A cause-and-effect diagram is a tool that shows the relationship between a quality characteristic (effect) and possible sources of variation for this quality characteristic (causes) (Ishikawa, 1982). Figure 1 is an example of a cause-and-effect diagram, which shows what makes an effective quality development program. The effect in this diagram is "quality development." The "causes" of variation in this characteristic are categorized into main factors: vendor quality support, employee involvement, awareness and publicity, definition of customer needs, measurement systems, education and training, and management leadership. Each of these main factors is divided into detailed causes. For example, a detailed cause under the "Education and Training" main factor is "statistical methods."

The causes are most commonly categorized in terms of Shewhart's concept of a process (machines, methods, environment, materials, measurement, and people). This is particularly true for manufacturing applications. Figure 2 is an example of a cause-and-effect diagram using this approach. The effect in this example is the color of a polyvinyl chloride dryblend. The main factors are simply the six elements in Shewhart's concept

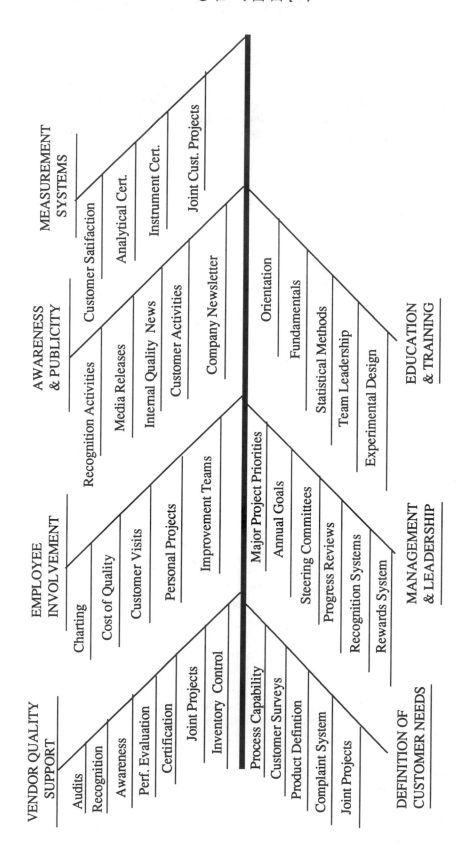

Figure 1

CAUSE-AND-EFFECT DIAGRAM ON QUALITY DEVELOPMENT

Figure 2

PRODUCT COLOR CAUSE-AND-EFFECT DIAGRAM

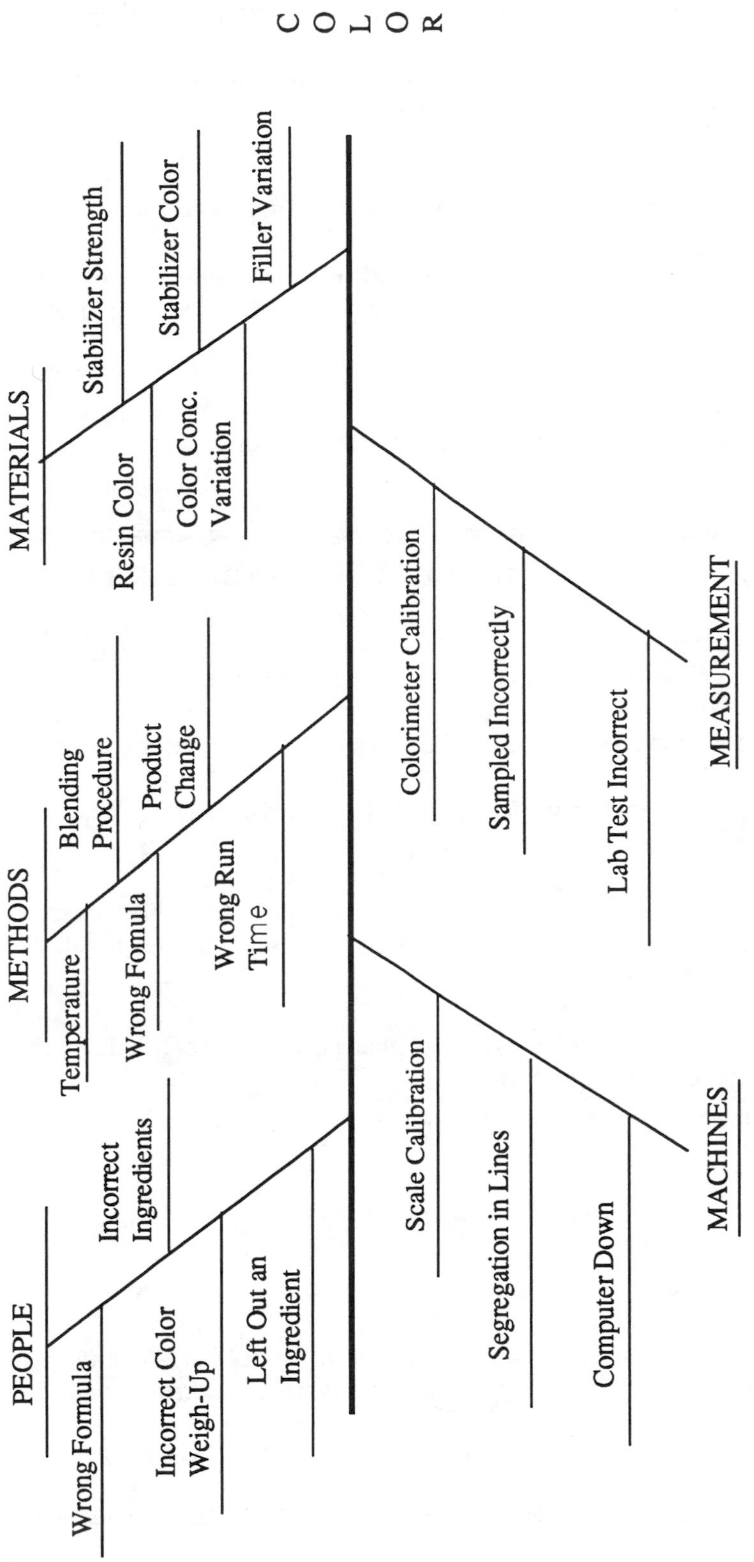

(Note: Environment is often added as a main factor)

of a process. Each main factor is broken down into detailed causes. For example, a detailed cause of variation under the main factor "Methods" is "Wrong Blending Procedure."

Cause-and-effect diagrams are also called fishbone diagrams (because of their appearance) and Ishikawa diagrams (because of their developer).

WHEN TO USE A CAUSE AND EFFECT DIAGRAM

The cause-and-effect diagram can be used in the "Determine Causes" step of the problem solving model. In this step of the problem solving model, we are trying to determine why our process behaves as it does.

Cause-and-effect diagrams can be used for any problem in any department (manufacturing, administration, controllers, supply and transportation, marketing, etc.).

STEPS IN CONSTRUCTION OF A CAUSE-AND-EFFECT DIAGRAM

Below are the steps which should be followed in the construction of a cause-and-effect diagram. A process flow diagram of these steps is shown in Figure 3.

1. Pinpoint the quality characteristic you want to improve or control.

2. Write the quality characteristic (effect) selected on the right hand side and draw a horizontal line to the left.

3. Select the categories for the causes of variation in the quality characteristic. These categories will be the main factors of the cause-and-effect diagram. Each main factor forms a branch off the horizontal line.

4. Brainstorm detailed causes for each main factor. These detailed causes are written on branches off those of the main factors.

   Use the brainstorming rules shown below to maximize the number of causes your group develops.

BRAINSTORMING RULES

1. Go for quantity of ideas.
2. Have everyone participate.
3. Encourage creativity and mind-set breaking.
4. Don't discuss/criticize/evaluate ideas.
5. Build off other ideas.
6. Pass if you don't have an idea to contribute.

Figure 3

CONSTRUCTION OF CAUSE-AND-EFFECT DIAGRAMS

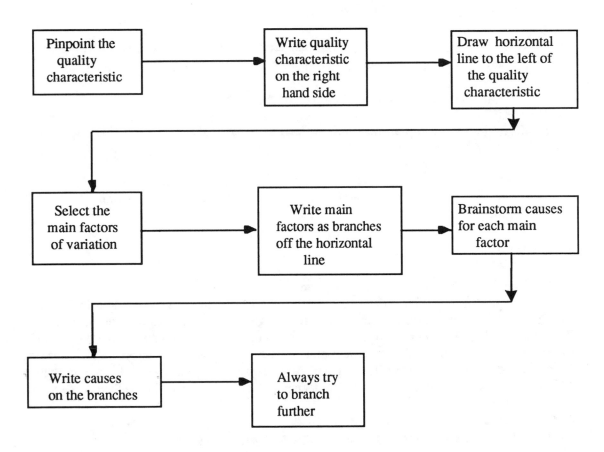

It is sometimes helpful for the group to brainstorm causes onto a blank piece of paper first and then to write the causes onto the diagram instead of writing the causes onto the diagram as they are mentioned. In this way the flow of ideas during brainstorming is not cut off.

5. Always try to branch further by continuously asking "What causes variation in this branch?" In this way, you can add to the cause-and-effect diagram until it fully shows all the possible causes of variation.

EXERCISE: CAUSE-AND-EFFECT DIAGRAM

Think about a problem that is familiar to all members of the group (e.g. late arrival at work). Draw a cause-and-effect diagram for this problem using the following steps.

1. Pinpoint the quality characteristic you want to improve or control.

2. Write the quality characteristic (effect) selected on the right hand side and draw a horizontal line to the left.

3. Select categories for causes of variation in the quality/characteristic. These categories will be the main factors of the cause-and-effect diagram. Each main factor forms a branch off the horizontal line.

SUMMARY

Use cause-and-effect diagrams when you are looking at why your process is behaving the way it does. They will give you an organized look at the causes of problems or variation. Developing a cause-and-effect diagram is educational for the group, helps the group think of causes of variation it may have never thought of, and it gives the group a good indication of how well it knows its process. In developing a cause-and-effect diagram, brainstorm the potential causes and write them onto the branches of the diagram. Always try to branch the diagram further. Maintain the cause-and-effect diagram as a living document by continually adding to the diagram new causes that are discovered.

APPLICATIONS OF CAUSE-AND-EFFECT DIAGRAMS

Think about two instances where you might be able to use cause-and-effect diagrams at work and write them down below.

1.

2.

27

SCATTER DIAGRAMS

Suppose you are faced with a problem. You have followed the steps in the problem solving model. You have defined the problem using Pareto diagrams and pinpointing. In addition, you have analyzed how the process is behaving through the use of process flow diagrams, histograms and control charts. The process is in control, but the results are not acceptable. There is too much variation in the process (or perhaps it is operating at the wrong level or average). You need to find out what is causing the process to behave as it does. A cause-and-effect diagram has been constructed. This diagram lists all the possible causes of the problem. How do you determine what causes are really responsible for the variation. For example, is reaction yield influenced more by run time or pressure? One method of doing this is to use a scatter diagram. The scatter diagram is introduced in this chapter.

OBJECTIVES

In this chapter you will learn:

1. What a scatter diagram is
2. When to use a scatter diagram
3. How to construct a scatter diagram

The scatter diagram provides a method of determining the degree of association between two variables. The objective of constructing a scatter diagram is to determine if one variable (y) can be controlled by controlling another variable (x). This is a very valuable tool in determining why a process behaves as it does. The scatter diagram provides a method of determining what factors in a process may affect a product response or a process parameter.

INTRODUCTION TO SCATTER DIAGRAMS

A scatter diagram is used to show the relationship between two kinds of data. It could be the relationship between a cause and an effect, between one cause and another, or even between one cause and two others. To understand how scatter diagrams work, consider the following example. Suppose you have been working on the process of getting to work within a certain time period. The control chart you constructed on the process shows that, on average, it takes you 25 minutes to get to work. The process is in control. You would like to decrease this average to 20 minutes. What causes in the process affect the time it takes you to get to work? There are many possible causes, including traffic, the speed at which you drive, the time you leave for work, weather conditions, etc. Suppose you have decided that the speed at which you drive is the most important cause. A scatter diagram can help you determine if this is true.

In this case, the scatter diagram would be showing the relationship between a "cause" and an "effect." The cause is the speed at which you drive and the effect is the time it takes to get to work. You can examine this cause-and-effect relationship by varying the speed at which you drive to work and measuring the time it takes to get to work. For example, on one day you might drive at 40 mph and measure the time it takes to get to work. The next day, you might drive at 50 mph. After collecting enough data, you can then plot the speed at which you drive versus the time it takes to get to work. Figure 1 is an example of a scatter diagram for this case. The cause (speed) is on the x axis. The effect (time it takes to get to work) is on the y axis. Each set of points is plotted on the scatter diagram.

Figure 1

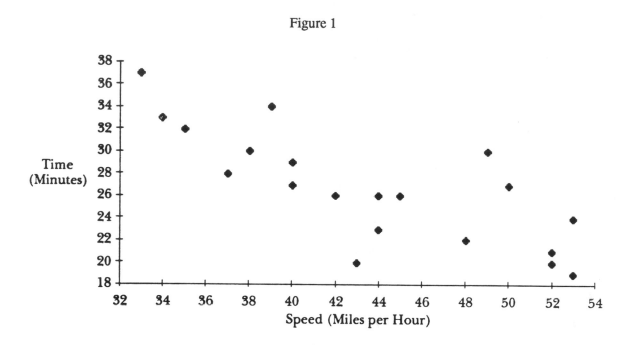

The next step is to determine if there is a relationship between speed and the time it takes to get to work. Figure 2 shows the general types of relationships that can exist. Figure 2a shows a strong positive correlation between x and y. This means if x increases, then so will y. If x is the speed at which you drive and y the time it takes to get to work, a strong positive correlation would mean that the faster you drive (increasing x), the longer it takes to get to work (increasing y). Figure 2b shows a situation where a positive correlation may be present. This means if x increases, y will increase somewhat. However, there are probably other factors that are affecting y.

Figure 2c shows an example of no relationship or correlation between x and y. In other words, y is affected by causes other than x. For the driving-to-work example, this would mean that the speed at which you drive has no effect on the time it takes to get to work.

Figure 2d is an example of a possible negative relationship between x and y. Increasing x (the speed) decreases y (the time) somewhat, but there appear to be other causes that affect y. Figure 2e shows a strong negative relationship between x and y. This means that an increase in x causes a decrease in y. For example, the faster you drive, the quicker you get to work.

Figure 2

GENERAL TYPE OF CORRELATIONS

2a. Positive Correlation

2b. Possible Positive Correlation

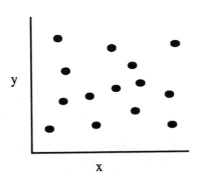

2c. No correlation

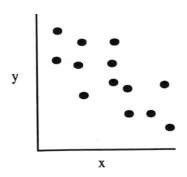

2d. Possible negative correlation

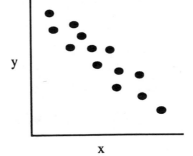

2e. Negative Correlation

Comparing Figure 1 with Figure 2 appears to indicate that there might be a negative correlation present between the speed at which you drive and the time it takes to get to work. By driving faster, you can decrease the time it takes to get to work a little. However, other causes (such as traffic, time you leave to get to work, etc.) are evidently affecting the time it takes to get to work.

Exact methods of determining if a correlation exists between x and y are given below. A measure of the correlation between x and y is the linear correlation coefficient (R_{yx}). If R_{yx} is close to +1, there is a strong positive correlation. If R_{yx} is close to -1, there is a strong negative correlation. If R_{yx} is close to 0, there is no correlation.

There is one item to be cautious about. It is possible that a scatter plot can show a relationship between x and y, but x might not be the cause of y. There may be some other factor that affects both x and y in the same manner. For example, a study during the 1930's showed a relationship between the number of babies born in one part of England and the stork population. As the stork population increased, so did the number of babies. We realize that the reason for the increase in babies born was not caused by the increase in stork population.

In statistical terms, x is called the independent variable and y is called the dependent variable. This terminology is not very useful. It is helpful to think of x as the process variable and y as the response variable.

WHEN TO USE A SCATTER DIAGRAM

A scatter diagram can be used in the "Determine Causes" step of the problem solving model. In this step, we are trying to determine why the process is behaving as it does.

STEPS IN CONSTRUCTION OF A SCATTER DIAGRAM

The steps in constructing a scatter diagram are given below. A process flow diagram of the steps is shown in Figure 3.

1. Gather the data.

   a. Collect 50 to 100 paired samples of data (x and y values) whose relationship you wish to investigate and record the data. Less data can be used if necessary.

2. Plot the data.

   a. Select the scales for the x and y axes.

   b. Plot each paired value of sample data on the chart.

3. Determine if a correlation exists between x and y.

    a. Calculate the linear correlation coefficient, R_{yx}

        $$R_{yx} = \Sigma[(X_i - \overline{X})(Y_i - \overline{Y})]/\sqrt{\Sigma(X_i - \overline{X})^2 \Sigma(Y_i - \overline{Y})^2}$$

        or

        $$R_{yx} = (\Sigma X_i Y_i - n\overline{X}\overline{Y})/(n - 1)s_x s_y$$

        where s_x and s_y are the standard deviations of the x and y values, respectively, and n is the number of data points.

    b. Determine the degrees of freedom. The degrees of freedom (df) is the number of paired sample points (n) minus 2.

        $df = n - 2$

    c. Compare the value of R_{yx} to the value in Appendix Table 11 (p. 496) for the degrees of freedom. If R_{yx} is larger than or equal to the table value, there is a statistically significant correlation between x and y.

Figure 3

STEPS IN CONSTRUCTION OF A SCATTER DIAGRAM

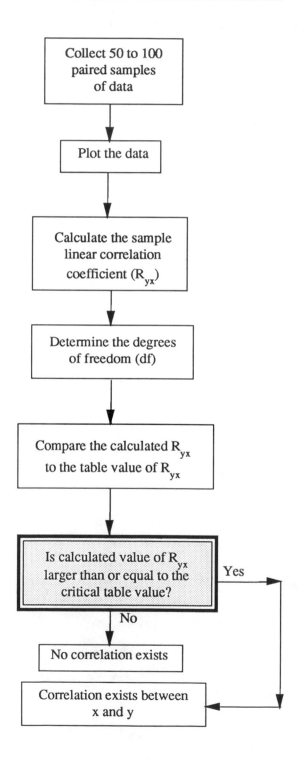

SCATTER DIAGRAM EXAMPLE

Run Time and Purity

An operating unit contains a batch reactor used to make a liquid product from various raw materials. The reactor is controlled on temperature. The reaction is considered finished when the pressure begins to drop. This normally takes about three hours. Purity of the liquid product is one critical parameter for downstream operations. A team was working on how to improve the purity by either increasing the purity or making it more consistent. Front line personnel on the team indicated that they felt the run time had an effect on purity. They felt that the longer the run time (i.e., the time from the start of the reaction to pressure drop), the higher the purity. The team decided to construct a scatter diagram to determine if there was a correlation between run time (minutes) and product purity (percent). Data from the last 12 batches are given below.

Run Time (x)	Purity (y)	xy
185	90.3	16705.5
176	89.8	15804.8
189	91.8	17350.2
195	92.5	18037.5
174	86.4	15033.6
198	93.4	18493.2
190	90.1	17119.0
176	87.5	15400.0
179	86.7	15519.3
175	85.0	14875.0
186	89.6	16665.6
205	93.1	19085.5

	Run Time (x)	Purity (y)	xy
Sum	2228	1076.2	200089.2
Average	185.67	89.68	
Standard Deviation	10.1115	2.7774525	

Use the data above to determine if there is correlation between run time and purity. When you have completed the calculations, answer the following questions. A plot of the data is given on the page 382.

1. What is the value of R_{yx}?

2. What is the critical value for R_{yx}?

3. Is there a correlation between run time and purity?

4. If there is a correlation between run time and purity, what type of correlation is it?

5. What would you suggest doing next?

RUN TIME VERSUS PURITY

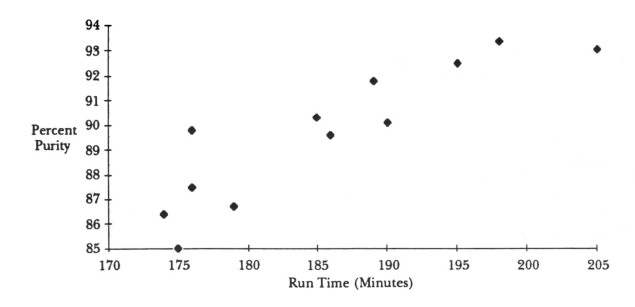

SCATTER DIAGRAM EXAMPLE

Color and Specific Gravity

A customer of a company had complained recently about the color of a granular product. A team was formed to work on the problem. One person suggested that color may vary with the specific gravity of the product. Twelve samples were pulled from the process and tested for color and specific gravity. The data are given below.

Color (x)	Specific Gravity (y)	xy
10.6	1.201	12.7306
13.9	1.218	16.9302
11.9	1.210	14.3990
17.2	1.213	20.8636
14.9	1.213	18.0737
14.7	1.209	17.7723
12.0	1.210	14.5200
14.1	1.206	17.0046
13.0	1.209	15.7170
12.4	1.203	14.9172
16.3	1.215	19.8045
10.4	1.204	12.5216
Sum	161.4	14.511	195.2543
Average	13.45	1.20925	
Standard Deviation	2.1258154	0.0050834	

Use the data above to determine if there is a correlation between color and specific gravity. When you have completed the calculations, answer the following questions. A plot of the data is given on page 384.

1. What is the value of R_{yx}?

2. What is the critical value for R_{yx}?

3. Is there a correlation between color and specific gravity?

4. If there is a correlation between color and specific gravity, what type of correlation is it?

5. What would you suggest doing next?

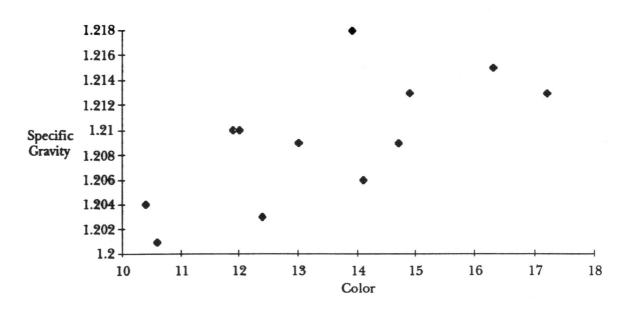

SPECIFIC GRAVITY VERSUS COLOR

DISCUSSION OF SCATTER DIAGRAMS

The linear correlation coefficient, R_{yx}, is a measure of the association between two variables. The maximum value for R_{yx} is +1. The minimum value for R_{yx} is - 1. In both these cases, all sample points fall on a straight line. As R_{yx} approaches +1 or -1, the stronger the correlation between x and y. Note that there may be no engineering basis for considering one variable to be x and the other variable to be y. The square of this coefficient (R^2) indicates the fraction of variation in y that is associated with x.

If R_{yx} is significant, this means that there is a statistically significant correlation between x and y. It does not say that this correlation is important in terms of engineering or a management system. It may or may not be a strong enough correlation to use in an engineering fashion. With large sample sizes, "small" values of R_{yx} may be statistically significant.

It is important to remember that there are many things associated with a process result. Only a subset of these are true causes of the process result. Even if R_{yx} is significant, the cause of y might not be x. However, if R_{yx} is not significant, we are sure that x is not the cause of y.

One question not addressed by scatter diagrams is the stability of the process. If the process is not in statistical control, the correlation between two variables might not be consistent over time.

To study variation, there must be enough variation in x to detect changes in y. Figure 4 shows an example of this. The response variable (y) might not be affected by changes in the process variable (x) if x does not change over a large enough range.

Figure 4

VARIATION IN y DUE TO x

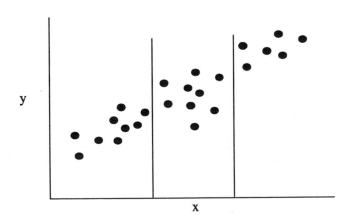

If R_{yx} is not significant, it does not mean that there is no relationship between x and y. It is possible that a nonlinear relationship exists.

If two variables have control charts that look the same, it is probable that the two variables are positively correlated. If two variable have control charts that are the inverse of one another (i.e., when one variable goes up, the other tends to go down), it is probable that the two variables are negatively correlated.

SUMMARY

A scatter diagram provides a method of determining the degree of association between two variables. This is accomplished by calculating the linear correlation coefficient, R_{yx}, for paired data values of x and y. The calculated value of R_{yx} is then compared to a table value. If the calculated value is larger than or equal to the table value, there is a statistically significant correlation between x and y. The data values are plotted to form the scatter diagram. Scatter diagrams are useful tools to determine why a process behaves as it does.

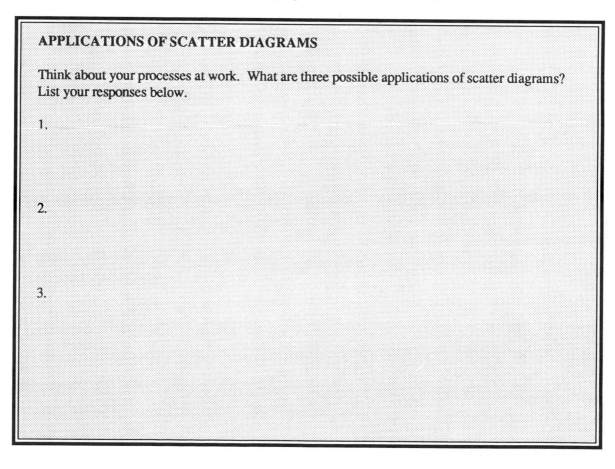

APPLICATIONS OF SCATTER DIAGRAMS

Think about your processes at work. What are three possible applications of scatter diagrams? List your responses below.

1.

2.

3.

28
MEASUREMENT SYSTEMS

Boss: This railcar shipment must go out today. The customer needs it or they will have to shut down.

Supervisor: No problem. It's loaded and the lab is testing it now.

(Thirty minutes later.)

Supervisor: Oh NO! This railcar sample is out of spec. Hurry, pull another sample and get it to the laboratory to test.

(Twenty minutes later.)

Supervisor: [Expletive Deleted]! This sample is still out of spec. Tell the lab to retest that sample.

(Twenty minutes later.)

Supervisor: Finally! The sample is in spec. It took the lab long enough to run the analysis right. I knew we made it right. Ship the railcar!

Does this sound familiar to you? Many operating units still do this, i.e., continuously test the material until a "suitable" result is obtained. What is wrong with this approach? This type of approach to operating indicates a lack of understanding of variation and a lack of confidence in the accuracy and precision of the measurement system. It also leads to increased costs due to more sampling and testing. What can be done to increase the confidence people have in measurement systems? What methods are there to determine when a measurement system should be calibrated or when it should be left alone? This module introduces methods to monitor and determine the accuracy and precision of measurement systems. The effect that measurement systems have on the process is also discussed.

OBJECTIVES

In this chapter you will learn:

1. How measurement systems cause variability in processes
2. How to view the measurement system itself as a process
3. The definition of accuracy and precision
4. How to determine and monitor the accuracy of measurement systems
5. How to determine and monitor the precision of measurement systems
6. How to determine how much of the total process variation is due to the measurement system

Collecting and analyzing data are a vital part of process improvement. It is important to

---

be sure that the data we are collecting are accurate and precise. If not, we may not be able to see the effect of improvements we are trying to make in the process. In addition, we may be looking for an assignable cause in an operating unit when the real cause is in the measurement system.

THE IMPORTANCE OF MEASUREMENT SYSTEMS

There are many types of measurement systems. Statistical process control (SPC) was originally developed for the production of widgets. Measurements in this case often involve dimensions. This type of measurements can be very precise and accurate. However, this is not true for all types of measurements. For example, in the process industries, there are many types of measurement systems. These include temperature and pressure gauges/indicators, on-line analyzers, wet chemical techniques and physical test methods. Some of these measurement systems are not very accurate or precise.

We rely on measurements to tell us when our process is operating correctly, when there is a problem, or when we have made an improvement. When we take a measurement, we often assume that the measurement result is the "true" value of our sample. Unfortunately, this is not true. Measurement systems, like any other process, are subject to variation. You will not always get the same result, for example, when you run a sample again or take another temperature measurement. The following examples illustrate two ways the measurement system can impact a process.

Suppose you are an operator in a processing unit. You have started a control chart on a certain product parameter. This parameter is determined by taking a sample from a process stream and sending it to the laboratory to be analyzed. The lab reports the result to you. You in turn plot the point on the control chart. Suppose the last point you plotted is above the upper control limit. This means there is an assignable cause present in the process. Your responsibility is to begin looking for the cause. Where do you start? There are many sources of variation present in a process. A control strategy helps you determine what to look for. The cause could be due, for example, to a change in raw materials or a sudden drop in the outside temperature. You begin to look for the cause. This takes time. What about the test method the lab used? Is it possible that the assignable cause was due to something occurring in the lab and not in the processing unit? The answer to this is yes; it is possible that the lab test did not reflect an accurate value for the product parameter. This could have occurred, for example, due to using a new reagent or to a miscalculation. If this is the case, you will spend a lot of time looking for the assignable cause in the processing unit when the real cause was in the laboratory. Thus, errors in measurement can cause assignable causes in the process.

Now suppose you are an engineer. You have determined some process changes that should increase the color of a certain product. The operating unit makes the process changes and samples are taken. The samples are sent to the laboratory for testing. Is the test method used by the laboratory good enough to "see" the color improvement caused by the process changes? If not, you will not be able to determine what impact the process changes had. Worse, you may make a wrong judgement about what impact the process changes had. Thus, test methods with a lot of variability may mask process improvements you try to make.

The two preceding examples illustrate how measurement systems can affect a process, in general. The impact that measurement systems have on a process is discussed below.

What is a process? A process refers to how we do things, such as making chemical or oil products, typing a memo, processing an invoice, etc. There may be a few steps or many steps in a process. In general, a process may be divided into six major elements. These elements are people, materials, methods, machines, measurement and environment. Walter Shewhart, the father of SPC, thought of processes in these terms. The output of a process was due to the manner in which these six elements worked together. Shewhart's concept of a process is shown in Figure 1. Variation in any of the six elements causes variation in the process output. Thus, the variation in the measurement system will cause variation in the process output.

Figure 1

SHEWHART'S CONCEPT OF A PROCESS

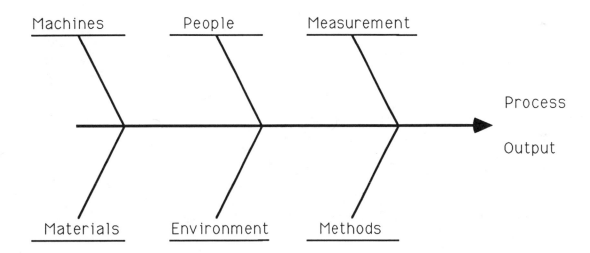

There is a certain amount of inherent or natural variation in the six elements. For example, no two people can do a job exactly alike. In fact, one person cannot do the very same job exactly the same every time. The same is true for the measurement system. No matter how accurate and precise the measurement system is, the same result will not be obtained time after time. This natural variation in the six elements is what is called common cause variation. Common cause variation is always present, and the measurement system will always contribute to the common cause variation in the process output. Measurement systems, like the other elements, can also cause assignable cause variation. For a measurement system to be useful, the reasons for assignable causes due to the measurement system must be eliminated. In addition, the measurement system's contribution to the process's common cause variation must be minimized.

Control charts are used to monitor various process or product parameters. These charts indicate when there is assignable cause variation present or when there is only common cause variation present. The measurement system itself can be viewed as a process. The output from the measurement system, as shown in Figure 2, is due to the way people, methods, machines, materials, and the environment interact. Thus measurement systems are subject to assignable as well as common causes. Control charts can then be used to monitor the measurement system.

Figure 2

VIEWING A MEASUREMENT SYSTEM AS A PROCESS

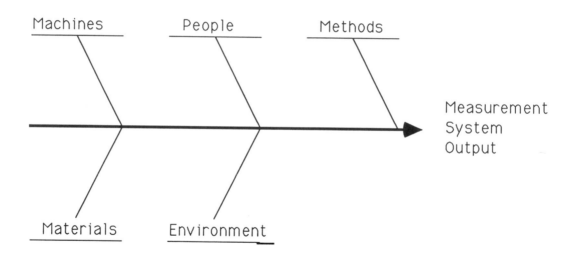

ACCURACY AND PRECISION DEFINITIONS

Accuracy refers to the absolute correctness of the measurement system relative to a standard. There are many standards. For example, one method of checking the accuracy of a laboratory weigh scale is to put a standard 10 gram weight on the scale. On average, the scale should read 10 grams if it is accurate. Another example involves an on-line gas analyzer. To check the accuracy of this analyzer, a standard gas is injected into the analyzer. On average, the analyzer should indicate the standard gas composition if the analyzer is accurate. Obviously, we want test methods which reflect the true value of a standard (or whatever we are measuring).

In some instances, standards do not exist. For example, polyvinyl chloride samples with accurately known particle size distributions or heat stabilities do not exist. In these instances, one must resort to some artificial or secondary standards.

Precision of a test method refers to how reproducible or repeatable the results are. How close will the results be if a sample is rerun or a temperature is remeasured? Consider the laboratory weigh scale example. If the standard weighs 10 grams and is weighed multiple times, how different will the results be? This variation in reproducing the result is called the precision of the test method. The smaller this difference, the more precise the test method.

Figure 3 presents another method of developing an understanding of accuracy and precision.

Figure 3

DEFINITION OF ACCURACY AND PRECISION

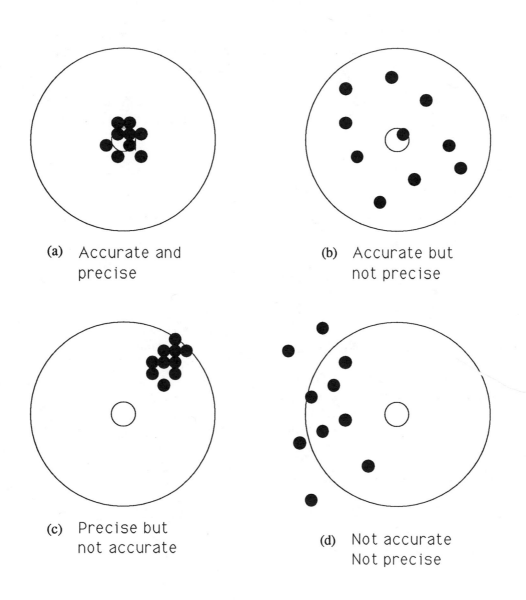

(a) Accurate and
precise

(b) Accurate but
not precise

(c) Precise but
not accurate

(d) Not accurate
Not precise

In Figure 3, the objective was to hit the target's bullseye. Figure 3a shows a marksman who is precise and accurate. Each shot hits near or in the bullseye (i.e., the marksman is accurate). The difference between successive shots is small (i.e., the marksman is precise). Figure 3b shows a marksman who is accurate but not precise. If you average all his shots, the average would be close to the bullseye. However, the difference between shots is very large. Figure 3c shows a marksmen who is precise but not accurate. His average shot is not near the bullseye. However, the differences between consecutive shots is very low. Figure 3d is an example of a marksman who is neither accurate or precise. There is no telling where the next shot will end up. He is off the bullseye and the differences between shots are very large.

ACCURACY AND PRECISION OF MEASUREMENT SYSTEMS

As mentioned before, measurement systems themselves can be viewed as a process. As such, measurement systems have common and assignable causes of variation. Control charts can be used to monitor the accuracy and precision of measurement systems.

The procedure for determining the accuracy and precision of a measurement is given below. A process flow diagram of the steps is given in Figure 4. A detailed example is then given.

Procedure for Determining the Accuracy and Precision of Measurement Systems

1. *Run a standard on a regular basis.*

2. *Plot the results using an individuals chart.*

3. *Bring the measurement system into control by finding and eliminating assignable causes.*

4. *Determine the accuracy of the measurement system by comparing the center line on the X chart to the the true value of the standard.*

5. *Determine the precision of the measurement system by calculating the standard deviation of the measurement system from the range chart.*

6. *Determine the percent of total variance due to the measurement system.*

7. *Continue to monitor the accuracy and precision of the measurement system over time.*

Figure 4

DETERMINING THE ACCURACY AND PRECISION OF A MEASUREMENT SYSTEM

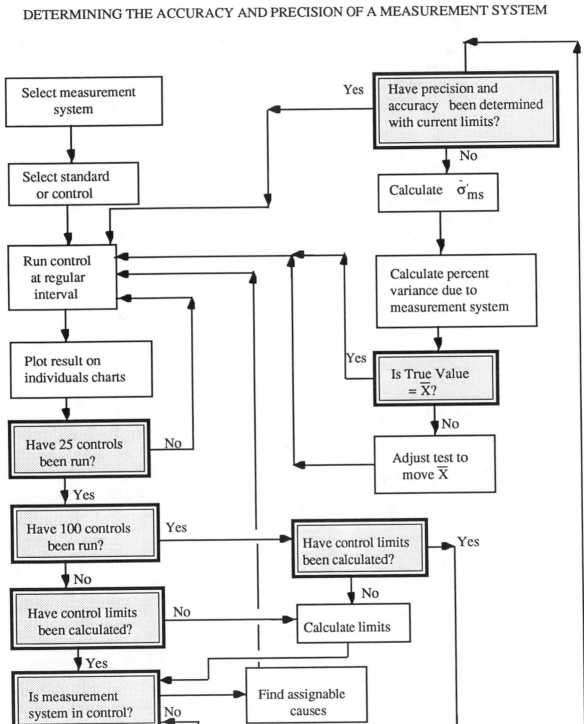

The following example gives each step in detail. A critical processing parameter in the production of a solid material is the moisture content of the solid. The moisture is measured using an infrared device. The laboratory wanted to determine if the moisture measurement was accurate and precise. In the past the instrument was calibrated at the start of each shift. (Calibration should be done only when there is evidence (statistical) that the instrument needs to be calibrated. Recalibrating without evidence of the need is the same as overcontrolling an operating process.) The instrument was calibrated at the start of this study. It was not calibrated again.

Step 1: Run a standard on a regular basis.

Since there was no standard solid with a known moisture content, laboratory technicians set aside a large sample of solid to use as a control. The sample was kept in a sealed plastic container to minimize any changes in moisture content. The moisture content of this control was measured once per shift. The moisture results from running the control for the last 25 shifts are given below.

Shift Number	$\%H_2O$	Shift Number	$\%H_2O$
1	0.10	14	0.12
2	0.14	15	0.11
3	0.10	16	0.11
4	0.12	17	0.12
5	0.12	18	0.11
6	0.12	19	0.13
7	0.11	20	0.12
8	0.12	21	0.12
9	0.10	22	0.13
10	0.12	23	0.14
11	0.12	24	0.12
12	0.10	25	0.13
13	0.11		

Step 2: Plot the results using an individuals control chart.

Since data are available on a limited basis only (once per shift) and measurement error is normally distributed, an individuals control chart can be used to monitor the results from the control samples. The moving ranges and the calculation of overall averages and control limits are given below. The completed control charts are shown in Figure 5.

Shift Number	%H$_2$O	Moving Range
1	0.10	
2	0.14	0.04
3	0.10	0.04
4	0.12	0.02
5	0.12	0.00
6	0.12	0.00
7	0.11	0.01
8	0.12	0.01
9	0.10	0.02
10	0.12	0.02
11	0.12	0.00
12	0.10	0.02
13	0.11	0.01
14	0.12	0.01
15	0.11	0.01
16	0.11	0.00
17	0.12	0.01
18	0.11	0.01
19	0.13	0.02
20	0.12	0.01
21	0.12	0.00
22	0.13	0.01
23	0.14	0.01
24	0.12	0.02
25	0.13	0.01
Sum	2.94	0.31

$\overline{X} = \Sigma X/k = 2.94/25 = 0.118; \qquad \overline{R} = \Sigma R/(k - 1) = 0.31/24 = 0.013$

$UCLx = \overline{X} + 2.66\overline{R} = 0.118 + (2.66)(0.013) = 0.152$

$LCLx = \overline{X} - 2.66\overline{R} = 0.118 - (2.66)(0.13) = 0.083$

$UCLr = 3.27\overline{R} = 3.27(0.013) = 0.042$

Figure 5

MOISTURE CONTENT OF CONTROL CHART SAMPLE
INDIVISUALS CHART

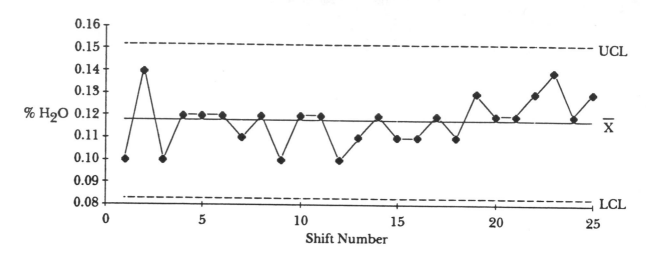

MOVING RANGE CHART (n = 2)

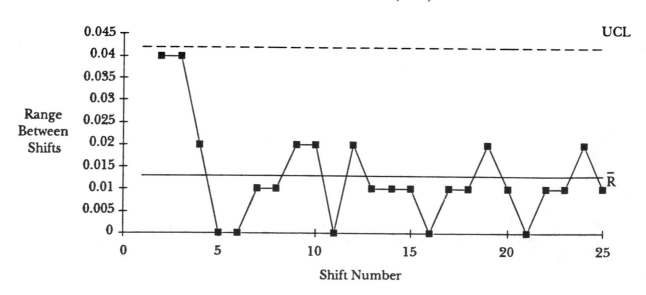

Step 3: *Bring the measurement system into statistical control by finding and eliminating assignable causes.*

As can be seen in Figure 5, the range chart is in statistical control. However, the X chart is not in statistical control. The last seven points on the X chart are above the average. This means that there is an assignable cause present. Something has happened that caused the average to increase. When this occurs, the measurement system should be treated as if something was mechanically broken. No more samples should be run until the reason for the out-of-control point has been found and eliminated. Control strategies should be used to help find the reasons for out-of-control points. It may be that the instrument needs to be recalibrated.

Step 4: *Determine the accuracy of the measurement system.*

To determine the accuracy of the measurement, the center line ($\overline{X}$) on the X chart is compared to the true value of the standard. The measurement system must be in control before this can be done. If $\overline{X}$ does not equal the true value of the standard, adjustments must be made to the measurement system to shift the average to the true value. In this example, there is no "true" standard. In this situation, you assume that the center line is the true value and continually monitor the measurement system versus this center line. The-out-of control situation means that the moisture measurement is not consistent with respect to the average.

An accurate measurement system is defined as a measurement system that is in statistical control with respect to the average of the standard (i.e., the X chart) and whose center line ($\overline{X}$) on the X chart is equal to the true value of the standard.

Step 5: *Determine the precision of the measurement system.*

To determine the precision of the measurement system, calculate the measurement system standard deviation, $\hat{\sigma}'_{ms}$, from the average range on the range chart.

$$\hat{\sigma}'_{ms} = \overline{R}/d_2$$

This is a measure of the amount of variation among the individual results. What causes variation among the individual results? The same material is being used all the time so this does not cause variation. The variation is caused by the test method itself and the laboratory technicians who run the test. The range chart must be in statistical control before you can calculate this standard deviation. For the moisture control example, the range chart is in control. The estimate of the measurement system standard deviation is then:

$$\hat{\sigma}'_{ms} = \overline{R}/d_2 = 0.013/1.128 = 0.0115$$

Step 6: Determine the percent of total process variation due to the measurement system.

The total process variation can be divided into three major components: the process variation itself, the sampling variation and the measurement system variation. These three components of variation are related by the following:

$$\sigma_t'^2 = \sigma_p'^2 + \sigma_s'^2 + \sigma_{ms}'^2$$

where $\sigma_t'^2$ is the total process variance, $\sigma_p'^2$ is the process variance, $\sigma_s'^2$ is the sampling variance and $\sigma_{ms}'^2$ is the measurement system variance. Note that the relationship is linear in terms of the variance, not the standard deviation.

The percent variance due to the measurement system is then given by:

$$\% \text{ Variance due to measurement system} = 100(\sigma_{ms}'^2)/\sigma_t'^2$$

To determine this percent, some estimate of the total variance is required. The best estimate would be from a range chart kept on the quality parameter (in this case, moisture content) in the operating unit. If the operating unit is not keeping a chart on the quality parameter, you can take a month's worth of data from the unit and calculate a variance.

Assume that a control chart has been kept in the operating unit. The standard deviation, $\hat{\sigma}_t'$, has been calculated from the average range to be 0.025. In this case, the percent of total variance due to the measurement system is:

$$100(\hat{\sigma}_{ms}'^2)/\hat{\sigma}_t'^2 = 100(.0115)^2/(.025)^2 = 21.2\%$$

There is no hard-and-fast rule that says a measurement system should be responsible for some percent of the total variance. In general, to help improve processes, the measurement system should account for no more than 10% of the total variance.

A precise measurement system is defined as a measurement system that is in statistical control with respect to variation (i.e., the range chart) and that is responsible for less than 10% of the total process variance.

Step 7: Continue to monitor the accuracy and precision of the measurement system over time.

The control limits on the chart should be extended into the future. Controls should continue to be run on a regular basis. If any out-of-control situations occur, the reasons should be found and the cause eliminated. Continuous monitoring of the measurement system will ensure that test results are accurate and precise. After 100 controls have been run, the control limits should be recalculated to give better estimates of the averages.

MEASUREMENT SYSTEM EXAMPLE

Gas Analyzer

A processing plant is using an on-line gas analyzer to continuously monitor one stream's content of a certain gas. The analyzer is checked weekly by the maintenance department. A maintenance worker injects a standard containing 100 ppm of the gas into the analyzer. Depending on what the resulting reading is, the analyzer is recalibrated. Several people wondered if they were "overcontrolling" this calibration process. The maintenance department decided to inject a standard 100 ppm sample into the analyzer on a daily basis. An individuals chart would be used to monitor the resulst. There were two objectives of this study. One was to determine how often the analyzer should be recalibrated. The other was to develop a method of monitoring the accuracy and precision of the analyzer. The results from injecting the standard for the first 25 days are given below.

Standard Number	Conc. ppm	Moving Range
1	105	
2	97	8
3	103	6
4	102	1
5	99	3
6	100	1
7	96	4
8	102	6
9	102	0
10	100	2
11	98	2
12	100	2
13	103	3
14	101	2
15	96	5
16	102	
17	98	
18	100	
19	98	
20	103	
21	106	
22	102	
23	99	
24	100	
25	97	

Use the data to construct an individuals chart to monitor the accuracy and precision of this test method. The first 15 moving ranges have been calculated. The first 15 results have been plotted and ranges have been plotted for you (p. 401). When you have finished the chart, answer the following questions.

1. Is the gas analyzer in statistical control with respect to the average and variation? What does this mean?

2. Is the gas analyzer accurate?

3. What is a measure of the precision of the gas analyzer?

4. Do you have enough information to determine if the gas analyzer is precise? If yes, is the gas analyzer precise?

5. Has there been any indication that the gas analyzer should be recalibrated? If yes, when did this occur? If no, what could this mean to the maintenance department?

6. What should be done next?

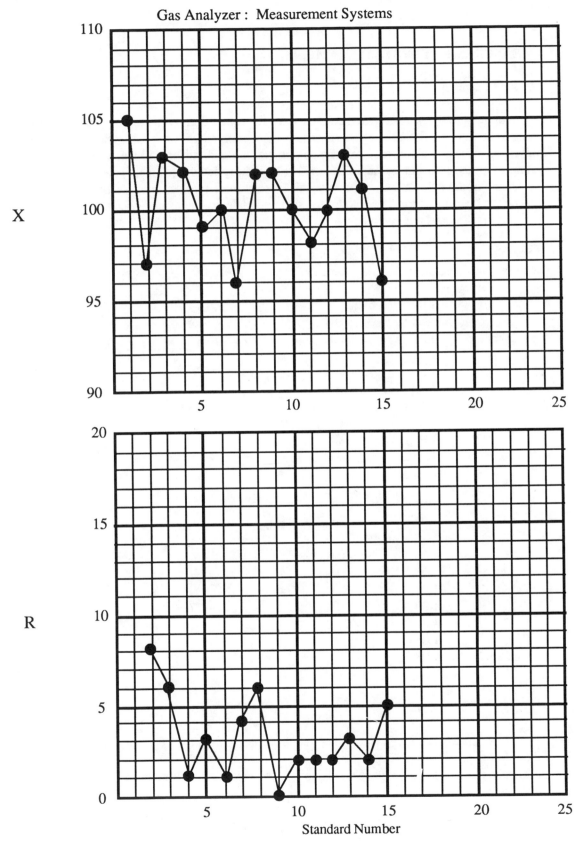

Gas Analyzer : Measurement Systems

MEASUREMENT SYSTEM EXAMPLE

Color Measurement

A company produces a laminated colored sheet. Specifications on the color of the product are very tight. Several methods are used to ensure that the color is within specifications. One method involves the use of a colorimeter. The colorimeter is standardized daily using a standard plate supplied by the colorimeter manufacturer. The laboratory decided to investigate the accuracy and precision of the colorimeter. The colorimeter reads three types of color. Only one color was chosen to be monitored. The color of the standard plate was measured once a day. The results for the first 25 days are given below.

Sample Number	Color	Moving Range
1	1.09	
2	1.08	0.01
3	1.07	0.01
4	1.08	0.01
5	1.06	0.02
6	1.10	0.04
7	1.10	0.00
8	1.07	0.03
9	1.05	0.02
10	1.05	0.00
11	1.06	0.01
12	1.08	0.02
13	1.05	0.03
14	1.03	0.02
15	1.10	0.07
16	1.05	0.05
17	1.05	0.00
18	1.07	0.02
19	1.03	0.04
20	1.05	0.02
21	1.06	
22	1.03	
23	1.02	
24	1.04	
25	1.07	

Use these data to construct an individuals control chart to monitor the accuracy and precision of the colorimeter. The first 20 moving ranges have been calculated for you. The first 20 points have also been plotted (p. 404). Once you have finished the individuals chart, answer the following questions.

1. Is the colorimeter in statistical control with respect to the average and the variation? What does this mean?

2. The colorimeter manufacturer lists the true value of the standard plate color as 1.06. Is the colorimeter accurate?

3. What is a measure of the precision of the colorimeter?

4. What is the problem with determining if the colorimeter is precise?

5. What would have to done to determine if the colorimeter is precise?

6. What should be done next?

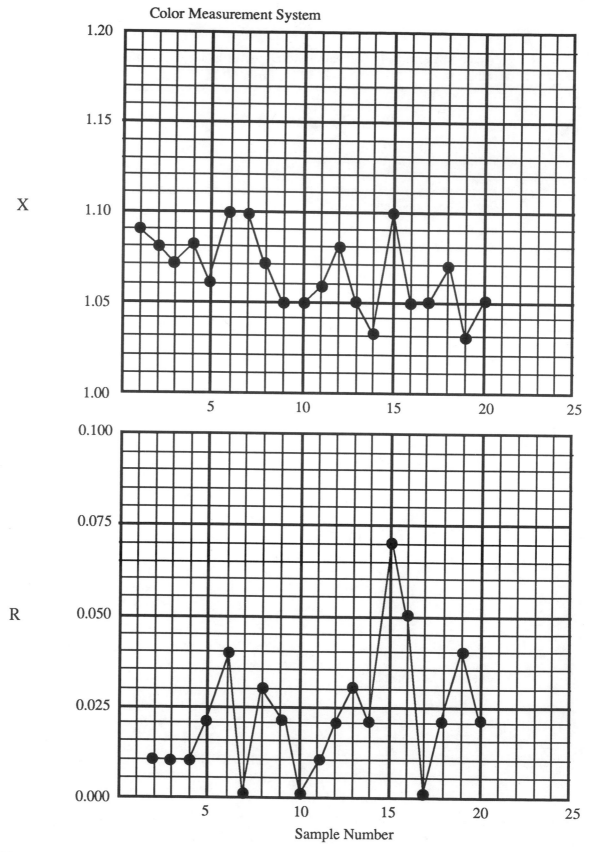

MEASUREMENT SYSTEM PRECISION BY DUPLICATE SAMPLE ANALYSIS

The method described above for determining the precision of measurement systems involves the running of standards or controls on an on-going basis. Often you may be interested in getting a quick look at the precision of a measurement system that is not being monitored on a regular basis. You might also be considering a new measurement system and want to determine if its precision is better than the method you are currently using. An alternate method of determining the precision of a measurement system exists. This method involves the duplicate analysis of a number of different samples. The alternate procedure for determining precision is given below (AT&T, 1956). A process flow diagram of the steps is given in Figure 6. A detailed example is then given.

Determining Measurement System Precision Through Duplicate Sample Analysis

1. Collect 30 - 50 different samples that are representative of the process output and divide each sample in half.

2. One technician runs the test method on one-half of each of the 30 - 50 samples.

3. At a later time, the same technician runs the other half of each sample in the same order.

4. Analyze the results using an $\overline{X}$-R chart with a subgroup size of two. The subgroups are formed from the results obtained from running the same sample twice.

5. Determine if the range chart is in statistical control. If it is not, the reasons for the assignable causes must be found and eliminated. Once this done, the procedure is started over.

6. Determine the precision of the measurement system by calculating the standard deviation of the the measurement system from the range chart.

7. Determine the percent of total variance due to the measurement system.

8. Examine the $\overline{X}$ chart to qualitatively determine the discriminating power of measurement system.

The following example gives each step in detail. One parameter of interest in the production of a powdered product is the bulk density. The bulk density is measured by weighing the amount of material held in a given volume and then calculating the bulk density in grams/cc. This measurement system was not being monitored by running a control on a regular basis. Some questions had arisen about just how precise this measurement was. To get a quick look at the precision, it was decided to run duplicate sample analysis to determine the precision.

Figure 6

DETERMINING THE PRECISION OF A MEASUREMENT SYSTEM BY
DUPLICATE SAMPLE ANALYSIS

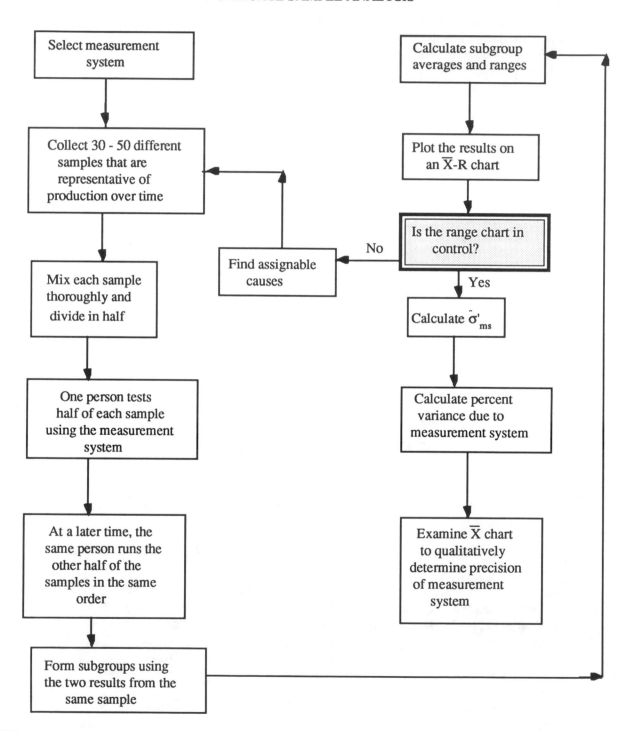

Step 1: *Collect 30 - 50 different samples that are representative of the process output and divide each sample in half.*

Laboratory personnel pulled 50 samples at random from shipment retains that had been stored over the last month. Each sample was numbered. Each sample was thoroughly mixed and divided in half. Each half was labeled "A" or "B."

Step 2: *One technician runs the test method on one-half of each of the 30 - 50 samples.*

One technician ran the bulk density test for each of the 50 samples labeled "A." The samples were run in numerical order starting with 1A, then 2A, etc.

Step 3: *At a later time, the same technician runs the other half of each sample in the same order.*

The next day, the same technician ran the bulk density test on each of the 50 samples labeled "B." The samples were run in numerical order starting with 1B, then 2B, etc.

Step 4: *Analyze the results using an $\overline{X}$-R chart with a subgroup size of two. The subgroups are formed from the results obtained from running the same sample twice.*

The data from the 50 samples are shown in Table 1. Subgroup number 1 is composed of the two sample results for sample 1 (i.e., the bulk density for sample 1A was 0.587; for sample 1B, it was 0.585). The subgroup average is the average of these two results. The subgroup range is the difference between these two results. The subgroup averages and subgroup ranges were calculated for all 50 subgroups. The results of these calculation are also shown in Table 1.

Overall averages and control limits are then calculated as shown below.

$$\overline{\overline{X}} = \Sigma\overline{X}/k = 29.247/50 = 0.5849$$

$$\overline{R} = \Sigma R/k = 0.050/50 = 0.001$$

$$UCLx = \overline{\overline{X}} + A_2\overline{R} = 0.5849 + (1.88)(0.001) = 0.5868$$

$$LCLx = \overline{\overline{X}} - A_2\overline{R} = 0.5849 - (1.88)(0.001) = 0.5831$$

$$UCLr = D_4\overline{R} = 3.27(0.001) = 0.0033$$

$$LCLr = D_3\overline{R} = None$$

The finished control charts are shown in Figure 7.

Table 1
BULK DENSITY PRECISION TEST RESULTS

Sample Number	Bulk Density (g/cc) A	Bulk Density (g/cc) B	Subgroup Average	Subgroup Range
1	0.587	0.585	0.5860	0.002
2	0.593	0.594	0.5935	0.001
3	0.593	0.594	0.5935	0.001
4	0.585	0.585	0.5850	0.000
5	0.591	0.592	0.5915	0.001
6	0.587	0.585	0.5860	0.002
7	0.586	0.587	0.5865	0.001
8	0.576	0.575	0.5755	0.001
9	0.594	0.593	0.5935	0.001
10	0.581	0.582	0.5815	0.001
11	0.589	0.587	0.5880	0.002
12	0.595	0.595	0.5950	0.000
13	0.593	0.593	0.5930	0.000
14	0.592	0.593	0.5925	0.001
15	0.595	0.595	0.5950	0.000
16	0.579	0.578	0.5785	0.001
17	0.558	0.558	0.5580	0.000
18	0.593	0.594	0.5935	0.001
19	0.582	0.581	0.5815	0.001
20	0.592	0.593	0.5925	0.001
21	0.586	0.586	0.5860	0.000
22	0.586	0.587	0.5865	0.001
23	0.589	0.589	0.5890	0.000
24	0.574	0.574	0.5740	0.000
25	0.584	0.585	0.5845	0.001
26	0.591	0.593	0.5920	0.002
27	0.575	0.575	0.5750	0.000
28	0.586	0.587	0.5865	0.001
29	0.577	0.578	0.5775	0.001
30	0.592	0.593	0.5925	0.001
31	0.557	0.557	0.5570	0.000
32	0.586	0.587	0.5865	0.001
33	0.585	0.587	0.5860	0.002
34	0.574	0.576	0.5750	0.002
35	0.595	0.593	0.5940	0.002
36	0.602	0.602	0.6020	0.000
37	0.589	0.590	0.5895	0.001
38	0.582	0.582	0.5820	0.000
39	0.566	0.567	0.5665	0.001
40	0.593	0.595	0.5940	0.002
41	0.580	0.584	0.5820	0.004
42	0.589	0.588	0.5885	0.001
43	0.597	0.598	0.5975	0.001
44	0.558	0.557	0.5575	0.001
45	0.588	0.590	0.5890	0.002
46	0.557	0.556	0.5565	0.001
47	0.588	0.590	0.5890	0.002
48	0.592	0.593	0.5925	0.001
49	0.576	0.576	0.5760	0.000
50	0.602	0.603	0.6025	0.000

Figure 7

BULK DENSITY
$\overline{X}$ CHART (n = 2)

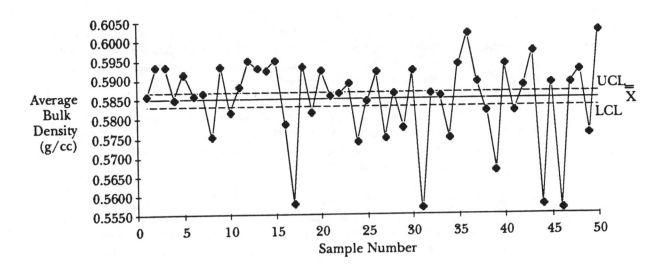

RANGE CHART (n = 2)

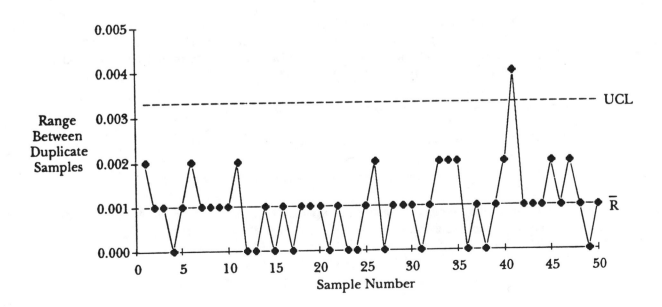

Step 5: Determine if the range chart is in statistical control. If it is not, the reasons for the assignable cause must be found and eliminated. Once this is done, the procedure is started over.

As can be seen from the range chart in Figure 7, there is one out-of-control point. This means that there was an assignable cause present when one part of that sample was run. In this case, it is not necessary to be worried about this one point. Out of the 100 samples run, there is only one point out of control. The estimate of $\overline{R}$ will still be very good since this one point has minimal impact on the results.

The range chart provides an indication of how well the technician runs the test. If it is in control, the technician is capable of reproducing the results. If it is not in control, the technician is not capable of reproducing the results. If this is true, the reasons for the poor performance should be found and corrected.

If the range chart is out of control, the value calculated for the average range will not be accurate. As a result, any values calculated for the standard deviation or the control limits on the $\overline{X}$ chart will be invalid. In this case, the range chart must be brought into statistical control by eliminating assignable causes, and the entire procedure must be started over.

Step 6: Determine the precision of the measurement system by calculating the standard deviation of the measurement system from the range chart.

Since the range chart shows stability, the standard deviation can be calculated. This calculation is shown below.

$$\hat{\sigma}'_{ms} = \overline{R}/d_2 = 0.001/1.128 = 0.000886$$

What are the sources of variation in this number? This number represents the variation obtained in running the bulk density test on the same sample. The sources of variation are primarily the way the technician runs the test and the methods used. If the test method is good, one would expect this number to be small.

Step 7: Determine the percent of total variance due to the measurement system.

The percent variance due to the measurement system is given by:

$$\% \text{ Variance due to measurement system} = 100(\hat{\sigma}'^2_{ms})/\hat{\sigma}'^2_t$$

To determine this percent, some estimate of the total variance is required. A control chart had been kept in the operating unit on bulk density. The average range on this chart was 0.02. The total process standard deviation can then be calculated as:

$$\hat{\sigma}'_t = \overline{R}/d_2 = 0.02/1.128 = 0.0177$$

The percent variance due to the test method is then:

$$100(\hat{\sigma'}_{ms}{}^2)/\hat{\sigma'}_t{}^2 = 100(.00088/.01777)^2 = 0.25\%$$

This indicates that the bulk density measurement method is responsible for only 0.25% of the total variance seen in the process. This is a very small percentage and means that the bulk density test is very precise.

Step 8: *Examine the X chart to qualitatively determine the discriminating power of the measurement system.*

The $\overline{X}$ chart is shown in Figure 7. Note that most of the points are out of control. Why is this so? Remember the way the subgroups were formed. The two results from running the same sample formed a subgroup. Since the range is based on the difference between results on the same sample, one would expect this range to be small if the test method is precise. The average range is used in calculating control limits on the $\overline{X}$ chart. Since the average range should be small, the control limits on the $\overline{X}$ chart should be extremely narrow compared to the range of subgroup averages plotted. Remember that samples representative of the process were selected. Thus, the $\overline{X}$ chart contains process variation. This technique of determining the precision of a measurement system is unique in that you want points to be out of control. That means that the measurement system can distinguish between samples. The $\overline{X}$ chart provides a qualitative method of looking at the percent of total process variance due to the test method.

MEASUREMENT SYSTEM EXAMPLE

Stream Contaminant

A company sells a by-product stream to a certain customer. The customer will only buy the stream if a certain contaminant is below 100 ppm. Recently the customer requested that the stream be supplied with a maximum of 60 ppm contaminant. The overall process average of the stream is currently close to that maximum. Some part of the production will meet that specification; other parts will not. The question of how accurate and precise the measurement system is has been raised. Since this was not an important stream, this test method was not being monitored on a regular basis by running a control. It was decided to use the duplicate sample analysis technique to determine the precision of the test method. Thirty production samples were randomly collected. Each sample was thoroughly mixed and divided in half. Each half was labeled #1 or #2. Each sample was analyzed using this wet chemical test method. The results are shown below.

Subgroup Number	Contaminant Level (ppm) #1	Contaminant Level (ppm) #2	Subgroup Average	Subgroup Range
1	61	62	61.5	1
2	60	56	58.0	4
3	60	57	58.5	3
4	62	57	59.5	5
5	55	55	55.0	0
6	59	58	58.5	1
7	60	57	58.5	3
8	66	59	62.5	7
9	56	55	55.5	1
10	60	61	60.5	1
11	57	62	59.5	5
12	60	57	58.5	3
13	58	60	59.0	2
14	61	56	58.5	5
15	56	59	57.5	3
16	60	62	61.0	2
17	57	57	57.0	0
18	60	61	60.5	1
19	58	60	59.0	2
20	59	62	60.5	3
21	53	56	54.5	3
22	63	63	63.0	0
23	55	53	54.0	2
24	60	61	60.5	1
25	57	60	58.5	3
26	60	62		
27	56	54		
28	59	59		
29	61	61		
30	65	62		

Use these data to determine the precision of this measurement system by using an $\overline{X}$-R chart. The overall process standard deviation has been estimated from one month's worth of data to be 2.50. The first 25 subgroup averages and ranges have calculated and plotted for you (p. 414). Once you have completed the $\overline{X}$-R chart, answer the following questions.

1. Is the range chart in or out of statistical control? What does this mean?

2. What is a measure of the precision of the test method?

3. How much of the total variance is due to the test method?

4. Is the $\overline{X}$ chart in or out of statistical control? What does this mean?

5. What can be said about the accuracy of this test method?

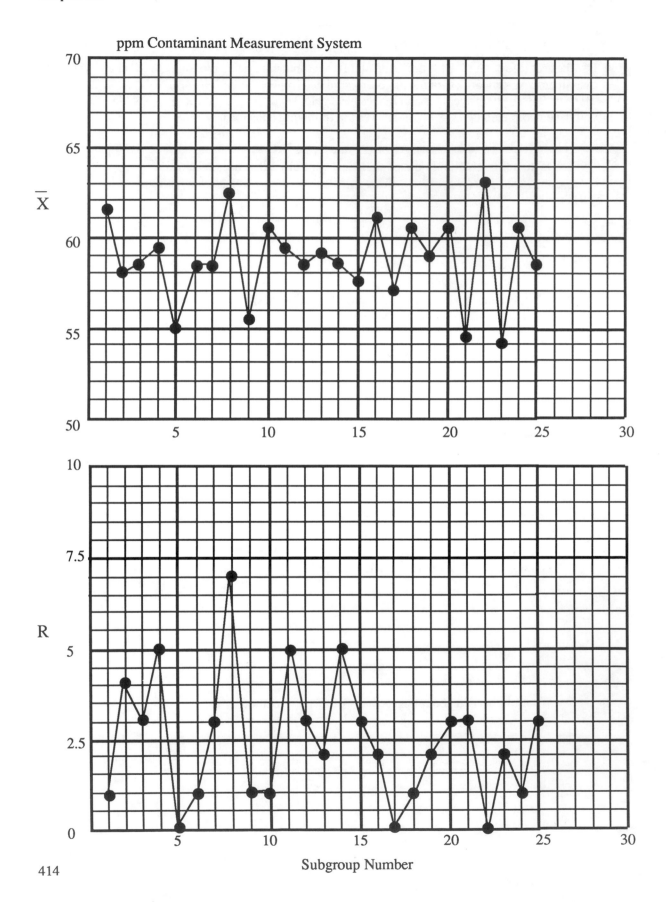

ppm Contaminant Measurement System

MEASUREMENT SYSTEM EXAMPLE

Particle Size Analyzer

A plant produces a product in powdered form. One important parameter is the average particle size of the powder. The plant currently uses a set of screening sieves to measure the particle size distribution. The average particle size is then calculated. A new particle size analyzer has been suggested. This analyzer examines a solution of the powder and automatically calculates the particle size distribution and average particle size. The analyzer is much faster than the sieve method. The sieve method has a standard deviation of 4.5 microns for average particle size. Thirty production samples were selected at random for testing in the new particle size analyzer. Each sample was thoroughly mixed and divided in half. Each half of a sample was labeled #1 or #2. The samples were analyzed for average particle size (in microns) using the procedure for determining precision by duplicate analysis. The results are given below.

Subgroup Number	Particle Size (Microns) #1	#2	Subgroup Average	Subgroup Range
1	180	191	185.5	11
2	145	148	146.5	3
3	174	178	176.0	4
4	153	162	157.5	9
5	143	156	149.5	13
6	162	172	167.0	10
7	181	185	183.0	4
8	177	191	184.0	14
9	152	166	159.0	14
10	165	178	171.5	13
11	162	162	162.0	0
12	169	173	171.0	4
13	147	149	148.0	2
14	168	168	168.0	0
15	152	166	159.0	14
16	189	200	194.5	11
17	155	169	162.0	14
18	145	151	148.0	6
19	164	172	168.0	8
20	180	182	181.0	2
21	171	180	175.5	9
22	156	161	158.5	5
23	161	163	162.0	2
24	169	169	169.0	0
25	155	164	159.5	9
26	167	171		
27	160	161		
28	144	156		
29	174	177		
30	184	194		

Chapter 28

Use these data to to construct an $\overline{X}$-R chart and determine if the particle size analyzer is more precise than the sieve method. The first 25 subgroup averages and ranges have been calculated and plotted for you (p. 417).. When you have completed the $\overline{X}$-R chart, answer the following questions.

1. Is the range chart in or out of statistical control? What does this mean?

2. What is a measure of the precision of the particle size analyzer?

3. Is the particle size analyzer more or less precise than the sieve analysis?

4. Is there enough information to determine quantitatively how much of the process variance would be due to the particle size analyzer? What about qualitatively?

5. What can be said about the accuracy of the particle size analyzer?

6. Reexamine the raw data from the duplicate sample analysis. Is there anything peculiar about these data? If so, what does it mean?

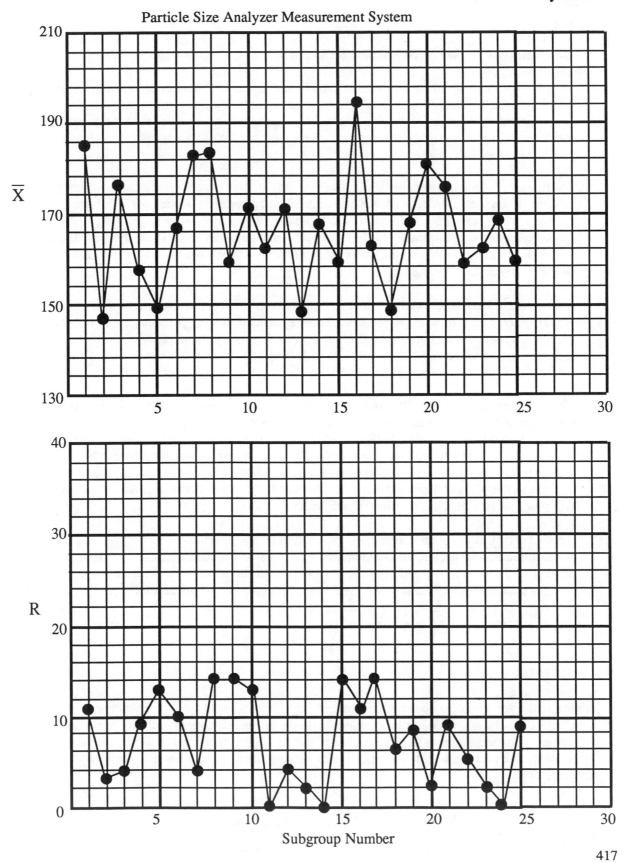

Particle Size Analyzer Measurement System

DISCUSSION OF MEASUREMENT SYSTEMS

In general, all measurement systems that exhibit random variation should be monitored on an ongoing basis. The best method of doing this is to use an individuals control chart to monitor the accuracy and precision over time. It is impossible to begin monitoring all measurement systems at one time. A plan should be developed to determine when the various measurement systems will be monitored.

The first measurement systems to be monitored should be the important systems that have a major impact on customer-related parameters or process-related parameters. Once a control or standard has been run for some time, the percent of total process variance due to the measurement system can be calculated. If this number is less than 10%, the frequency with which the standard or control is run can be reduced if the system has remained in statistical control for some period of time (usually at least one month).

The percent of total process variance due to the measurement system can also be used to determine which systems need to be improved first. Those systems with large percentages are not very precise and need to be improved. One method of decreasing the percentage is to run the measurement system more than one time on each sample. However, this should not be a long-term solution to the problem of lack of precision.

The measurement system precision obtained by running a standard or control over time is a better estimate of the true precision than that obtained by the duplicate sample analysis technique. When the duplicate sample analysis technique is used, only one person runs the samples and the time frame is short compared with the time usually involved in running a control or standard on a regular basis. Thus, there are fewer opportunities for variation when running the duplicate sample analysis technique, and the value obtained for the standard deviation of the measurement system will probably be a little smaller than the true value.

Another disadvantage of the duplicate sample analysis technique is that it does not provide a very good estimate of the accuracy of the measurement system without additional work. The overall average on the $\overline{X}$ chart obtained from this method could be compared to the overall average from process data (assuming the samples used were representative of production). The biggest disadvantage of the duplicate sample analysis technique is that it does not provide a method for ongoing monitoring of the measurement system.

SUMMARY

In this chapter, methods of determining the accuracy and precision of a measurement system were introduced. It was shown that measurement systems can and do cause variability in a process. The measurement system is itself a process and, thus, is subject to common cause variation as well as assignable cause variation. The objective is to develop measurement systems that are accurate and precise.

One method of determining and monitoring the accuracy and precision of a measurement system is to run a control on a regular basis. The results are analyzed on an individuals chart. The X chart is a measure of the accuracy of the measurement system. As long as the X chart stays in control and the centerline is equal to the true value of the standard or control, the measurement is accurate. The range chart is used to determine the precision of the measurement system. This precision is then compared to the overall process variance. The measurement system is precise as long as the range chart remains in statistical control and the measurement system is responsible for less than 10% of the total process variance.

An alternative method to determine the precision of a measurement system is to run duplicate sample analysis. This method provides a quick look at how precise the measurement system is.

APPLICATIONS OF MEASUREMENT SYSTEMS

Think about your own work situation. What are three important measurement systems? For each of these systems, how would you determine the accuracy and precision? What would you use for a standard, control or different samples? Write down your responses below.

1.

2.

3.

29
PROCESS CAPABILITY

Suppose you are monitoring a quality characteristic, X. Based on successive samples from the process, you have constructed an $\overline{X}$-R chart. The R chart is in statistical control. Since it is in control, you have established limits on the $\overline{X}$ chart and calculated $\hat{\sigma}' = \overline{R}/d_2$. The $\overline{X}$ chart is in control. This gives you an estimate of $\overline{X}'$. The specifications for the quality characteristic are USL (upper specification limit) and LSL (the lower specification limit). Specifications are based on individual measurements. You would like all the individual measurements to fall between USL and LSL. The question is: Is the process capable of meeting specifications? This chapter introduces process capability.

OBJECTIVES

In this chapter you will learn:

1. The definition of process capability
2. How to calculate the process capability
3. How to handle situations where the population is not normally distributed

Bringing a process into statistical control is not process improvement. Bringing a process into statistical control is putting the process where it should be. Once the process is in statistical control, real efforts at process improvement can begin. Process capability is one method of measuring the effectiveness of a process in meeting customer specifications as well as measuring process improvement efforts.

INTRODUCTION TO PROCESS CAPABILITY

Assume that the quality characteristic in a process is in statistical control and that the individual measurements are normally distributed. The same result will not be obtained each time a measurement is taken of the quality characteristic. This quality characteristic has an inherent or normal amount of variation (common cause variation). This variation is called the natural tolerance (NT) and is defined as $6\hat{\sigma}'$. The natural tolerance is sometimes called the process capability. It describes what the process is capable of producing when the process is in statistical control.

Since the individual measurements are normally distributed, the shape of the population can be easily drawn as shown in Figure 1. $\overline{\overline{X}}$ is the estimate of population average ($\overline{X}'$) and is obtained from the $\overline{X}$ chart. $\hat{\sigma}'$ is the estimate of the population standard deviation (σ') and is obtained from the R chart ($\hat{\sigma}' = \overline{R}/d_2$). Nearly all the data (99.7%) will fall between $\overline{\overline{X}} \pm 3\hat{\sigma}'$. The natural tolerance is simply the distance from $-3\hat{\sigma}'$ to $+3\hat{\sigma}'$. If the specifications are outside the distribution, the process is capable of meeting specifications. The upper specification limit (USL) minus the lower specification limit (LSL) is called the engineering tolerance (ET).

Figure 1

NATURAL AND ENGINEERING TOLERANCES

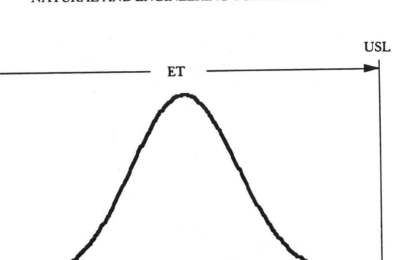

PROCESS CAPABILITY INDICES

There are several methods of measuring process capability. Unless specifically stated otherwise, the process capability indices calculations require that the process be in statistical control and that the individual measurements are normally distributed.

Cp

Cp is the ratio of the engineering tolerance to the natural tolerance:

$$Cp = ET/NT = (USL - LSL)/6\hat{\sigma}'$$

If the engineering tolerance is less than the natural tolerance (i.e., Cp < 1.0), the process is not capable of meeting specifications. If the engineering tolerance is equal to or larger than the natural tolerance (Cp ≥ 1.0), the process is capable of meeting specifications. A process can be capable of meeting specifications but not be meeting specifications if the process is not centered relative to the specifications. Examples of processes that are capable and are not capable are shown in Figure 2 (University of Tennesee, 1986).

Figure 2

EXAMPLES OF PROCESS CAPABILITIES

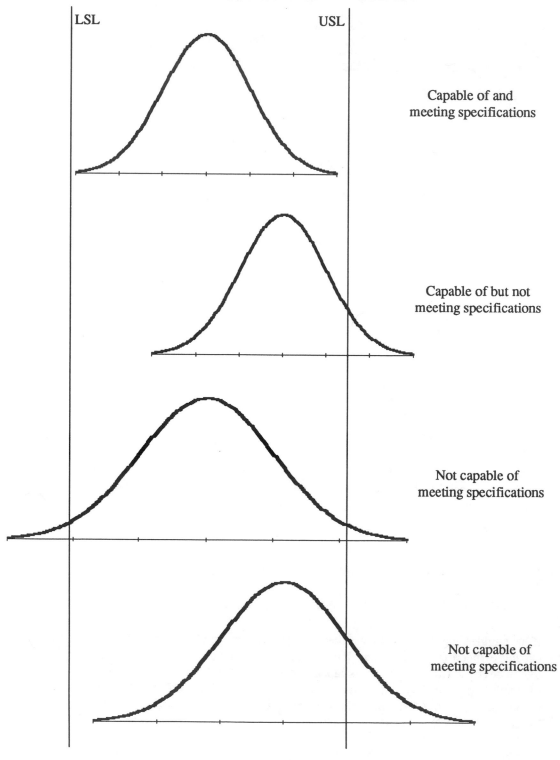

LSL   USL

Capable of and
meeting specifications

Capable of but not
meeting specifications

Not capable of
meeting specifications

Not capable of
meeting specifications

The z table (Appendix Table 7, p. 491) can be used to determine the percentage of product or measurements that is out of specification. To determine the percentage of product above the upper specification limit, use the z value determined from:

$$z = (USL - \overline{\overline{X}})/\hat{\sigma}'$$

The z table will give the percentage of measurements above the USL for the calculated z value. To determine the percentage of product below the lower specification limit, use the z value determined from:

$$z = (\overline{\overline{X}} - LSL)/\hat{\sigma}'$$

The z table will give the percentage of measurements below the LSL for the calculated z value.

Cp EXAMPLES

1. A team in an operating unit has been working on improving the yield from a batch reactor. An $\overline{X}$-R chart has been constructed. The process is in statistical control with respect to $\overline{X}$ and to R. The control charts indicate that $\overline{\overline{X}}$ = 80.2% yield and that $\overline{R}$ = 6.49. A subgroup size of five was used to construct the control chart. The individual measurements are known to be normally distributed. Internal specifications for yield from the reactor are LSL = 75% and USL = 85%. Is the process capable of meeting these internal specifications? What is the value of Cp? What percentage, if any, of the batches are out of specification?

Chapter 29

2. A company produces a product in granular form. One product response of interest to a customer is the brightness of the product. An $\overline{X}$-R chart with a subgroup size of 4 has been used to monitor the product brightness. After 20 subgroups, the process was in control with respect to $\overline{X}$ and to R. The data indicated that $\Sigma\overline{X} = 2107.725$ and $\Sigma R = 366.8$. The individual measurements are known to be normally distributed. Specifications on product brightness are LSL = 70.0 and USL = 125.0. Is the process capable of meeting specifications? What is the value of Cp? What percentage, if any, of the product is out of specification?

3. A plant produces plastic pellets for several customers. One customer is interested in the hardness of the plastic. The plant has begun monitoring hardness using an individuals chart. After 25 samples, $\Sigma X = 1037$ and $\Sigma R = 102$. The process is in statistical control with respect to X and to R. The individual measurements are known to be normally distributed. The specifications for hardness are LSL = 30.0 and USL = 55.0. Is the process capable of meeting specifications? What is the value of Cp? What percentage, if any, of the product is out of specification?

Cpk

The Cp values calculated above are not the best indicators of process capability. The value of Cp does not take into account where the process is centered. Just knowing that a process is capable (Cp ≥ 1.0) does not ensure that all the product being received is within the specifications. A process can have a Cp ≥ 1.0 and produce no product within specifications. In addition, Cp values can't be calculated for one-sided specifications. A better measure of process capability is Cpk.

Cpk takes into account where the process is centered. The value of Cpk is the minimum of two process capability indices. One process capability is Cpu, which is the process capability based on the upper specification limit. The other is Cpl, which is the process capability based on the lower specification limit. Algebraically, Cpk is defined as:

$$Cpk = \text{Minimum (Cpu, Cpl)}$$

where

$$Cpu = (USL - \overline{\overline{X}})/3\hat{\sigma}'$$

$$Cpl = (\overline{\overline{X}} - LSL)/3\hat{\sigma}'$$

Both Cpu and Cpl take into account where the process is centered. The value of Cpk is the difference between the process average ($\overline{\overline{X}}$) and the nearest specification limit divided by three times the standard deviation ($\hat{\sigma}'$). Cpk values above 1.0 are desired. This means that essentially no product is being produced above USL or below LSL.

If Cpk is less than 1.0, this means that there is some product being produced out of specification. To determine how much product is being produced out of specification, the individual values of Cpu and Cpl must be considered. Note that when Cpk is less than 1.0, you don't know if product is out of specification on the high side, the low side, or both. To determine the amount out of specification, use the z table. Note that the values of z, Cpu, and Cpl are related by the following:

For the USL: z = 3Cpu

For the LSL: z = 3Cpl

Thus, to obtain the z value to look up in the table, simply multiply the value of Cpu or Cpl by 3.

Cpk EXAMPLES

1. An operating unit has been monitoring reactor run time since it is known to affect several final product parameters. An $\overline{X}$-s chart with a subgroup size of four has been used to monitor the run time. The process is in control with respect to $\overline{X}$ and to s. After 20 subgroups, $\Sigma\overline{X} = 7274.5$ and $\Sigma s = 333.7$. The individual measurements are

known to be normally distributed. The internal specifications are LSL = 330 minutes and USL = 390 minutes. What is the value of Cpk? What percentage, if any, of the reactor run times is out of the internal specifications?

2. A company produces a liquid product that contains small amounts of iron. The USL for iron is 100 ppm. There is no LSL. A moving average/moving range chart has been used to monitor the iron data. The process is in statistical control with respect to $\overline{X}$ and R. The control charts indicate that $\overline{\overline{X}}$ = 26.72 and $\overline{R}$ = 16.23. A subgroup size of four was used in the moving average/moving range chart. The individual measurements are known to be normally distributed. What is the value of Cpk? What percentage, if any, of the product is out of specification?

3. An operating unit is monitoring the purity of an overhead stream from a distillation column using a moving average/moving range chart with a subgroup size of four. The process is in statistical control with respect to $\overline{X}$ and R. The control charts indicate that $\overline{\overline{X}}$ = 97.17% and $\overline{R}$ = 3.39. The LSL is 95%. There is no USL. The individual measurements are known to be normally distributed. What is the value of Cpk? What percentage, if any, of the production is out of specification?

PROCEDURE FOR DEVELOPING Cpk VALUES

Figure 3 is a process flow diagram of one procedure for determining process capability for a product response. It assumes that a product response has been chosen by some method, i.e., dialogue with customers, largest netback product, etc. Major steps in the process flow diagram are described below.

Is the product response currently monitored? This question simply asks if you currently test the product for the selected response. If the answer is no, you must determine if laboratory test methods exist. If not, laboratory test methods must be developed. If laboratory tests do exist, you must begin taking samples to measure the product response.

Review sampling plan and select type of control chart. The purpose of this step is to review the sampling plan to ensure the control chart is examining the variation in which you are interested. The type of chart you select and the way you subgroup the data affect your interpretation of the charts. For example, suppose your current sampling plan calls for one sample per day to be tested for the product response. If you select an individuals control chart, the range chart will be examining the short-term variation in day-to-day results. The X chart will be examining the long-term variation in day-to-day results. If you select an $\overline{X}$-R chart with a subgroup size of 7 (one week), the range chart will be examining the within-week (day-to-day) variation, and the $\overline{X}$ chart will be examining the week-to-week variation in the average daily product response. Thus, the chart you select and the way you subgroup the data affect the interpretation of the chart. You may have to change the sampling plan. For example, if you are interested in the variation within each day, you will have to sample more than once per day.

Begin the control chart on the product response. The next step is to start the control chart you have selected. Before calculating control limits, 25 subgroups are normally required. Historical data can be used if available.

Is the product response in statistical control? Being in statistical control is not the natural state. A process is brought into statistical control by finding and permanently eliminating assignable causes. As a rule of thumb, a process can be considered in statistical control if less than 5% of the points are out of control. This does not mean that front-line personnel can avoid looking for reasons for the out-of-control points. The time frame used in determining if a product response is in control is critical. Enough time must pass so that all possible sources of variation have had an opportunity to be included in the process. These sources of variation include raw material variations, different shifts, environmental changes, etc. The time period for determining whether a product response is in statistical control should normally be the results obtained over a 30-day period.

If the product response is not in statistical control, it must be brought into control by eliminating assignable causes. Once 25 subgroups in a row are in statistical control, control limits should be recalculated. These limits should not be recalculated unless there is a reason to do so (e.g., a process change).

Is the distribution of individual measurements normal? If the product response is in statistical control, the distribution of the individual measurement must be checked. If the distribution is normal, the Cpk value can be calculated. If the distribution is not normal, an alternate method of determining process capability must be used.

Calculate Cpk. If the product response is in statistical control and the individual measurements are normally distributed, the process capability may be calculated. The procedure for calculating Cpk is given below.

1. Determine the process standard deviation by using the range chart:

$$\hat{\sigma}' = \bar{R}/d_2$$

   where $\hat{\sigma}'$ is the estimate of the process standard deviation, $\bar{R}$ is the average range and d_2 is the control chart constant that depends on subgroup size.

2. Calculate the process capability based on the upper specification limit (Cpu) and the process capability based on the lower specification limit (Cpl) (if both exist):

$$Cpu = (USL - \bar{\bar{X}})/3\hat{\sigma}'$$

$$Cpl = (\bar{\bar{X}} - LSL)/3\hat{\sigma}'$$

   where $\bar{\bar{X}}$ is the overall process average, USL is the upper specification limit and LSL is the lower specification limit.

4. Determine Cpk, which is the smaller of Cpu and Cpl.

Is Cpk greater than or equal to 1? If Cpk is greater than or equal to 1, this means that the product response is capable of meeting specifications. If Cpk is less than 1, this means the product response is not capable of meeting specifications. There are only three ways to increase the value of Cpk. These are to reduce process variation, to widen the specifications or to center the process average relative to the specifications. Since most customers will not widen the specifications, the process variation must be reduced or the process average centered to improve the Cpk value.

Does Cpk value meet the customer's requirement? The calculated Cpk must meet the customer's requirement. Most customers consider a Cpk value of 1.0 to be sufficient. However, more and more customers are beginning to demand higher values of Cpk, such as 1.33. If the process Cpk value does not meet the customer's requirements, the process variability must be reduced further.

Figure 3

PROCEDURE FOR DEVELOPING Cpk VALUES

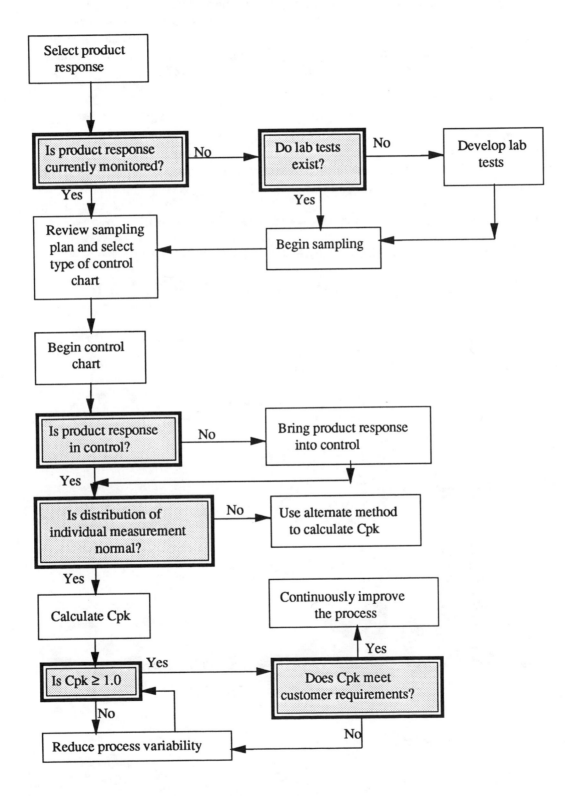

HOW TO HANDLE SITUATIONS WITH NONNORMAL DISTRIBUTIONS

The calculation of Cp and Cpk assumes that the population is normally distributed. In the process industries, this often is not true. There are many situations where the population is not normally distributed. This is particularly true for situations when a quality characteristic has a one-sided specification (either a maximum or a minimum). If the distribution is nonnormal, alternative methods of determining process capability must be used (ASQC, 1987).

Transformation of Data

One alternate method is to transform the data so that the distribution of the transformed data is normal. Consider the following example. A company is attempting to minimize the amount of iron in one of its product streams. A control chart has been used to monitor the iron content. The process is in control with respect to the average and to the variation. $\overline{\overline{X}}$ is 37.9 ppm and $\hat{\sigma}'$ is 30.4. The upper specification limit is 130 ppm. There is no lower specification limit. A value of Cpk can be calculated from this information (Cpk = Cpu since there is no Cpl):

$$Cpk = Cpu = (USL - \overline{\overline{X}})/3\hat{\sigma}' = (130 - 37.9)/(3)(30.4) = 1.01$$

Based on this calculation, the process appears capable of meeting specifications. To calculate Cpk, the individual measurements must be normally distributed. A histogram of the individual measurements is given in Figure 4.

Figure 4

IRON CONTENT HISTOGRAM

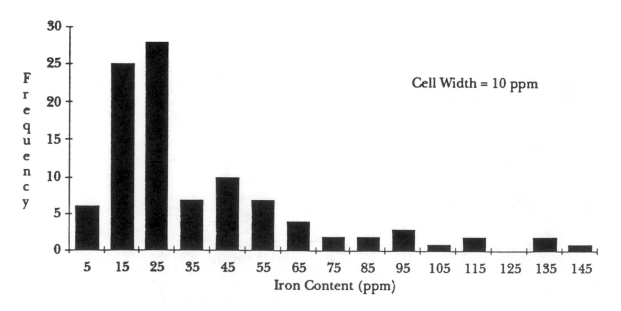

As can be seen, the individual measurements are not normally distributed. In addition, there are some points above 130 ppm (3 points out of 100). Since the individual measurements are not normally distributed, Cpk values should not be calculated on the actual data. Skewed distributed like the one above can often be transformed into a normal distribution by using natural logarithms. Each individual measurement is transformed using the following:

$$y = \ln(x)$$

where x is the individual iron content result and y is the transformed variable. A histogram of the transformed data is given in Figure 5.

Figure 5

IRON CONTENT: HISTOGRAM OF TRANSFORMED DATA

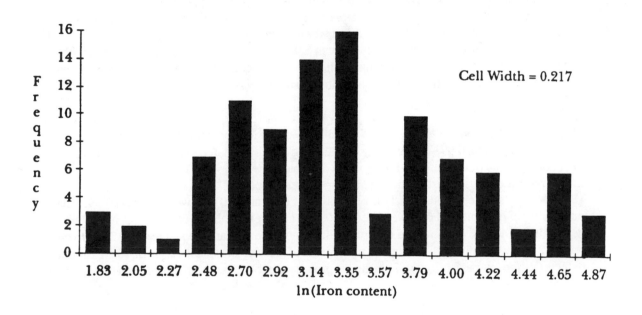

As can be seen, the data are now normally distributed. A Cpk value can now be calculated using these transformed data:

$$Cpk = [\ln(USL) - \overline{\overline{Y}}]/3\hat{\sigma}'_y$$

where $\overline{\overline{Y}}$ is the average of the transformed data and $\hat{\sigma}'_y$ is the standard deviation of the transformed data. The Cpk value is now:

$$Cpk = [\ln(USL) - \overline{\overline{Y}}]/3\hat{\sigma}'_y = (4.87 - 3.37)/(3)(.731) = 0.68$$

Using the transformed data, the process is not capable of meeting specifications. The calculated value of Cpk corresponds to a z value of 2.04. This implies that 2.07% of the production is out of specification. This corresponds closely with the actual percentage out of specification (3%).

Known Distributions

On occasion you will know what the underlying distribution is. You can make use of this knowledge to determine the process capability. In this case, you simply determine what the values are for the underlying distribution corresponding to a Cpk of 1 (i.e., Cpl = 1 and Cpu = 1). To do this, determine the value for the distribution where 0.13% of the points are less than that value (this corresponds to a Cpl = 1) and determine the value for the distribution where 0.13% of the points are above this value (this corresponds to a Cpu = 1). This procedure doesn't allow you to calculate a certain value of Cpk. However, you can determine if the value of Cpk is greater than or less than 1.

To understand how this works, consider the following example. Suppose you are interested in the number of surface imperfections on a plastic sheet. A few surface imperfections are not a problem. However, too many cause problems for a customer. The specification is a maximum of 10 surface imperfections in a 10 square inch area. A c chart has been used to monitor the process. The c chart is in statistical control with c = 4.6. Since the distribution of the population is not normal (in this case, the underlying distribution is the Poisson distribution), Cpk values can't be calculated.

To determine if the process is capable of meeting specifications, you must determine the c value that corresponds to a 0.9987 probability of obtaining c or fewer defects (or a 0.0013 probability of obtaining more than c defects) for a Poisson distribution with c = 4.6. The probability function that governs the Poisson distribution or Molina's tables (Appendix Table 13, p. 505) may be used. Using Molina's tables indicate that the probability of obtaining 10 or fewer defects is 0.992, the probability of obtaining 11 or fewer defects is 0.997, and the probability of obtaining 12 or fewer defects is 0.999 for $\bar{c}$ = 4.6. Since the probability is not greater than or equal to 0.9987 until you reach 12 or fewer defects, the process is not capable of meeting the specification of 10 maximum defects (surface imperfections). To make the process capable, $\bar{c}$ must be reduced to a maximum of 3.6.

Omitting the Calculation of Cpk

The third option is to omit the calculation of Cpk. The percent out of specifications can be determined directly from the histogram of individual measurements. This is not the preferred method and should be used only as a last resort. If this method is used, sufficient data should be used to construct the histogram. These data should cover at least a 30-day period.

DISCUSSION OF PROCESS CAPABILITY

There are numerous ways to determine process capability. It is important that you understand which method has been used to determine process capability and the time period covered during the capability study. For example, if the time period covered is small (hours or several days), the process capability reported will probably be greater than the actual performance of the process over a longer time period (such as a month).

In some instances, companies are subtracting out the variation due to the measurement system and/or sampling before calculating the process capability. The total process variance is the sum of three components:

$$\sigma'^2_t = \sigma'^2_p + \sigma'^2_s + \sigma'^2_{ms}$$

where σ'^2_t is the total process variance, σ'^2_p is the process variance, σ'^2_s is the sampling variance and σ'^2_{ms} is the measurement system variance. The total process variance is the variation being measured on the range chart kept in an operating unit. Since the measurement system variability (and in some cases, the sampling variability) is often large, it may be subtracted out of the total process variance. If this done, it should be clearly stated that this is the case. If not, misleading interpretation of the process capability may occur.

The process capability can be used as a guideline for how often the process should be sampled. If Cpk $\geq$ 2.0, sampling can be done infrequently. A control chart should be kept on an ongoing basis. If Cpk < 1.0, you should sample as frequently as you can afford. If Cpk is between 1.0 and 2.0, you should consider the following in determining how often to sample:

- What is the probability of the process making a major shift between samples?
- Can you detect this shift without sampling?
- What is the cost of a large process shift going undetected?

The frequency of sampling will also affect the value of Cpk. The more frequently you sample the process, the smaller the variation in range values. This will lead to larger values of Cpk. However, more points will be out of control on the $\overline{X}$ chart. The less frequently you sample, the larger the variation in range values. This will lead to smaller values of Cpk.

Chapter 29

SUMMARY

This chapter has discussed the concept of process capability. Two questions were addressed. One question was "Is the process capable of meeting specifications?" The second question was "Are the individual measurements normally distributed?" If the process is in statistical control and the individual measurements are normally distributed, the process capability can be calculated by Cp or Cpk. If the individual measurements are not normally distributed, one of three alternate methods must be used. One method involves transforming the data into a normal distribution. Another method involves using the underlying distribution if it is known. The third method involves not calculating a process capability. Percent out of specification is determined from a histogram. Regardless of the method used, the time period over which the process capability is determined is important. Better estimates of true process capability are obtained from longer time periods.

APPLICATIONS OF PROCESS CAPABILITY

Think about your processes at work. What do you think are the three most important product responses to calculate process capabilities on? Do you expect the individual measurements to be normally distributed? Record your responses below.

30
COMPARING PROCESSES

It is usually not possible to keep control charts on all product responses and process parameters. What can one say about the average or variation in one process if control charts aren't being kept? How can two processes, such as two reactors, be compared if control charts aren't being used? How can multiple processes be compared? What about comparing two test methods or two different analysts? The techniques introduced in this chapter will handle these types of situations. These techniques are "snapshots" in time of the process or processes. You can't be sure of obtaining similar results in the future if the process is not in statistical control.

OBJECTIVES

In this chapter you will learn:

1. What random samples and variables are
2. What statistical inference is
3. What a point estimator is
4. What a confidence interval is
5. How to construct a confidence interval around one average and one variance
6. How to construct a confidence interval around the difference in two averages (comparing two processes)
7. How to construct a confidence interval around multiple averages (comparing more than two processes)
8. How to construct a confidence interval around the difference in paired sample averages

The techniques presented in this chapter provide a method of comparing processes when control charts are not in use.

NOMENCLATURE

In working with control charts, we have said that a subgroup was composed of n *independent samples*. If the process was in statistical control, the center line of the $\overline{X}$ chart, $\overline{\overline{X}}$, was an estimate of the population average, $\overline{X}'$. The center line on the R chart was used to estimate the population standard deviation, σ'.

The nomenclature for this chapter will be different in order to be consistent with most nomenclature in statistical textbooks. A sample will now be composed of n *observations*. In addition, the population average will be referred to as μ instead of $\overline{X}'$.

RANDOM SAMPLES AND VARIABLES

The major objective of this chapter is to learn how to summarize information obtained from a sample of n observations so as to infer something about the population parameters. Instead of using a large number of samples or subgroups as is done in control charts to estimate these population parameters, we are going to try to estimate them based on one sample only.

Sample	Population
Sample average	Average
Sample variance	Variance
Sample standard deviation	Standard deviation
Sample range	Standard deviation
Sample histogram	Distribution

Suppose X is a product response that is in statistical control. Since it is in control, there are estimates of the process average (μ), standard deviation (σ'), and shape of the distribution.

Suppose a sample is drawn at random from the process. A sample is drawn at random from a population if every possible sample result in the population has an equal chance of being chosen.

What is known about this randomly chosen sample without analyzing it? The exact value of X is not known. However, something is known about X because the shape of distribution is known. Knowing this distribution, you can determine the probability that the sample result will be above or below a given value. For example, if the distribution is normal, the z table (Appendix Table 7, p. 491) could be used to determine these probabilities.

The quantity, X, which is not exactly known but for which the probability distribution is known, is called a random variable (Box, Hunter and Hunter, 1978).

STATISTICAL INFERENCE: POINT AND INTERVAL ESTIMATION

Statistical inference involves estimating information about an unknown population (the average, the variability, and the shape of the distribution are not known) from a small amount of sample data. There are two types of statistical inference of interest: point estimation and interval estimation.

Point Estimation

The use of a sample statistic (such as the sample average) to determine a single value to estimate a population parameter (such as the population average) is called point estimation. The statistic is a point estimator of the population parameter. Point estimators are random variables and have sampling distributions. A sampling distribution is a theoretical probability distribution that shows the relationship between the possible values of a sample statistic and the probability density associated with each value of the sample statistic over all possible samples of a particular size from a particular population.

For example, if the population is normally distributed with known variance, the sampling distribution of sample averages will be normal. Even if the population is not normally distributed, the sampling distribution of sample averages will tend to be normal, provided n is large enough (central limit theorem).

The sample represents only a small subset of observations drawn from a much larger set of potential observations. This makes it risky to say that any estimate is exactly like the population parameter. Many different factors can make the sample a poor representation of the population. Lumping these factors together as random effects should mean that randomly drawn samples reflect the population parameters. What you want to do is use the available information in the best possible manner to infer information about the population.

You want to choose statistics to use as parameter estimators that have the following desirable properties:

1. *Unbiasedness* - the expected value of the statistic equals the value of the parameter.
2. *Consistency* - the sample estimator should be closer to the population parameter as the sample size increases.
3. *Efficiency* - the variance of the sample estimator should be no larger than the variance of any other estimator.
4. *Sufficiency* - the sample estimator should use all the data.

All estimators examined below will have these properties.

Point Estimation Example

In the production of an alcohol mixture, the total alcohol content is of interest. Ten samples were pulled from a process and analyzed for % alcohol content. The results are given below.

97.5 98.5 95.5 97.3 97.8
96.2 97.0 97.1 96.4 98.0

Determine the following:

n =   Maximum =

$\overline{X}$ =   Minimum =

s^2 =   R =

Note that there is not enough data for a histogram. This will often be the case for techniques used in this chapter.

Interval Estimation

Interval estimation is the other type of statistical inference of interest. Interval estimation uses knowledge about the sampling distribution to place a confidence interval around a point estimator. It is clear that a point estimator will not exactly equal the unknown population parameter. In addition, the point estimator alone does not give any indication of the magnitude of the process variation.

For example, it is known that the sampling distribution of sample averages with a known variance will be normal. This information can be used to infer the likelihood that the population average lies within a certain interval. This interval is called a confidence interval. It is a random interval of values with a given probability (usually 95%) of containing the true population average (University of Tennessee, 1986 and 1987).

The following confidence intervals will be covered in thisc hapter:

1. Around the average and variance for one process

2. Around the difference in two averages (comparing two processes)

3. Around the difference in multiple averages (comparing more than two processes)

4. Around the difference in paired sample averages

CONFIDENCE INTERVAL AROUND ONE AVERAGE

Here we will be concerned with only one process. Based on a sample of n independent observations, an interval will be determined that has a certain probability of containing the true population average. Two cases will be examined.

1. The population is normal and has a known variance. The normal distribution will be used to construct a confidence interval around the unknown average. This case is introduced to demonstrate how confidence intervals are derived.

2. The population is normal and the variance is not known. The t distribution will be used to construct a confidence interval around the unknown average.

Case 1: Normal Population, Known Variance

The z table can be used to find the percentage of product out of or in spec as well as the probability of getting a sample out of or in spec. An example for a process with an average of 100 and a standard deviation of 10 is shown below. The upper specification limit is 120, which corresponds to a z value of 2.0. The area (from the z table) above z = 2.0 is 0.0228. This means that 2.28% of the product will be out of spec on the high side. The lower specification limit is 75, which corresponds to a z value of -2.50. The area (from the z table) below z = -2.50 is 0.0062. This means that 0.62% of the product is out of spec on the low side.

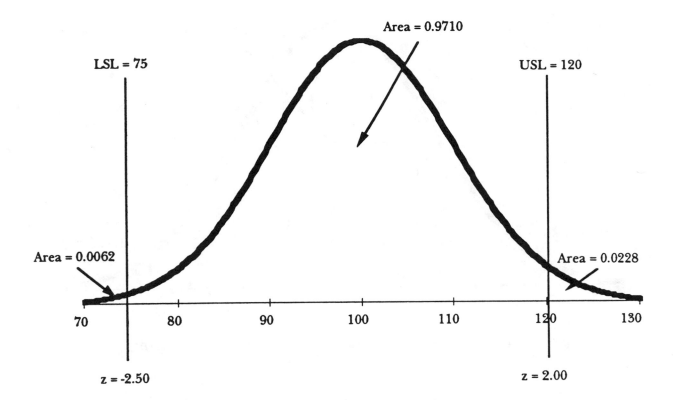

These areas can be viewed as probabilities. For example, the probability of pulling a sample from the process with a value less than 75 is 0.0062. The probabilities are written as follows:

$$P\ (z < -2.50) = .0062$$

$$P\ (-2.50 < z < 2.0) = .9710$$

$$P\ (z > 2.0) = .0228$$

This idea of probabilities can be used to construct confidence intervals around averages. Suppose the process is normally distributed with known variance, σ'^2. Each sample pulled from the process is an estimate of the unknown population average, μ. The z value will be used to construct a confidence interval around μ.

$$z = (X - \mu)/\sigma'$$

439

In general, $P(-a < z < a) = 1 - \alpha$, i.e., the probability of z being between -a and a is equal to $1 - \alpha$ as shown in the figure below. $1 - \alpha$ is called the confidence coefficient.

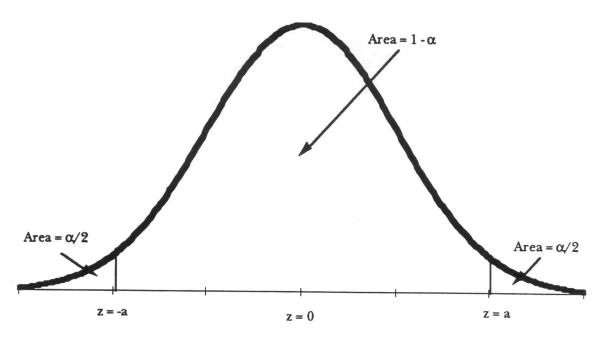

Suppose 95% confidence limits around m are desired. Thus $1 - \alpha = .95$ or $\alpha = .05$.

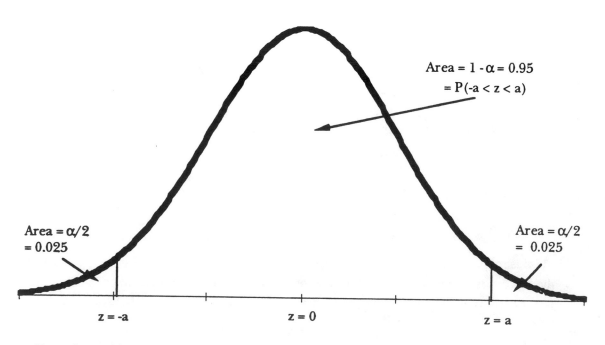

From the z tables, what values of a and -a will give the areas above?

The expression for z can be solved for m. The result is $\mu = X - z\sigma'$. z can have the values of +a or -a, thus:

$$\mu = X \pm a\sigma'$$

This is the confidence interval for a sample of one observation that is likely to contain the true average.

In most cases, one sample with one observation will not be taken to construct confidence intervals. Usually a sample with n observations will be taken. In this case, the standard normal variable, z, is

$$z = (\overline{X} - \mu)/(\sigma'/\sqrt{n})$$

The confidence interval around the population average is then given by

$$\mu = \overline{X} \pm a(\sigma'/\sqrt{n})$$

For confidence limits of 95%, a = 1.96.

CONFIDENCE INTERVAL AROUND AN AVERAGE EXAMPLE--NORMAL DISTRIBUTION WITH KNOWN VARIANCE

Insulation Resistance

In the production of a certain plastic, the insulation resistance is important. Ten samples of the plastic are taken and measured for insulation resistance. Determine a confidence interval that has a 95% chance of containing the true average based on the results obtained below.

Sample Number	Insulation Resistance
1	697
2	698
3	673
4	657
5	710
6	702
7	680
8	709
9	670
10	671

The population is normally distributed with a variance of 400.

Solution:

The expression for the confidence inteval is

$$\mu = \overline{X} \pm a(\sigma'/\sqrt{n})$$

For 95% confidence interval, a = 1.96, n = number of observations = 10, $\overline{X}$ = average of n observations = 686.7, σ' = $\sqrt{400}$ = 20.

$$\mu = \overline{X} \pm a(\sigma'/\sqrt{n})$$

$$\mu = 686.7 \pm 1.96(20/\sqrt{10})$$

$$\mu = 686.7 \pm 12.4$$

The confidence interval with a 95% probability of containing the average is

$$674.3 < \mu < 699.1$$

CONFIDENCE INTERVAL AROUND AN AVERAGE EXAMPLE--NORMAL DISTRIBUTION WITH KNOWN VARIANCE

Determining If a Process Change Has Affected the Average

In a chemical plant, methyl chloride (MeCl) content of the final product is important. From control charts kept in the field, it is known that the process average is 22.5 and the process standard deviation is 2.333. The process is in control. It is felt that a certain process change will lower the MeCl content. The process change is made. Fifteen samples are then collected and analyzed for MeCl content. The results are given below. Is there any evidence that the process change has lowered the MeCl content of the final product?

To solve this problem, a 95% confidence interval around the unknown average is constructed based on the sample results. If the interval does not contain the process average of 22.5, one will conclude that the process change did affect the MeCl content. If the interval contains the process average, one will conclude that the process change had no effect.

Sample	ppm MeCl
1	25
2	24
3	23
4	25
5	19
6	21
7	23
8	20
9	18
10	20
11	23
12	17
13	21
14	22
15	24

Suppose a confidence interval of 98% is desired. What is a in this case and what is the interval?

If it is felt that the interval is too wide, what can be done to reduce it?

Case 2: Normal Distribution, Unknown Variance

In real life, the average and variance of the population will often not be known. When σ' is not known, a z value can't be calculated and the z table can't be used. However, there will be an estimate of σ'. This estimate is s, the sample standard deviation,

$$s = \sqrt{(\Sigma X_i - \overline{X})^2)/(n - 1)}$$

s can be substituted for s' in the expression for z. A new quantity, t, can be calculated

$$t = (\overline{X} - \mu)/(s/\sqrt{n})$$

Under certain assumptions, such as when X is normally distributed, t has a known distribution. Note that the same value of X will always give the same z value but could give different t values since s is different for different samples.

The precise form of the t distribution depends on one parameter only, ν, the degrees of freedom used to calculate s. The t distribution has a shape similar to the normal distribution except that there is more area in the tails. As n approaches infinity, there is no uncertainty in s, and the t distribution becomes the standard normal distribution. The smaller ν is, the larger the area in the tails of the t distribution. A table of t values for various degrees of freedom and confidence coefficients is given in Appendix Table 8 (p. 492).

The confidence interval around the average can be developed in a manner similar to that when the normal distribution was used. The confidence interval for a normal population with unknown variance is

$$\mu = \overline{X} \pm t(s/\sqrt{n})$$

where

  n = number of independent observations
  $\overline{X}$ = average of the observations
  s = standard deviation of the observations
  t = value form the t table (based on ν and α)
  ν = degrees of freedom = n - 1
  1 - α = confidence coefficient (α is evenly distributed in the tails)

The t distribution can be thought of as a distribution that represents chance variations in the values of averages. $s/\sqrt{n}$ measures this chance variation and is called the standard error of the mean. It is a standard deviation since it measures variability.

CONFIDENCE INTERVAL AROUND AN AVERAGE EXAMPLE--NORMAL DISTRIBUTION WITH UNKNOWN VARIANCE

MeCl Revisited

Consider the MeCl data again. In this case, assume that control charts are not being kept, but it is desired to find an interval that should contain the true average. The sample statistics are calculated first.

$\overline{X} = 21.667$

$s = 2.498$

$n = 15$

$\nu = n - 1 = 15 - 1 = 14$

For $\nu = 14$ and $\alpha = 0.05$, $t = 2.145$ (from Appendix Table 8).

These values are then substituted into the equation for the confidence interval around an average.

$$\mu = \overline{X} \pm t(s/\sqrt{n})$$

$$\mu = 21.667 \pm 2.145(2.498/\sqrt{15})$$

$$\mu = 21.667 \pm 1.383$$

$$20.283 < \mu < 23.05$$

CONFIDENCE INTERVAL AROUND AN AVERAGE EXAMPLE--NORMAL DISTRIBUTION WITH UNKNOWN VARIANCE

Calibration Check of an Analyzer

Ethylene concentration is measured by an on-stream analyzer. Some question has arisen about whether or not the analyzer needs to be recalibrated. A sample containing 50% ethylene was run eight times. The results obtained are given below. Is there any indication that the analyzer needs recalibrating?

Sample	% Ethylene
1	47
2	52
3	53
4	49
5	63
6	46
7	50
8	57

Technique to use: make several, independent measurements on the same material, determine the average and standard deviation, and determine a 95% confidence interval. If the interval contains 50, no recalibration is needed.

Based on the confidence interval calculated for the ethylene analyzer, are the following statements true or false? Why?

1. The analyzer needs to be recalibrated.

2. 95% of the values on the test material will fall in this interval.

3. The probability is 0.95 that the next sample of size 8 on this test material will have an average that falls in this interval.

CONFIDENCE INTERVAL AROUND AN AVERAGE EXAMPLE--NORMAL DISTRIBUTION WITH UNKNOWN VARIANCE

Use of Confidence Interval To Check a Claim

A raw material used in a process contains an impurity which can be detrimental to the final product if the concentration of the impurity is too large. A supplier of the raw material claims his product contains 50 ppm, on average, of the impurity. The first 20 lots received from the supplier were tested for the impurity level. Results are given below. Is the supplier's claim valid?

59	52	54	48
55	59	57	44
57	68	47	53
50	60	52	51
50	47	52	54

CONFIDENCE INTERVAL AROUND THE POPULATION VARIANCE

The procedure for constructing a confidence interval around the average for a process has been introduced. A confidence interval that will likely contain the population variance can also be constructed. To do this requires the use of the chi-square distribution. The chi-square distribution is related to the square of the standard normal variable, z. In general it can be shown that

$$\chi^2_{(n-1)} = (n - 1)s^2/\sigma'^2$$

This can be used to set confidence limits around σ'^2. Suppose there are n observations, and a 100(1 - α)% confidence interval for σ'^2 is desired. This interval is given by

$$(n - 1)s^2/b < \sigma'^2 < (n - 1)s^2/a$$

where a is the $\alpha/2$ fractile and b is the 1 - ($\alpha/2$) of the chi-square distribution with n - 1 degrees of freedom. Values of the chi-square distribution are given in Appendix Table 9, p. 493.

CONFIDENCE INTERVAL AROUND A VARIANCE EXAMPLE

Insulation Resistance

In the production of a certain plastic, the insulation resistance is important. Ten samples of the plastic are taken and measured for insulation resistance. We do not know the average or standard deviation of our process. Based on these samples, determine the 95% confidence limits for the variance using the chi-square distribution.

Sample	Insulation Resistance
1	697
2	698
3	673
4	657
5	710
6	702
7	680
8	709
9	670
10	671

Solution:

The confidence interval for the unknown population variance is given by

$$(n - 1)s^2/b < \sigma'^2 < (n - 1)s^2/a$$

where a is the $\alpha/2$ fractile and b is the $1 - (\alpha/2)$ of the chi-square distribution with n - 1 degrees of freedom.

$n = 10$
$s = 18.7$
$s^2 = 349.8$
$v = n - 1 = 10 - 1 = 9$
$a = 2.70$ (from chi-square distribution, fractile for $\alpha/2 = .025$)
$b = 19.02$ (from chi-square distribution, fractile for $1 - \alpha/2 = .975$)

$$(10 - 1)(349.8)/19.02 < \sigma'^2 < (10 - 1)(349.8)/2.70$$

$$165.5 < \sigma'^2 < 1166$$

CONFIDENCE INTERVAL AROUND AN AVERAGE AND A VARIANCE EXAMPLE

Solids Content

In the production of a product, the percent solid content is important. To determine 95% confidence intervals for the population average and variance, 12 samples were taken from the process and analyzed. The results are given below. Determine the 95% confidence intervals for the average and variance.

Sample	% Solid
1	70.7
2	69.2
3	73.1
4	70.3
5	68.8
6	70.9
7	71.2
8	69.4
9	69.6
10	70.8
11	72.0
12	68.9

CONFIDENCE INTERVAL FOR THE DIFFERENCE BETWEEN TWO AVERAGES

This section describes how to compare two processes. The question is "Are the two processes making the same product?" If asked this question, one needs to know in what respect are the processes making the same product. With respect to part-to-part variation or with respect to the average value? Both questions will be addressed here.

Sample averages and variances differ from one another. That is not the issue. The question is "What is the difference in the average or the variance if the product is made under the same conditions in these two processes?" We don't know. However, the variances of the two processes can be compared to determine if there is any difference. In addition, a confidence interval for the difference between the two averages based on the sample results can be constructed.

Suppose there are two processes (such as two different reactors) making the same product. Suppose n_1 samples are taken from process 1 and n_2 samples are taken from process 2. Some sample statistics can be calculated.

	Process 1	Process 2
Sample size	n_1	n_2
Sample average	$\overline{X}_1$	$\overline{X}_2$
Sample variance	s_1^2	s_2^2

The approach used to construct a confidence interval for the difference in two averages assumes that the variation is consistent between the two processes, i.e., each process has the same variance. This must be checked. To determine if two variances are the same, the F distribution is used. The F distribution is related to the ratio of chi-square for both sample variances. To calculate the F value based on sample results, the larger sample variance is divided by the smaller sample variance.

$$F = \text{Larger } s^2/\text{Smaller } s^2$$

To determine if there is a difference in the variances of the two processes, this calculated value of F is compared to an F value from a table for the F distribution. The table values of F depend on two degrees of freedom. v_1 is the degrees of freedom in the variance in the numerator (n - 1). v_2 is the degrees of freedom in the variance in the denominator (n - 1). The other item that the table values of F depend upon is the confidence coefficient. Table 10 (p. 495) lists the values of the F distribution for various values of v_1 and v_2 for $\alpha = 0.05$.

If the calculated value of F is larger than the table value, it is concluded that there is a difference in the two variances. If the calculated value is less than the table value, it is concluded that there are no differences in the two variances.

Once it has been determined that the variation is consistent between the two processes, the confidence interal for the difference between the two averages can be constructed.

If the variances are not the same between the two processes, you cannot use this method.

The next step is to estimate the variance. There are two estimates of the variance from the sample variances. These are pooled to get a single estimate of the variance. The pooled variance is given by:

$$s_p^2 = [(n_1 - 1)s_1^2 + (n_2 - 1)s_2^2]/(n_1 + n_2 - 2)$$

The pooled variance is just a weighted average of the sample variances based on degrees of freedom for each sample. The square root of the pooled variance is a measure of the part-to-part variation or the process standard deviation using information from both samples.

The pooled variance is used to help estimate the standard deviation, s_{diff}, which estimates the variation of differences in two sample averages.

$$s_{diff} = \sqrt{s_p^2((1/n_1)+(1/n_2))}$$

This standard deviation is then used to construct a confidence interval around the difference in the two averages.

The $100(1 - \alpha)\%$ confidence interval for $\mu_1 - \mu_2$ is given by

$$\overline{X}_1 - \overline{X}_2 \pm t\sqrt{s_p^2((1/n_1) + (1/n_2))}$$

$$\overline{X}_1 - \overline{X}_2 \pm ts_{diff}$$

where t is the value for the t distribution corresponding to $n_1 + n_2 - 2$ degrees of freedom.

If the confidence interval does not contain zero, it is concluded that the two processes are operating at different averages.

COMPARING TWO PROCESSES EXAMPLE

Railcar Wheels

The weight of railcar wheels is determined by a certain procedure (procedure 1). A new procedure (procedure 2) has been suggested. The weight of the railcar wheels was measured using one procedure or the other. Results are given below. Are there any differences in the variances of the two procedures?

(Results are in thousand pounds)

PROCEDURE 1	PROCEDURE 2
0.37	0.33
0.32	0.31
0.42	0.36
0.38	0.26
0.34	0.34
0.33	

Solution:

The first step is to calculate the sample variances.

$$s_1^2 = 0.0014 \qquad s_2^2 = 0.00145$$

The next step is to calculate the F value by dividing the larger variance by the smaller variance.

$$F = s_2^2/s_1^2 = .00145/.0014 = 1.04$$

The next step is to compare this F value with the table value of F.

$$v_1 = \text{degrees of freedom in the numerator} = 5 - 1 = 4$$
$$v_2 = \text{degrees of freedom in the denominator} = 6 - 1 = 5$$
$$\alpha = 0.05$$

The table value of F (from Table 10) is 5.19.

Since the table value of F is greater than the value of F calculated from the sample variances, it is concluded that the variation is consistent between the two processes.

Since the variation is consistent, the $100(1 - \alpha)\%$ confidence interval for $\mu_1 - \mu_2$ is given by:

$$\overline{X}_1 - \overline{X}_2 \pm t\sqrt{s_p^2((1/n_1)+(1/n_2))}$$

The next step is to calculate the pooled variance.

$$s_p^2 = [(n_1 - 1)s_1^2 + (n_2 - 1)s_2^2]/(n_1 + n_2 - 2)$$

$$s_p^2 = [(6 - 1)(.0014) + (5 - 1)(.00145)]/(6 + 5 - 2)$$

$$s_p^2 = 0.001422$$

The next step is to calculate s_{diff}.

$$s_{diff} = \sqrt{s_p^2((1/n_1)+(1/n_2))}$$

$$s_{diff} = \sqrt{(.001422)((1/6) + (1/5))}$$

$$s_{diff} = 0.0228$$

The next step is to find the t value.

$$v = n_1 + n_2 - 2 = 9$$

From the t table (Appendix Table 8), for $\alpha = 0.05$, $t = 2.262$

The 95% confidence interval can now be determined.

$$\overline{X}_1 - \overline{X}_2 \pm ts_{diff}$$

$$0.36 - 0.32 \pm 2.262(0.0228)$$

$$0.04 \pm 0.0516$$

The confidence interval for the difference in the two averages is

$$-0.0116 < \mu_1 - \mu_2 < 0.0916$$

Since this interval contains zero, it is concluded that there is no difference in the averages of the two procedures.

COMPARING TWO PROCESSES EXAMPLE

Color Concentrates

Two color concentrates were being considered for use in a colored dryblend. We want to know if the color concentrates will produce the same color in the dryblend. To determine this 20 blenders of dryblend were produced using color concentrate 1. These 20 blenders had an average color of 3.01 with a standard deviation of 0.61. 20 blenders of dryblend using color concentrate 2 were also produced. These 20 blenders had an average color of 2.54, with a standard deviation of 0.54. Based on these results, would you conclude that the average color for the two color concentrates is different?

COMPARING TWO PROCESSES EXAMPLE

Techniques for Collecting Delinquent Accounts

Different techniques were tried to collect funds on extremely delinquent accounts. The techniques began when the accounts were 90 days overdue. The data below show the weeks it took on average to collect funds starting in nine successive months. Are there any differences between the techniques?

Month	Technique 1	Technique 2
1	30	28
2	26	28
3	20	31
4	9	13
5	2	18
6	23	10
7	7	19
8	15	17
9	28	

COMPARING TWO PROCESSES EXAMPLE

Comparing Reactors

Two different reactors are producing a polymer resin. Slurry samples are taken from each batch over three days' time. The samples are measured for inherent viscosity (IV), a measure of polymer molecular weight. The results are given below. Are there any differences in the inherent viscosity produced in the two reactors.

IV RESULTS

Reactor 1	Reactor 2
0.923	0.911
0.928	0.918
0.916	0.932
0.938	0.925
0.925	0.916
0.913	0.929
0.945	0.923
0.924	0.920
0.929	

COMPARING TWO PROCESSES WHEN THE VARIANCE IS NOT CONSISTENT

Occasionally the variance between two processes will not be consistent. The reasons for this should be investigated. One can still construct a confidence interval for the difference in two averages when the variance is not the same. The confidence interval is given by:

$$\overline{X}_1 - \overline{X}_2 \pm ts_{X1-X2}$$

where $s_{X1-X2} = \sqrt{((s_1^2/n_1) + (s_2^2/n_2))}$

The number of degrees of freedom (used to determine the value of t) is:

$$v = [(s_1^2/n_1) + (s_2^2/n_2)]^2/[\{(s_1^2/n_1)^2/(n_1 - 1)\} + \{(s_2^2/n_2)^2/(n_2 - 1)\}]$$

If the interval does not contain zero, it is concluded that the processes operate at different averages.

COMPARING TWO PROCESSES WHEN THE VARIANCE IS NOT CONSISTENT EXAMPLE

Purity

A material is made using two different processes. The purity of the material is important. Ten samples are taken from each process. Are there any differences in the average or variance between the two processes?

Process 1	Process 2
95	77
73	84
89	79
80	76
77	78
83	74
92	74
72	80
78	79
80	83

COMPARING MORE THAN TWO PROCESSES

Methods for comparing the average and variances of two processes have been introduced. In this section, a method of comparing more than two processes is introduced. This method is called Bonferroni's method. It involves the construction of multiple confidence intervals for differences in averages.

Assume that there are independent samples from k processes. The sample averages and variances can be calculated. An s chart is used to determine if there are any differences in variability between the k processes. If there are no differences, the pooled variance is calculated as:

$$s_p^2 = [\Sigma(n_i - 1)s_i^2]/\Sigma(n_i - 1)$$

where s_i^2 is the variance of process i and n_i is the number of samples taken from process i.

There will be $R = k(k-1)/2$ confidence intervals for differences of average that can be constructed. Each confidence interval is given by:

$$(\overline{X}_i - \overline{X}_j) \pm ts_{Xi\text{-}Xj}$$

where $\overline{X}_i$ and $\overline{X}_j$ are the averages of the samples taken from processes i and j, t is the value from the t distribution with $\Sigma(n_i - 1)$ degrees of freedom and a confidence coefficient of $1 - (\alpha/R)$, and

$$s_{Xi\text{-}Xj} = \sqrt{s_p^2((1/n_i) + (1/n_j))}$$

It is concluded that processes whose confidence intervals do not contain zero are operating at different averages.

Chapter 30

COMPARING MORE THAN TWO PROCESSES EXAMPLE

Comparing Four Shifts

Four different shifts make the same product for a company. Data from the last four times each shift operated is given below. Are there any differences between the shifts?

Sample	Shift 1	2	3	4
1	55.5	51.8	61.0	52.0
2	50.2	54.2	62.1	56.1
3	54.8	50.1	60.2	54.1
4	52.1	54.1	56.1	53.0
Average	53.15	52.55	59.85	53.80
Standard Dev.	2.4529	1.9740	2.6185	1.7569

The first step is to determine if there are any differences in the variability of the four shifts. This can be done by constructing an s chart on the standard deviations for the four shifts. The average standard deviation is $\bar{s} = 2.2006$. The control limits for the s chart are $UCL_s = B_4 \bar{s} = 2.266(2.2006)$ = 4.99 and LCL_s = none. Since none of the standard deviations are outside the control limits, it is assumed that the variation is consistent between the shifts.

The next step is to calculate the pooled variance. The pooled variance is

$$s_p^2 = [\Sigma(n_i - 1)s_i^2]/\Sigma(n_i - 1) = [3(2.4529)^2 + 3(1.9740)^2 + 3(2.6185)^2 + 3(1.7569)^2]/12$$

$$s_p^2 = 4.96415$$

There are 6 different intervals that can be constructed ($R = (k)(k - 1)/2 = 6$). Since the sample size is the same for all the processes, s_{Xi-Xj} will be the same for each interval.

$$s_{Xi-Xj} = \sqrt{s_p^2((1/n_i) + (1/n_j))} = \sqrt{4.96415((1/4) + (1/4))} = 1.5755$$

An overall confidence coefficient of 0.95 is desired. To obtain this, a t value with a confidence coefficient of $1 - \alpha/r = 0.99$ must be used for each interval. This will make the overall confidence coefficient be 0.94. There are 12 degrees of freedom ($\Sigma(n_i -1)$). The t value for 12 degrees of freedom and $\alpha = 0.01$ is 3.055. Each confidence interval will be:

$$(\bar{X}_i - \bar{X}_j) \pm ts_{Xi-Xj}$$

$$(\bar{X}_i - \bar{X}_j) \pm 3.055(1.5755)$$

$$(\bar{X}_i - \bar{X}_j) \pm 4.8132$$

458

The six intervals are given below:

$$\overline{X}_1 - \overline{X}_2 \pm 4.8132 = 0.60 \pm 4.8132 \qquad \text{or} \qquad -4.2132 < \overline{X}_1 - \overline{X}_2 < 5.4132$$

$$\overline{X}_1 - \overline{X}_3 \pm 4.8132 = -6.70 \pm 4.8132 \qquad \text{or} \qquad -11.5132 < \overline{X}_1 - \overline{X}_3 < -1.8868$$

$$\overline{X}_1 - \overline{X}_4 \pm 4.8132 = -0.65 \pm 4.8132 \qquad \text{or} \qquad -5.4632 < \overline{X}_1 - \overline{X}_4 < 4.1632$$

$$\overline{X}_2 - \overline{X}_3 \pm 4.8132 = -7.30 \pm 4.8132 \qquad \text{or} \qquad -12.1132 < \overline{X}_2 - \overline{X}_3 < -2.4868$$

$$\overline{X}_2 - \overline{X}_4 \pm 4.8132 = -1.25 \pm 4.8132 \qquad \text{or} \qquad -6.0632 < \overline{X}_2 - \overline{X}_4 < 3.5632$$

$$\overline{X}_3 - \overline{X}_4 \pm 4.8132 = -6.05 \pm 4.8132 \qquad \text{or} \qquad -10.8632 < \overline{X}_3 - \overline{X}_4 < -1.2368$$

The intervals that contain zero indicate no differences between the processes. The intervals that do not contain zero indicate differences between the processes. The data indicate that process 3 operates at a different average from processes 1, 2 or 4. There are no differences in the averages between processes 1, 2 and 4.

COMPARING MORE THAN TWO PROCESSES EXAMPLE

Comparing Furnaces

In an ethylene plant, three furnaces are used to crack ethane to eyhylene. The table below gives the % ethylene in the output stream from each of the three furnaces. Based on this data, is there a difference in the average or the variation in any of the three furnaces?

Sample	Furn. 1	Furn. 2	Furn. 3
1	66	65	67
2	61	64	65
3	61	66	68
4	62	64	70
5	61	65	67
6	62	63	68
7	60	64	64

PAIRED OBSERVATION COMPARISONS

So far, all the examples, in comparing processes, have involved the use of independent observations. There are times when the observations will not be independent. For example, suppose it is desired to compare two analytical test methods to see if there are any differences in the two methods. One might take a series of samples, divide each sample in half, and run each sample half in one the two test methods. In this case, the samples are not independent. The methods learned previously for comparing the difference between two averages can't be used. Instead, the procedure below for these paired sample comparisons must be used.

Consider the following example. Suppose there are two methods for determining nickel content. One method is a wet chemical method; the other method is atomic absorption. The same 10 samples were analyzed by each method. The results are given below.

Sample	ppm Nickel Chemical	A. Abs.	D
1	4.0	4.2	
2	4.0	3.9	
3	4.3	4.6	
4	4.1	4.2	
5	4.4	4.1	
6	4.3	4.3	
7	4.6	4.4	
8	4.7	4.6	
9	3.9	4.0	
10	4.5	4.4	

The first step in analyzing the results is to take the difference between the results for chemical and atomic absorption methods (for each sample, retaining the sign):

$$D = X_1 - X_2$$

The next step is to find the average difference, $\overline{D}$, and the standard deviation, s_D, of the D values.

$$\overline{D} = \qquad\qquad s_D =$$

The 100(1 - a)% confidence interval for the paired sample comparison is given by

$$\overline{D} \pm t s_D / \sqrt{n}$$

The number of degrees of freedom, v, in this case, is given by n - 1 = 10 - 1 = 9.

For $v = 9$ and $\alpha = 0.05$, t = 2.262.

The confidence interval becomes

$$D \pm ts_D/\sqrt{n}$$

$$0.01 \pm (2.262)(0.1853)/\sqrt{10}$$

$$0.01 \pm .1325$$

$$-0.1225 < \overline{D} < 0.1425$$

Since the interval contains zero, it is concluded that there are no differences between the two test methods.

PAIRED SAMPLE CONFIDENCE INTERVAL EXAMPLE

Comparing Two Technicians

An experiment was performed to see if two technicians have a tendency to obtain different results when determining the purity of a certain product. Each sample was split in half. Each technician made a determination of the purity on the half of the sample they were given. The results are given below. Is there any difference between the results obtained by the technicians?

Sample	Tech. 1	Tech. 2
1	74.0	73.0
2	73.1	71.3
3	73.5	73.2
4	73.9	71.1
5	71.2	70.3
6	72.5	71.5
7	73.0	73.4
8	74.3	72.4

SUMMARY

This chapter has introduced various methods of comparing processes. Methods of constructing confidence intervals around the average or variance of one process were introduced. A method of comparing two processes was introduced. This method involved comparing the variances of the two process to determine if they were consistent. The averages of the two processes could then be compared. A method of comparing the averages of two processes if the variances were not consistent was also introduced. Bonferroni's method of comparing multiple processes was introduced. Paired sample comparisons were also introduced. It should be remembered that all of these techniques are "snapshots" in time of the processes. The same results may not be obtained at a later date if the processes are not in statistical control.

APPLICATIONS OF COMPARING PROCESSES

Think about the material covered in this chapter. Which methods do you think will be most useful to you and why?

31

NESTED EXPERIMENTAL DESIGNS

There are many sources of variation in a process. What do I begin to work on first? Where is the greatest opportunity for improvement? A method of determining what you should work on first is introduced in this chapter. This method allows you to make this decision based on data. It essentially involves designing an experiment to examine multiple sources of variation. These sources of variation include batch-to-batch variability, within-batch variability, measurement variability, sampling variability, shift-to-shift variability, etc.

OBJECTIVES

In this chapter you will learn:

> 1. A method of determining what to work on first in process improvement

The objective of this chapter is to learn how to design an experiment to determine what the different sources of variation contribute to the overall process variation. This permits you to determine what to work on first. This designed experiment is called a nested experimental design. This method represents a "snapshot in time" of the process. If the process is not in statistical control, you can't be sure of obtaining similar results in the future.

INTRODUCTION TO NESTED EXPERIMENTAL DESIGNS

This section describes how to develop a nested experimental design (Box, Hunter and Hunter, 1978). This design allows you to examine multiple sources of variation. Once the experiment is designed and completed, the data are analyzed. This analysis permits you to determine what percent of the total variation is due to the specific sources of variation you selected. The source of variation responsible for the largest percentage is the one you will normally try to reduce first.

One advantage of this method is that the experiment is carried out under normal plant operating conditions. There is no need to change levels of various operating parameters at this stage. The approach simply involves collecting samples at appropriate places and times.

The concept of a nested experimental design will be introduced through the use of an example. The example is for a batch process, although the concept is easily expanded to a continuous process.

Suppose the process has a population average of 100 and a population standard deviation of 10. Suppose an individuals control chart has been used to monitor the process. Figure 1 shows the control chart for the last 100 batches. As can be seen from the figure, the process is in statistical control with a process average ($\overline{X}$) of 100.82 and an average range ($\overline{R}$) of 12.36. Since the process is in control, an estimate of the total process standard deviation can be obtained from the average

Figure 1

CONTROL CHARTS ON BATCH PROCESS

INDIVIDUALS CHART FOR PRODUCT RESPONSE

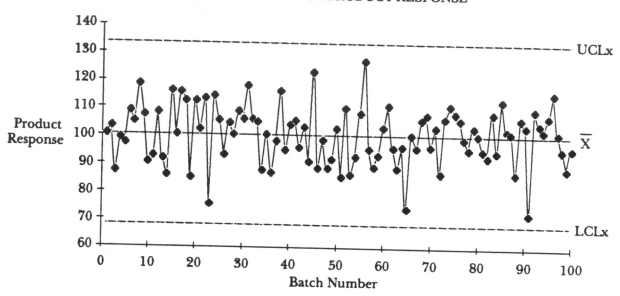

MOVING RANGE CHART (n = 2)

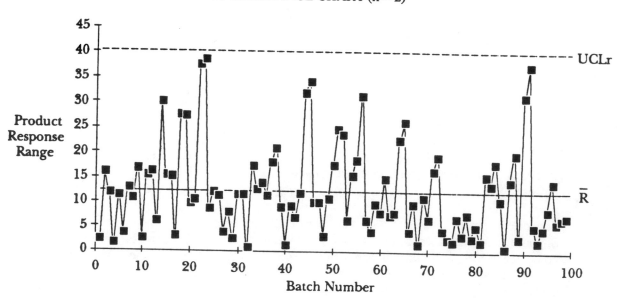

range as shown below:

$$\hat{\sigma}'_t = \overline{R}/d_2 = 12.36/1.128 = 10.95$$

Note that the overall process average and the standard deviation obtained from the control charts are very close to the population average and standard deviation.

Each result plotted on the individuals chart represents one sample taken from the batch. This sample is taken to the laboratory and tested for the product response. The objective is to decrease the amount of variation in the product response.

What are the sources of variation in the product response, i.e., why do the results vary? There are essentially three major sources of variation in this process. These are the batch-to-batch variability, the within-batch or sampling variability, and the measurement variability. Numerous factors may affect each of these sources of variation.

Each of the three sources of variability adds variation or error to the results. Assume that $\overline{X}'$ is the unknown population average. The sample pulled from the batch process and tested in the laboratory will probably not be equal to this unknown population average. The difference between the laboratory result and the unknown population average is the overall error, ε. Thus,

$$\varepsilon = X - \overline{X}'$$

where X is the laboratory result. This overall error has three components corresponding to the three sources of variation. Thus,

$$\varepsilon = \varepsilon_b + \varepsilon_s + \varepsilon_{ms}$$

where ε_b is the error cause by the batch-to-batch variability, ε_s is the error caused by the sampling variability, and ε_{ms} is the error caused by the measurement system. By definition, these three components of error have zero averages and can be assumed to be represented as normal distributions with fixed variances. Figure 2 shows the relationship among the three components of error.

The standard deviation, $\hat{\sigma}'_t$, calculated from the moving range chart represents an estimate of the combination of these three error components. There is a linear relationship among the variances. The total process variance (σ'^2_t) is given by:

$$\sigma'^2_t = \sigma'^2_b + \sigma'^2_s + \sigma'^2_{ms}$$

where σ'^2_b is the batch-to-batch variance, σ'^2_s is the sampling or within-batch variance, and σ'^2_{ms} is the measurement system variance. Note that the linear relationship is between the variances, not the standard deviations.

If one can estimate each of the individual variances, one can determine where to start working

Figure 2

COMPONENTS OF VARIATION

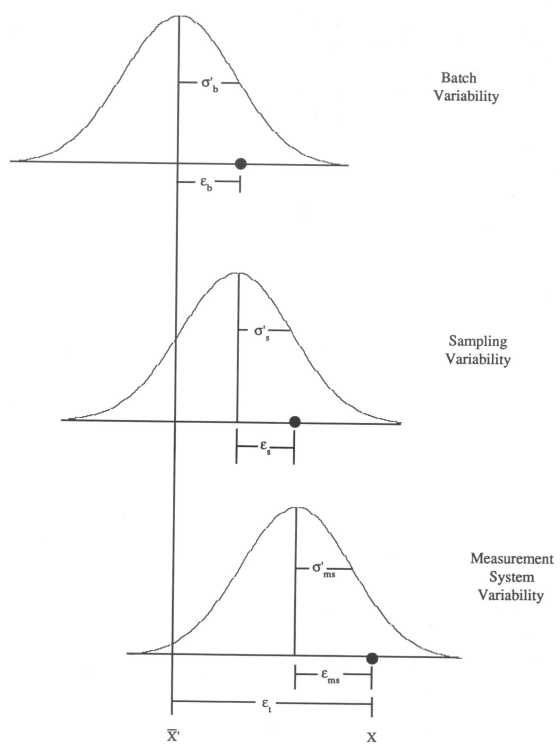

on decreasing the variation in the product response. This is accomplished using a nested experimental design.

For a batch process, a nested design involves looking at a multiple number of batches, sampling each batch multiple times, and running each sample in the laboratory multiple times. Assume that you decide to look at 15 batches of material. Each batch will be sampled twice. In addition, each sample will be tested twice in the laboratory. Figure 3 is a schematic of the design for one batch. The results from the sampling program are shown in Table 1. The next step is to analyze the data.

Table 1

RESULTS FROM NESTED DESIGN FOR A BATCH PROCESS

Batch Number	Sample Number	Test 1 Result	Test 2 Result	Var. of two Test Results	Avg. of two Test Results	Var. of two Batch Samples	Avg. of four Batch Samples
1	1	92.32	97.38	12.802	94.850		
	2	120.12	111.36	38.369	115.740	218.196	105.295
2	1	123.21	121.86	0.911	122.535		
	2	111.58	115.66	8.323	113.620	39.739	118.078
3	1	99.43	96.73	3.645	98.080		
	2	102.26	99.90	2.785	101.080	4.500	99.580
4	1	100.73	98.67	2.122	99.700		
	2	89.90	91.71	1.638	90.805	39.561	95.253
5	1	101.43	103.61	2.376	102.520		
	2	94.47	97.96	6.090	96.215	19.877	99.368
6	1	84.51	75.41	41.405	79.960		
	2	93.67	99.71	18.241	96.690	139.946	88.325
7	1	90.06	94.31	9.031	92.185		
	2	114.43	113.99	0.097	114.210	242.550	103.198
8	1	98.57	96.00	3.302	97.285		
	2	87.87	78.37	45.125	83.120	100.324	90.203
9	1	91.40	97.02	15.792	94.210		
	2	98.73	88.39	53.458	93.560	0.211	93.885
10	1	105.29	103.95	0.898	104.620		
	2	100.67	97.39	5.379	99.030	15.624	101.825
11	1	93.33	95.07	1.514	94.200		
	2	108.10	93.28	109.816	100.690	21.060	97.445
12	1	103.64	104.23	0.174	103.935		
	2	114.81	119.29	10.035	117.050	86.002	110.493
13	1	81.22	81.57	0.061	81.395		
	2	92.68	87.72	12.301	90.200	38.764	85.798
14	1	105.31	107.38	2.142	106.345		
	2	95.86	104.17	34.528	100.015	20.034	103.180
15	1	100.81	102.88	2.142	101.845		
	2	101.01	98.91	2.205	99.960	1.777	100.903

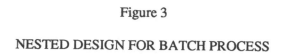

Figure 3

NESTED DESIGN FOR BATCH PROCESS

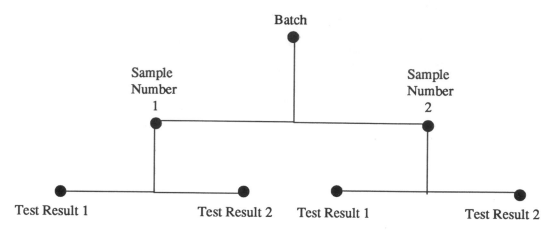

Consider the measurement variability first. As can be seen from the table, batch 1 was sampled twice. Each batch sample was divided in two and tested in the laboratory. For batch sample number 1, test result 1 was 92.32. Test result 2 was 97.38. Since this is the same material, the difference in these two test results is due to measurement error only. This will be used to estimate the measurement system variance ($\hat{\sigma}'_{ms}{}^2$). The steps in estimating $\hat{\sigma}'_{ms}{}^2$ are given below.

Calculating $\hat{\sigma}'_{ms}{}^2$

1. Calculate the variance of each set of test results. For example, for sample number 1 of batch number 1, the two test results were 92.32 and 97.38. The variance of these two numbers is 12.802. This is done for all sets of samples.

2. An estimate of the variance of the measurement system is the pooled variance of the sets of samples. Since each batch sample was tested the same number of times, the pooled variance is the average of the variances of the two test results. Thus,

$$\hat{\sigma}'_{ms}{}^2 = 14.890$$

Next consider the sampling variability. The sampling variance will be calculated by using the batch sample averages. The steps in estimating the sample variance ($\hat{\sigma}'_s{}^2$) are given as follows.

Calculating $\hat{\sigma}'^2_s$

1. Take the average of the two test results for each batch sample. For example, the average of the two test results (92.32 and 97.38) for sample number of 1 batch number 1 is 94.85.

2. Calculate the variance of the two sample averages for each batch. For example, for batch number 1, the averages of the two sample results are 94.850 and 115.740. The variance of these two sample averages is 218.196.

3. Calculate the pooled (average) variance of the batch sample variances. The average variance is 65.878. Note that this is not the sampling variance. Why do the sample averages vary? One reason is the sampling variation. However, measurement variability is also present in the sample averages. To determine the sampling variance, you must subtract out the effect due to measurement variability. If V_s equals the average of the variances of the sample averages, then

$$V_s = \hat{\sigma}'^2_s + (\hat{\sigma}'^2_{ms}/n_{m1})$$

where n_{m1} equals the number of duplicate samples run. In this example, $n_{m1} = 2$. The sampling variance is then calculated as:

$$\hat{\sigma}'^2_s = V_s - (\hat{\sigma}'^2_{ms}/n_{m1}) = 65.878 - (14.890/2) = 58.432$$

Next consider the batch-to-batch variability. The batch-to-batch variance is determined from the batch averages. The following procedure is used to estimate the batch-to-batch variance ($\hat{\sigma}'^2_b$).

Calculating $\hat{\sigma}'^2_b$

1. Calculate the batch average for each batch from the individual test results. For example, for batch number 1 there are four test results: 92.32, 97.38, 120.12, and 11.36. The average of these four test results is 105.295.

2. Calculate the variance of the batch averages. The variance is 70.501. This is not the batch-to-batch variance. The batch averages do include variability due to the batches, but they also include the variability due to sampling and measurement. To calculate the batch-to-batch variance, you must subtract out the effects of the measurement and sampling variance. If V_p is the variance of the batch averages, then

$$V_p = \hat{\sigma}'^2_b + (\hat{\sigma}'^2_s/n_s) + (\hat{\sigma}'^2_{ms}/n_{m2})$$

where n_s equals the number of samples taken per batch and n_{m2} equals the number of test results per batch. In this example, $n_s = 2$ and $n_{m2} = 4$. The process variance is then calculated as follows:

$$\hat{\sigma}'^2_b = V_p - (\hat{\sigma}'^2_s/n_s) - (\hat{\sigma}'^2_{ms}/n_{m2}) = 70.501 - (58.432/2) - (14.890/4) = 37.563$$

Determining Total Variance and Percentages

The total variance and the percentages due to each component can now be determined as follows:

1. Calculate the total variance using the following:

$$\hat{\sigma}'^2_t = \hat{\sigma}'^2_b + \hat{\sigma}'^2_s + \hat{\sigma}'^2_{ms}$$

$$\hat{\sigma}'^2_t = 37.563 + 58.432 + 14.890 = 110.885$$

The square root of the total variance should be close to the standard deviation calculated from the moving range chart. The square root of $\hat{\sigma}'^2_t$ is 10.04, which is close to the 10.95 determined from the range chart; however, it will usually not be the same.

2. Determine the percentage of total variance caused by each component as shown below.

% of total variance due to the batch-to-batch variability:

$$100(\hat{\sigma}'^2_b/\hat{\sigma}'^2_t) = 100(37.563/110.885) = 33.9\%$$

% of total variance due to the sampling variability:

$$100(\hat{\sigma}'^2_s/\hat{\sigma}'^2_t) = 100(58.432/110.885) = 52.7\%$$

% of total variance due to the measurement system variability:

$$100(\hat{\sigma}'^2_{ms}/\hat{\sigma}'^2_t) = 100(14.890/110.885) = 13.4\%$$

The largest percentage is usually the one you will select to work on improving first. In this example, it is the sampling variability.

The preceding example dealt with a batch process. The same technique can be used with a continuous process. For example, suppose you are interested in determining what the sources of variation are in a continuous process. You currently sample the process once a day. You can consider all the material you produced during that day a "batch." If you wanted to determine shift-to-shift variability, you would take samples on each shift. To determine sampling variability, you would take multiple samples on each shift. You could also determine measurement variability by running the same sample multiple times in the laboratory. The analysis procedure is similar to the procedure above.

The procedure for analyzing a nested design always begins with the innermost nested part of the design. For the above batch process example, the innermost nest was the measurement system variability. The procedure continues until you reach the outer-most nested part of the design. It is important to remember to subtract out the variability of contained nests (such as subtracting out the measurement system variance to determine the sampling variance). The denominator in these equations is always the number of samples used to calculate the average value (for example, n_{m1} = 2 in the preceding example since each average contained two samples).

It should be pointed out that a nested design is really a study of rational subgrouping. The same information could be obtained over time by using a series of control charts. The subgroups used in the control charts would examine the source of variation that you are interested in.

NESTED DESIGN EXAMPLE

Comparison of Shift-to-Shift, Sampling, and Measurement Variability

A company produces a liquid product using a continuous process. A certain product response is measured once a shift (8-hour shifts). The results are being monitored using an individuals chart. The control charts are shown Figure 4. The process is in statistical control with $\overline{X} = 59.6$ and $\overline{R}$ = 6.9. The upper specification limit is 75. The lower specification limit is 45. The process is not capable of meeting specifications. It was decided to run a nested experimental design to determine where the greatest source of variability in the process was. After some discussion, the sources of variability to be studied were selected as the shift-to-shift variance, the sampling variance, and the measurement system variance. Each shift was sampled ten times. Each sample was tested twice in the laboratory. The data are given in Table 2. Use these data to determine what source of variation causes the most process variability.

Figure 4

CONTROL CHARTS FOR NESTED DESIGN EXAMPLE

INDIVIDUALS CHART FOR PRODUCT RESPONSE

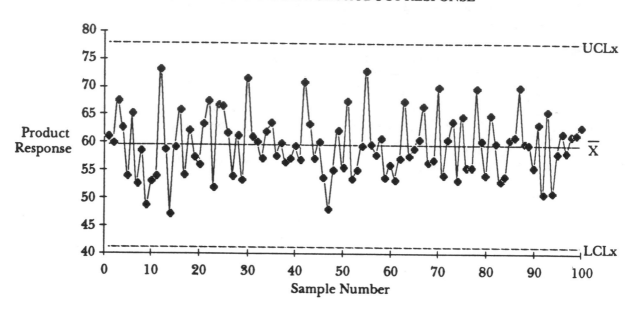

MOVING RANGE CHART (n = 2)

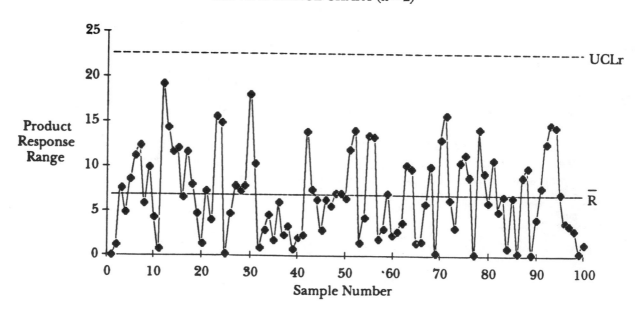

Table 2

DATA FOR NESTED DESIGN EXAMPLE

Shift Number	Sample Number	Test Result 1	Test Result 2	Variance of Two Tests	Average of Two Tests	Variance of Samples	Average of Samples
1	1	59.6	62.7	4.82	61.14		
	2	58.4	55.3	4.62	56.85		
	3	55.7	51.2	10.25	53.43		
	4	59.0	57.5	1.08	58.25		
	5	63.0	59.3	6.83	61.10		
	6	56.1	56.3	0.02	56.24		
	7	68.5	64.1	9.46	66.30		
	8	57.4	59.7	2.66	58.52		
	9	57.9	59.9	2.04	58.89		
	10	58.7	56.7	1.91	57.69	11.98	58.8
2	1	52.7	53.0	0.04	52.87		
	2	57.5	55.7	1.66	56.62		
	3	58.0	62.6	10.71	60.33		
	4	54.3	55.4	0.53	54.86		
	5	59.3	62.5	5.36	60.90		
	6	55.9	53.9	1.89	54.91		
	7	50.9	53.4	3.19	52.16		
	8	59.4	58.9	0.12	59.18		
	9	52.4	50.9	1.10	51.66		
	10	55.8	55.9	0.01	55.86	11.01	55.9
3	1	61.6	60.4	0.74	61.02		
	2	62.2	59.2	4.41	60.71		
	3	59.8	63.5	7.08	61.67		
	4	60.3	59.9	0.08	60.13		
	5	58.3	59.3	0.58	58.79		
	6	67.0	71.2	9.01	69.08		
	7	62.4	59.7	3.61	61.01		
	8	61.7	62.0	0.03	61.84		
	9	62.9	66.2	5.49	64.51		
	10	61.4	57.7	6.91	59.54	8.86	61.8

SUMMARY

This chapter has introduced nested experimental designs. These designs provide a method of determining what causes the most variation in a process. This permits you to determine, using data, what should be worked on first. A nested design examines multiple sources of variation by taking samples at appropriate places and times. Once the experiment is designed and completed, the data are analyzed. The analysis permits one to determine what percent of the total variation is due to the specific sources of variation selected. This type of design is carried out under normal operating conditions.

APPLICATION OF NESTED EXPERIMENTAL DESIGNS

In the space provided below, lay out a nested experimental design for a process at work, starting with the objective of the design.

32

THE DEMING PHILOSOPHY

Many things are required to be able to successfully implement Total Quality Management (TQM). The statistical methods taught in this book are important. Just as important, if not more so, is management's philosophy behind TQM. What will the goals and objectives of TQM be? How will the required resources be allocated? How will training be handled? Questions like these must be answered by management. Management needs guidance in answering these questions. The philosophy developed by Dr. W. Edwards Deming provides this guidance. Dr. Deming has developed fourteen points that tell management what must be done to remain competitive in the future. This chapter introduces the Deming philosophy. Deming's goal is the transformation of the western style of management. One part of the Deming philosophy involves the continuous reduction of variation. This reduction of variation is key to how the Japanese view customer satisfaction. Dr. Deming has also identified seven major diseases or obstacles to quality improvement. These seven diseases are also introduced in this chapter.

OBJECTIVES

In this module you will learn:

1. Who Dr. W. Edwards Deming is
2. How the United States and Japan's views of customer satisfaction differ
3. Deming's fourteen points
4. Seven deadly diseases that stand in the way of quality improvement

Deming's fourteen points are a road map that tell management what must be done to continuously improve quality and productivity. The fourteen points can be divided into three general categories. These are constancy of purpose, using objective data for decision-making, and a systems approach for the improvement of quality and productivity.

DR. W. EDWARDS DEMING

Dr. W. Edwards Deming is an internationally known consultant. He is best known for his work in Japan that revolutionized Japanese quality and productivity. His philosophy and methods helped Japan reach its current level of success.

Following World War II, Dr. Deming was invited by the Union of Japanese Scientists and Engineers to speak to leading industrialist about Japan's quality. After the war, Japan had a reputation for producing poor quality goods. Breaking into foreign markets was extremely difficult. Dr. Deming convinced the Japanese that they could break into foreign markets by improving quality and productivity using his methods. The Japanese listened to Deming and implemented his methods. The rest is history. The adoption of Dr. Deming's methods is now spreading across the United States.

Dr. Deming focuses on the improvement of products and services by reducing any uncertainty and variability in the design and manufacturing processes. This concept of reducing variation is key to the Japanese view of customer satisfaction. Deming emphasizes the PDCA cycle. This Plan, Do, Check, and Analyze or Act cycle is a never-ending cycle of improvement that occurs in all phases of a company, including product design, manufacturing, testing, sales, etc.

Deming emphasizes that top management has the overriding responsibility for improving quality. The distinction he makes is that employees work in the system while management works on the system. It is management's responsibility to improve the system and therefore the product or service being produced.

CUSTOMER SATISFACTION

In the past, there has been a definite distinction about how the Japanese and the United States view customer satisfaction. In the United States, if a product is within the specification limits, it is often assumed that the customer will be satisfied. This view is shown in Figure 1. If the product falls outside the specifications, the customer will not be satisfied.

Figure 1

A UNITED STATES VIEW OF CUSTOMER SATISFACTION

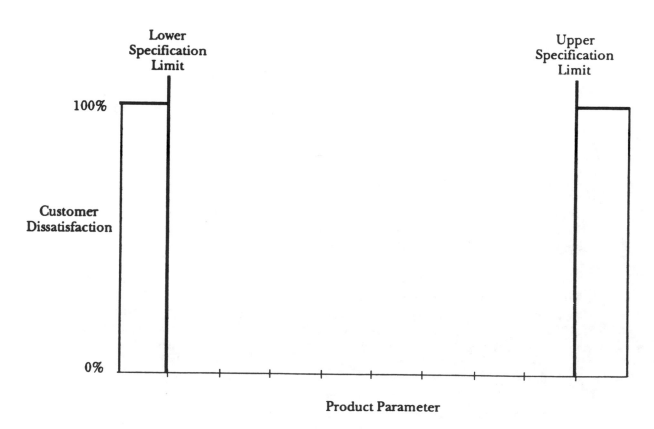

Product Parameter

The Japanese view customer satisfaction differently. They believe that customer satisfaction is highest when the product is precisely on the target value. Any departure from the target value is a cost. The loss increases exponentially as the product moves away from the target. This is demonstrated in Figure 2.

Figure 2

A JAPANESE VIEW OF CUSTOMER SATISFACTION

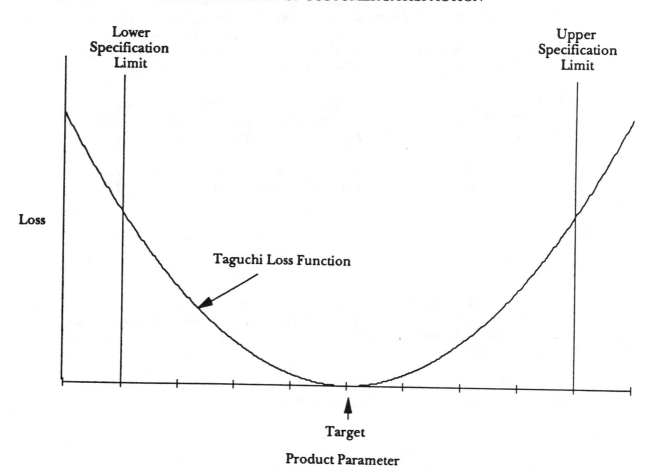

This loss can be measured. Genichi Taguchi has developed a loss function (a quadratic equation) to approximate this loss.

Customers are looking for consistency in the product or service. This feature has been recognized by the Japanese for many years. It is only now becoming recognized in the United States.

Chapter 32

DEMING'S FOURTEEN POINTS

Each of Deming's fourteen points is discussed below. For more information, see *Out of Crisis* by W. Edwards Deming.

1. *Create constancy of purpose toward improvement of product and service, with the aim to become competitive and to stay in business and provide jobs.*

   - Management is faced with two problems: (1) problems of today and (2) problems of tomorrow. Problems of today encompass maintenance of the quality of the product produced today. These problems must be handled efficiently. However, problems of the future command first and foremost constancy of purpose and dedication to improvement of competitive position to keep the company alive and to provide jobs for their employees.
   - Establishment of constancy of purpose means acceptance of obligations such as innovation, putting resources into research and education, improving designs of product and service, and putting resources into maintenance of equipment, furniture, fixtures, and aids to production.
   - If the management team talks about quality but applies pressure only for quantity, it becomes destructive to workers' morale. Communication is paramount between management and workers.

2. *Adopt the new philosophy. We are in a new economic age. Western management must awaken to the challenge, must learn its responsibilities, and must take on leadership for change.*

   - We can no longer tolerate commonly accepted levels of mistakes, defects, material not suited for the job, people on the job that do not know what the job is and are afraid to ask, handling damage, antiquated methods of training on the job, inadequate and ineffective supervision, management not rooted in the company, and job hopping in management.

3. *Cease dependence on inspection to achieve quality. Eliminate the need for inspection on a mass basis by building quality into the product in the first place.*

   - Quality comes from improvement of the process, rather than from inspection of a completed product. Routine 100 percent inspection to improve quality is equivalent to planning for defects and acknowledgment that the process does not have the capability required for the specifications. Routine inspection is in reality planning for defects, which means the process is out of statistical control or the specifications are impractical.
   - The problem is not the quality of the product, but the quality of the process. When the product leaves the door of a supplier, it is too late to do anything about its quality.
   - Inspection, scrap, downgrading, and rework are not corrective action on the process. Screening the good from bad (inspect and sort) does nothing to prevent the production of the bad part in the first place and is not 100% effective in detecting bad products.

- Shewhart developed the control chart in 1924. The concept of moving back into the production system to obtain statistical control of the manufacturing processes and continue to improve them until they meet requirements was integrated into Japan's production facilities. Only after this plateau was reached could just-in-time material control and automation be considered.

4. *End the practice of awarding business on the basis of price tag. Instead, minimize total cost. Move toward a single supplier for any one item, on a long-term relationship of loyalty and trust.*

   - It has become standard practice in America to obtain a number of competitive bids and buy from the lowest bidder without regard to the total cost of ownership and quality of the product.
   - We can no longer leave quality, service, and price to the forces of competition for price alone - not in today's requirements for uniformity and reliability.
   - The purchasing department must change its focus from lowest initial cost of material purchased to lowest total cost. This means education in purchasing.
   - It is also necessary to learn that specifications of incoming materials do not tell the whole story about performance. What problems does the material encounter in production? Materials and components may all be excellent, each by itself, yet not work well together in production or in the finished product. It is thus necessary to follow a sample of materials through the whole production process into complex assemblies, and onward, finally, to the customer.
   - A long-term relationship between purchaser and supplier is necessary for the best economy. How can a supplier be innovative and develop economy in his production processes when he can only look forward to short-term business with a purchaser?
   - There are also operational advantages. Even though two suppliers send excellent materials, there will be differences. Lot-to-lot variation from any one supplier is usually enough to give fits to manufacturing. It is reasonable to expect that variation between lots from two suppliers will give even more trouble.
   - Moreover, one should not overlook the simplification in the accounting and paperwork from a decrease in number of suppliers and fewer shipping points.
   - A worthy customer should expect his suppliers, if they are wise and looking ahead with constancy of purpose, to compete for selection as the single supplier. A supplier should himself work toward a single supplier for any one item.
   - A second source, for protection, in case ill luck puts one vendor out of business temporarily or forever, is a costly policy. There is lower investment and lower total inventory with a single vendor than with two.
   - Purchases of commodities and services should also move to the single supplier.

5. *Improve constantly and forever the system of production and service, to improve quality and productivity, and thus constantly decrease costs.*

   - A theme that appears over and over in Deming's philosophy is that quality must be built in at the design stage. It may be too late once plans are on their way.
   - Every product should be regarded as one of a kind; there is only one chance for optimum success.

- The employee's job is to produce consistent product or service within the capability of the process. Management's job is to constantly strive to improve the processes.
- Quality desired starts with the intent, which is fixed by management. The intent must be translated into plans, specifications, test, etc. in an attempt to deliver to the customer the quality intended, all of which are management's responsibility.
- Production workers can do little to improve the processes even when they are in an environment where management respects their recommendations.
- Deming credits Shewhart with the PDCA (plan, do, check, and analyze) cycle.
- Though no hard-and-fast rules exist, there seems to be a commonality of transformation for a typical organization. This transformation toward quality is an iterative process, progressing through each of the steps as describe below:

Plan:	Determine what can be and what changes are needed. List the obstacles to be overcome and decide what the more important results may be. Begin determining what new information is needed.
Do:	Begin with a small-scale implementation of the changes or test to provide the data for further implementation.
Check:	Measure and observe the effects of the change or test.
Analyze:	Determine whether the data confirmed the actions set forth in the plan. Determine the risks of proceeding with the plan on a larger scale.

- After completing these steps, modify the plan from the results of the analysis. Follow this process again and again, throughout the organization, expanding knowledge and applying further understanding with each cycle.
- Deming provides a simple yet effective method of looking at problems, and that is to distinguish between assignable causes and common causes. Using the inputs and transformations as well as the continuous improvement cycles of plan, do, check, and analyze, the customer's needs are considered to help produce a high quality and marketable product.
- This concept can finally be extended beyond the organization to include suppliers and distributors. Involving these people in your quality improvement effort is essential if quality is to be built in and carried all the way to the customer.

6. *Institute training on the job.*

- A total transformation in training philosophy must happen in most companies. Training must be totally reconstructed. Management needs training to learn about the company, all the way from incoming material to customer. A central problem is the need for appreciation of variation.
- Thorough training in statistical technique is useful in the job and will complement normal training and allow the employee to make greater contributions.
- With technology changing so rapidly, jobs are also changing. Retraining is an important component of keeping employees current with up-to-date skills for their jobs.

- It should be noted further that money spent on training, retraining, and education does not show on the balance sheet; it does not increase the tangible net worth of a company. In contrast, money spent for equipment is on the balance sheet and increases the present net worth of the company.
- Education in statistics is simple but powerful and will be needed for future jobs. Statistical techniques are required of all personnel in management and professional areas. People involved with quality and given training on the new management principles and statistical process control will be of great value to their companies. This training is not easy, and it will take time to train everyone.

7. *Institute leadership. The aim of supervision should be to help people and machines and gadgets to do a better job. Supervision of management is in need of overhaul, as well as supervision of production workers.*

- The job of management is not supervision, but leadership. Management must work on sources of improvement, the intent of quality of product and of service, and on the translation of the intent into design and actual product.
- The brainpower of all the employees of a company is its most valuable asset. It is the responsibility of the management team to use this asset. This can be done only with a proper organizational climate that asks for employee recommendations and is willing to act on them. Supervisors must be trained to bring out the workers' thoughts and report them to management without fear of reprisal.

8. *Drive out fear, so that everyone may work effectively for the company.*

- No one can put in his best performance unless he feels secure. Fear takes on many faces. A common denominator of fear in any form, anywhere, is loss from impaired performance and padded figures.
- The economic cost of fear in an organization is staggering. Many workers do not really understand what their job is and when they are doing it right. They are afraid to ask for more direction because they may appear to be slow learners. They are afraid to point out problems because they may be labeled as troublemakers or malcontents.
- Driving fear out of an organization takes time. Good management will solicit recommendations by employees, act on the good ones, and describe in detail reasons for not acting on bad ones and earn confidence of the workers. Mutual respect among employees and management is necessary to eliminate fear and promote constructive activity.

9. *Break down barriers between departments. People in research, design, sales, and production must work as a team, to foresee problems of production and in use that may be encountered with the product or service.*

- Lack of understanding of the problems in other departments can cause mistrust, delays and below par performance.
- Task forces, management implementation teams, statistics teams and other overt teamwork actions that communicate departmental needs, objectives, roadblocks and solve problems will increase the company performance.

10. *Eliminate slogans, exhortations, and targets for the work force asking for zero defects and new levels of productivity. Such exhortations only create adversarial relationships, as the bulk of the causes of low quality and low productivity belong to the system and thus lie beyond the power of the work force.*

   • Slogans, banners, posters and pledge cards do not help people do their jobs better. Employees already want to be proud of their work, their company and themselves. The only way to obtain long-term improvement in their work is to give them tools, methods and an organizational climate that helps them do their jobs better.
   • What is wrong with posters and exhortations? They are directed at the wrong people. They arise from management's supposition that the production workers could, by putting their backs into the job, accomplish zero defects, improve quality, improve productivity, and all else that is desirable. The charts and posters take no account of the fact that most of the trouble comes from the system. Calculations that indicate what proportion of defects and mistakes and high cost comes from the system (responsibility of management) and how much from the people on the job should be one of the chief tools of management, and certainly of leadership.
   • Exhortations and posters generate frustration and resentment. They advertise to the production worker that the management is unaware of the barriers to pride of workmanship.
   • The immediate effect of campaign of posters, exhortations, and pledges may well be some fleeting improvement of quality and productivity, the effect of elimination of some obvious assignable causes. In time, improvement ceases or even reverses. The campaign is eventually recognized as a hoax.
   • Management needs to learn that the main responsibility is theirs from now on to improve the system.

11. *Eliminate work standards (quotas) on the factory floor. Substitute leadership. Eliminate management by objective. Eliminate management by numbers, numerical goals.*

   • The job of management is to replace work standards by knowledgeable and intelligent leadership. Leaders must have some understanding of the job. Whenever work standards have been thrown out and replaced by leadership, quality and productivity have gone up substantially, and people are happier.
   • If you have a stable system, then there is no use to specify a goal. You will get whatever the system will deliver. A goal beyond the capability of the system will not be reached.
   • If you have an unstable system, then there is again no point in setting a goal. There is no way to know what the system will produce: it has no capability.
   • To manage, one must lead. To lead, one must understand the work that he and his people are responsible for. Who is the customer and how can we better serve the customer?

- An incoming manager, to lead, and to manage at the source of improvement, must learn. He must learn from his people what they are doing and must learn a lot of new subject matter. It is easier for an incoming manager to short-circuit his need for learning and his responsibilities, and instead to focus on the far end, to manage the outcome - get reports on quality, on failures, proportion defective, inventory, sales , people.
- Focusing on the outcome is not an effective way to improve a process or an activity.
- The job of supervision is to help workers perform better quality work and increase productivity on a continuing basis. To do this they have to have an understanding of the work and appreciation of the capability of the processes.
- When stability is reached with the processes in statistical control, management must continually strive to improve the processes to improve performance.

12. *Remove barriers that rob the hourly worker of his right to pride of workmanship. The responsibility of supervisors must be changed from sheer numbers to quality. Remove barriers that rob people in management and in engineering of their right to pride of workmanship. This means abolishment of the annual or merit rating and of management by objective.*

- Workers want to be proud of their work, but barriers exist that prevent them from producing the kind of work in which they take pride. Workers have a right to be proud of their work, and once these obstacles are removed, they begin to blossom in their work endeavors.
- These barriers must be removed from two groups of people. One group is management or people on salary. The barrier is annual rating of performance, or merit rating. The other group is hourly worker.
- The production worker in America is under handicaps that are taking a terrific toll in quality, productivity, and competitive position. Barriers and handicaps rob the hourly worker of his birthright, the right to be proud of his work, the right to do a good job. These barriers exist in almost every plan, factory, company, department store, and government office in the United States today.
- How can anyone on the factory floor take pride in his work when he is not sure what is acceptable workmanship, and what is not, and cannot find out? Right yesterday but wrong today. What is my job?
- For management to remove the barriers, they must get involved.

13. *Institute a vigorous program of education and self-improvement.*

- What an organization needs is not just good people, it needs people that are improving with education.
- With the coming changes in technology, the numbers of people needed for work will shift. Fewer people will be needed in some lines of work, while more will be needed in others.
- Present day adults made up 90% of the working population in 1990 and will still comprise 75% of the working population by the year 2000. Jobs will change dramatically in those 10 years, creating a need for education to keep workers on the job.

- In addition to providing workers with skills, it is also management's responsibility to educate workers so they are more flexible and adaptable and able to make contributions to the company.
- It takes management structure with vision for the long-range profitability of the company to commit to this kind of education.

14. *Put everyone in the company to work to accomplish the transformation. The transformation is everybody's job.*

- Management will require guidance from an experienced consultant, but the responsibility for promoting quality improvement on a continuous basis cannot be delegated to the consultant or other s further down the organization. The consultant can provide training on statistical process control and train teachers. The consultant cannot be a substitute for management in carrying out the transformation defined in these 14 Deming principles.
- Top management needs to seek obstacles to the 14 principles and remove them as they continually work to operate their organizations consistently with Deming's principles.
- The entire company must work to internalize Deming's philosophy into the daily operations of the company.

THE SEVEN DEADLY DISEASES

The seven deadly diseases are roadblocks to the transformation of the western style of management. These diseases are listed below.

1. Lack of constancy of purpose to plan product and service that will have a market and keep the company in business and provide jobs.

2. Emphasis on short-term profits: short-term thinking (just the opposite from constancy of purpose to stay in business), fed by fear of unfriendly takeover, and by push from bankers and owners for dividends.

3. Evaluation of performance, merit rating or annual review.

4. Mobility of management; job hopping.

5. Management by use only of visible figures, with little or no consideration of figures that are unknown or unknowable.

6. Excessive medical costs.

7. Excessive costs of liability, swelled by lawyers that work on contingency fees.

SUMMARY

This chapter has introduced the Deming philosophy. This philosophy includes fourteen points that tell management what their job is in improving quality and productivity. This module also introduced the seven deadly diseases. These diseases are road blocks to improving quality.

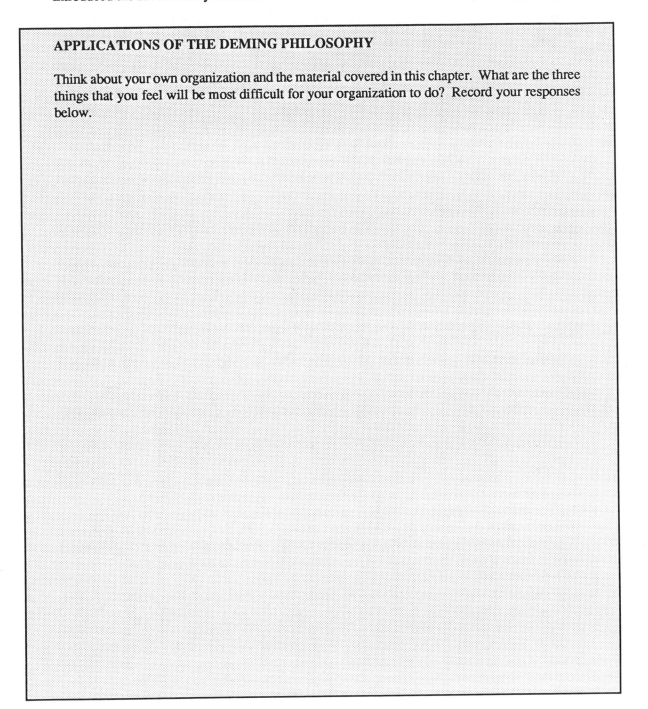

APPLICATIONS OF THE DEMING PHILOSOPHY

Think about your own organization and the material covered in this chapter. What are the three things that you feel will be most difficult for your organization to do? Record your responses below.

Appendix

TABLES

Table 1
CONTROL LIMITS SUMMARY FOR COMMONLY USED CONTROL CHARTS

Attributes Control Charts

p Charts: $\quad UCLp = \bar{p} + 3\sqrt{\bar{p}(1 - \bar{p})/\bar{n}} \qquad\qquad LCLp = \bar{p} - 3\sqrt{\bar{p}(1 - \bar{p})/\bar{n}}$

np Charts: $\quad UCLnp = n\bar{p} + 3\sqrt{n\bar{p}(1 - (n\bar{p}/n))} \qquad LCLnp = n\bar{p} - 3\sqrt{n\bar{p}(1 - (n\bar{p}/n))}$

c Charts: $\quad UCLc = \bar{c} + 3\sqrt{\bar{c}} \qquad\qquad\qquad LCLc = \bar{c} - 3\sqrt{\bar{c}}$

u Charts: $\quad UCLu = \bar{u} + 3\sqrt{\bar{u}/n} \qquad\qquad\quad LCLu = \bar{u} - 3\sqrt{\bar{u}/n}$

Variables Control Charts

$\overline{X}$-R Charts:
$$UCLx = \bar{\bar{X}} + A_2\bar{R} \qquad\qquad LCLx = \bar{\bar{X}} - A_2\bar{R}$$
$$UCLr = D_4\bar{R} \qquad\qquad\qquad LCLr = D_3\bar{R} \qquad\qquad \hat{\sigma}' = \bar{R}/d_2$$

X-R Charts:
$$UCLx = \bar{X} + 2.66\bar{R} \qquad\qquad LCLx = \bar{X} - 2.66\bar{R}$$
$$UCLr = 3.27R \qquad\qquad\qquad LCLr = None \qquad\qquad \hat{\sigma}' = \bar{R}/1.128$$

MA/MR
$$UCLx = \bar{\bar{X}} + A_2\bar{R} \qquad\qquad LCLx = \bar{\bar{X}} - A_2\bar{R}$$
$$UCLr = D_4\bar{R} \qquad\qquad\qquad LCLr = D_3\bar{R} \qquad\qquad \hat{\sigma}' = \bar{R}/d_2$$

$\overline{X}$-s Charts
$$UCLx = \bar{\bar{X}} + A_3\bar{s} \qquad\qquad LCLx = \bar{\bar{X}} - A_3\bar{s}$$
$$UCLs = B_4\bar{s} \qquad\qquad\qquad LCLs = B_3\bar{s} \qquad\qquad \hat{\sigma}' = \bar{R}/c_4$$

Table 2
FACTORS FOR USE WITH $\overline{X}$-R CHARTS

Subgroup Size (n)	A_2	D_3	D_4	d_2
2	1.880		3.267	1.128
3	1.023		2.574	1.693
4	0.729		2.282	2.059
5	0.577		2.114	2.326
6	0.483		2.004	2.534
7	0.419	0.076	1.924	2.704
8	0.373	0.136	1.864	2.847
9	0.337	0.184	1.816	2.970
10	0.308	0.223	1.777	3.078
11	0.285	0.256	1.774	3.173
12	0.266	0.284	1.716	3.258
13	0.249	0.308	1.692	3.336
14	0.235	0.329	1.671	3.407
15	0.223	0.348	1.652	3.472
16	0.212	0.364	1.636	3.532
17	0.203	0.379	1.621	3.588
18	0.194	0.392	1.608	3.640
19	0.187	0.404	1.596	3.689
20	0.180	0.414	1.586	3.735
21	0.173	0.425	1.575	3.778
22	0.167	0.434	1.566	3.819
23	0.162	0.443	1.557	3.858
24	0.157	0.452	1.548	3.895
25	0.153	0.459	1.541	3.931

Appendix

Table 3
FACTORS FOR USE WITH $\overline{X}$-s CHARTS

Subgroup Size (n)	A_3	B_3	B_4	c_4
2	2.659		3.267	0.7979
3	1.954		2.568	0.8862
4	1.628		2.266	0.9213
5	1.427		2.089	0.9400
6	1.287	0.030	1.970	0.9515
7	1.182	0.118	1.882	0.9594
8	1.099	0.185	1.815	0.9650
9	1.032	0.239	1.761	0.9693
10	0.975	0.284	1.716	0.9727
11	0.927	0.321	1.679	0.9754
12	0.886	0.354	1.646	0.9776
13	0.850	0.382	1.618	0.9794
14	0.817	0.406	1.594	0.9810
15	0.789	0.428	1.572	0.9823
16	0.763	0.448	1.552	0.9835
17	0.739	0.466	1.534	0.9845
18	0.718	0.482	1.518	0.9854
19	0.698	0.497	1.503	0.9862
20	0.680	0.510	1.490	0.9869
21	0.663	0.523	1.477	0.9876
22	0.647	0.534	1.466	0.9882
23	0.633	0.545	1.455	0.9887
24	0.619	0.555	1.445	0.9892
25	0.606	0.565	1.435	0.9896

Table 4
TESTS FOR PROCESS STABILITY

1. Points beyond the control limits
2. Too long a run length (see Table 5)
3. Too many runs or too few runs (see Table 6)
4. 2 out of 3 consecutive points in Zone A or beyond
5. 4 out of 5 consecutive points in Zone B or beyond
6. 8 consecutive points in Zone C or beyond
7. 15 consecutive points in Zone C on both sides of the aveerage
8. 8 consecutive points on both sides of the average with none in Zone C

Table 5
LENGTH-OF-RUN TEST

Number of Total Points on Chart	Critical Value (.05 Probability)	Critical Value (.01 Probability)
20 - 29	7	8
30 - 39	8	9
40 - 49	9	10
50 or over	10	11

Runs that are larger than the critical values listed above represent an out-of-control situation.

Table 6
NUMBER-OF-RUNS TEST

Total Number of Points on Chart	Expected Number of Runs	Small Critical Value	Large Critical Value
10 - 11	5	2	9
12 - 13	6	2	11
14 - 15	7	3	12
16 - 17	8	4	13
18 - 19	9	4	15
20 - 21	10	5	16
22 - 23	11	6	17
24 - 25	12	7	18
26 - 27	13	7	20
28 - 29	14	8	21
30 - 31	15	9	22
32 - 33	16	10	23
34 - 35	17	10	25
36 - 37	18	11	26
38 - 39	19	12	27
40 - 41	20	13	28
42 - 43	21	14	29
44 - 45	22	14	31
46 - 47	23	15	32
48 - 49	24	16	33
50 - 59	25	17	34
60 - 69	30	21	40
70 - 79	35	25	46
80 - 89	40	30	51
90 - 99	45	34	57
100 - 109	50	38	63
110 - 119	55	43	68
120 - 129	60	47	74

If the number of runs is smaller than or equal to the small critical value or larger than or equal to the large critical value, there is evidence of an assignable cause. The critical values are based on 0.01 probability. The number of runs is the number of times the centerline is crossed plus one.

Table 7
z TABLE: STANDARD NORMAL DISTRIBUTION

This table gives the proportion of process output that is beyond some value, x, that is z standard deviations away from the process average. The process must be normally distributed and in statistical control.

$|z|$	x.x0	x.x1	x.x2	x.x3	x.x4	x.x5	x.x6	x.x7	x.x8	x.x9
3.0	.00135									
2.9	.0019	.0018	.0018	.0017	.0016	.0016	.0015	.0015	.0014	.0014
2.8	.0026	.0025	.0024	.0023	.0023	.0022	.0021	.0021	.0020	.0019
2.7	.0035	.0034	.0033	.0032	.0031	.0030	.0029	.0028	.0027	.0026
2.6	.0047	.0045	.0044	.0043	.0041	.0040	.0039	.0038	.0037	.0036
2.5	.0062	.0060	.0059	.0057	.0055	.0054	.0052	.0051	.0049	.0048
2.4	.0082	.0080	.0078	.0075	.0073	.0071	.0069	.0068	.0066	.0064
2.3	.0107	.0104	.0102	.0099	.0096	.0094	.0091	.0089	.0087	.0084
2.2	.0139	.0136	.0132	.0129	.0125	.0122	.0119	.0116	.0113	.0110
2.1	.0179	.0174	.0170	.0166	.0162	.0158	.0154	.0150	.0146	.0143
2.0	.0228	.0222	.0217	.0212	.0207	.0202	.0197	.0192	.0188	.0183
1.9	.0287	.0281	.0274	.0268	.0262	.0256	.0250	.0244	.0239	.0233
1.8	.0359	.0351	.0344	.0336	.0329	.0322	.0314	.0307	.0301	.0294
1.7	.0446	.0436	.0427	.0418	.0409	.0401	.0392	.0384	.0375	.0367
1.6	.0548	.0537	.0526	.0516	.0505	.0495	.0485	.0475	.0465	.0455
1.5	.0668	.0655	.0643	.0630	.0618	.0606	.0594	.0582	.0571	.0559
1.4	.0808	.0793	.0778	.0764	.0749	.0735	.0721	.0708	.0694	.0681
1.3	.0968	.0951	.0934	.0918	.0901	.0885	.0869	.0853	.0838	.0823
1.2	.1151	.1131	.1112	.1093	.1075	.1056	.1038	.1020	.1003	.0985
1.1	.1357	.1335	.1314	.1292	.1271	.1251	.1230	.1210	.1190	.1170
1.0	.1587	.1562	.1539	.1515	.1492	.1469	.1446	.1423	.1401	.1379
0.9	.1841	.1814	.1788	.1762	.1736	.1711	.1685	.1660	.1635	.1611
0.8	.2119	.2090	.2061	.2033	.2005	.1977	.1949	.1922	.1894	.1867
0.7	.2420	.2389	.2358	.2327	.2297	.2266	.2236	.2206	.2177	.2148
0.6	.2743	.2709	.2676	.2643	.2611	.2578	.2546	.2514	.2483	.2451
0.5	.3085	.3050	.3015	.2981	.2946	.2912	.2877	.2843	.2810	.2776
0.4	.3446	.3409	.3372	.3336	.3300	.3264	.3228	.3192	.3156	.3121
0.3	.3821	.3783	.3745	.3707	.3669	.3632	.3594	.3557	.3520	.3483
0.2	.4207	.4168	.4129	.4090	.4052	.4013	.3974	.3936	.3897	.3859
0.1	.4602	.4562	.4522	.4483	.4443	.4404	.4364	.4325	.4286	.4247
0.0	.5000	.4960	.4920	.4880	.4840	.4801	.4761	.4721	.4681	.4641

Table 8
t DISTRIBUTION

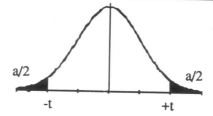

a/2   a/2

-t   +t

Probability = Area in the two tails of distribution outside ±t value in table

Probability

ν	0.8	0.5	0.2	0.1	0.05	0.02	0.01	0.005	0.002	0.001
1	0.325	1.000	3.078	6.314	12.706	31.821	63.657	127.32	318.31	636.62
2	0.289	0.816	1.886	2.920	4.303	6.965	9.925	14.089	22.326	31.598
3	0.277	0.765	1.638	2.353	3.182	4.541	5.841	7.453	10.213	12.924
4	0.271	0.741	1.533	2.132	2.776	3.747	4.604	5.598	7.173	8.610
5	0.267	0.727	1.476	2.015	2.571	3.365	4.032	4.773	5.893	6.869
6	0.265	0.718	1.440	1.943	2.447	3.143	3.707	4.317	5.208	5.959
7	0.263	0.711	1.415	1.895	2.365	2.998	3.499	4.029	4.785	5.408
8	0.262	0.706	1.367	1.860	2.306	2.896	3.355	3.833	4.501	5.041
9	0.261	0.703	1.383	1.833	2.262	2.821	3.250	3.690	4.297	4.781
10	0.260	0.700	1.372	1.812	2.228	2.764	3.169	3.581	4.144	4.587
11	0.260	0.697	1.363	1.796	2.201	2.718	3.106	3.497	4.025	4.437
12	0.259	0.695	1.356	1.782	2.179	2.681	3.055	3.428	3.930	4.318
13	0.259	0.694	1.350	1.771	2.160	2.650	3.012	3.372	3.852	4.221
14	0.258	0.692	1.345	1.761	2.145	2.624	2.977	3.326	3.787	4.140
15	0.258	0.691	1.341	1.753	2.131	2.602	2.947	3.286	3.733	4.073
16	0.258	0.690	1.337	1.746	2.120	2.583	2.921	3.252	3.686	4.015
17	0.257	0.689	1.333	1.740	2.110	2.567	2.898	3.222	3.646	3.965
18	0.257	0.688	1.330	1.734	2.101	2.552	2.878	3.197	3.610	3.922
19	0.257	0.688	1.328	1.729	2.093	2.539	2.861	3.174	3.579	3.883
20	0.257	0.687	1.325	1.725	2.086	2.528	2.845	3.153	3.552	3.850
21	0.257	0.686	1.323	1.721	2.080	2.518	2.831	3.135	3.527	3.819
22	0.256	0.686	1.321	1.717	2.074	2.508	2.819	3.119	3.505	3.792
23	0.256	0.685	1.319	1.714	2.069	2.500	2.807	3.104	3.485	3.767
24	0.256	0.685	1.318	1.711	2.064	2.492	2.797	3.091	3.467	3.745
25	0.256	0.684	1.316	1.708	2.060	2.485	2.787	3.078	3.450	3.725
26	0.256	0.684	1.315	1.706	2.056	2.479	2.779	3.067	3.435	3.707
27	0.256	0.684	1.314	1.703	2.052	2.473	2.771	3.057	3.421	3.690
28	0.256	0.683	1.313	1.701	2.048	2.467	2.763	3.047	3.408	3.674
29	0.256	0.683	1.311	1.699	2.045	2.462	2.756	3.038	3.396	3.659
30	0.256	0.683	1.310	1.687	2.042	2.457	2.750	3.030	3.385	3.646
40	0.255	0.681	1.303	1.684	2.021	2.423	2.704	2.971	3.307	3.551
60	0.254	0.679	1.296	1.671	2.000	2.390	2.660	2.915	3.232	3.460
120	0.254	0.677	1.289	1.658	1.980	2.358	2.617	2.860	3.160	3.373
∞	0.253	0.674	1.282	1.645	1.960	2.326	2.576	2.807	3.090	3.291

Table 9
THE CHI-SQUARE DISTRIBUTION

	Probability						
ν	0.005	0.01	0.025	0.05	0.1	0.25	0.5
1	3.92E-05	1.57E-04	9.82E-04	3.93E-03	0.016	0.102	0.455
2	0.010	0.020	0.051	0.103	0.211	0.575	1.386
3	0.072	0.115	0.216	0.352	0.584	1.213	2.366
4	0.207	0.297	0.484	0.711	1.064	1.923	3.357
5	0.412	0.554	0.831	1.145	1.610	2.675	4.351
6	0.676	0.872	1.237	1.635	2.204	3.455	5.348
7	0.989	1.239	1.699	2.167	2.833	4.255	6.346
8	1.344	1.646	2.180	2.733	3.490	5.071	7.344
9	1.735	2.088	2.700	3.325	4.168	5.899	8.343
10	2.156	2.558	3.247	3.940	4.865	6.737	9.342
11	2.603	3.053	3.816	4.575	5.578	7.584	10.341
12	3.074	3.571	4.404	5.226	6.304	8.438	11.340
13	3.565	4.107	5.009	5.892	7.042	9.299	12.340
14	4.075	4.660	5.629	6.571	7.790	10.165	13.339
15	4.601	5.229	6.262	7.261	8.547	11.037	14.339
16	5.142	5.812	6.908	7.962	9.312	11.912	15.339
17	5.697	6.408	7.564	8.672	10.085	12.792	16.338
18	6.265	7.015	8.231	9.390	10.865	13.675	17.338
19	6.844	7.633	8.907	10.117	11.651	14.562	18.338
20	7.434	8.260	9.591	10.851	12.443	15.452	19.337
21	8.034	8.897	10.283	11.501	13.240	16.344	20.337
22	8.643	9.542	10.982	12.338	14.042	17.240	21.337
23	9.260	10.196	11.689	13.091	14.848	18.137	22.337
24	9.886	10.856	12.401	13.848	15.659	19.037	23.337
25	10.520	11.524	13.120	14.611	16.473	19.939	24.337
26	11.160	12.198	13.844	15.379	17.292	20.843	25.336
27	11.808	12.879	14.573	16.151	18.114	21.749	26.336
28	12.461	13.565	15.308	16.928	18.939	22.657	27.336
29	13.121	14.257	16.047	17.708	19.768	23.567	28.336
30	13.787	14.954	16.791	18.493	20.599	24.478	29.336
40	20.707	22.164	24.433	26.509	29.051	33.660	39.335
50	27.991	29.707	32.357	34.764	37.689	42.942	49.335
60	35.535	37.485	40.482	43.188	46.459	52.294	59.335
70	43.275	45.442	48.758	51.739	55.329	61.698	69.334
80	51.172	53.540	57.153	60.392	64.278	71.145	79.334
90	59.196	61.754	65.647	69.126	73.291	80.625	89.334
100	57.328	70.065	74.222	77.930	82.358	90.133	99.334

Table 9
THE CHI-SQUARE DISTRIBUTION (continued)

	Probability						
ν	0.75	0.9	0.95	0.975	0.99	0.995	0.999
1	1.323	2.706	3.841	5.024	6.635	7.879	10.828
2	2.773	4.605	5.991	7.378	9.210	10.597	13.816
3	4.108	6.251	7.815	9.348	11.345	12.838	16.266
4	5.385	7.779	9.488	11.143	13.277	14.860	18.467
5	6.626	9.236	11.071	12.833	15.086	16.750	20.515
6	7.841	10.645	12.592	14.449	16.812	18.548	22.458
7	9.037	12.017	14.067	16.013	18.475	20.278	24.322
8	10.219	13.362	15.507	17.535	20.090	21.955	26.125
9	11.389	14.684	16.919	19.023	21.666	23.589	27.877
10	12.549	15.987	18.307	20.483	23.209	25.188	29.588
11	13.701	17.275	19.675	21.920	24.725	26.757	31.264
12	14.845	18.549	21.026	23.337	26.217	28.300	32.909
13	15.984	19.812	22.362	24.736	27.688	29.819	34.528
14	17.117	21.064	23.685	26.119	29.141	31.319	36.123
15	18.245	22.307	24.996	27.488	30.578	32.801	37.697
16	19.369	23.542	26.296	28.845	32.000	34.267	39.252
17	20.489	24.769	27.587	30.191	33.409	35.719	40.790
18	21.605	25.989	28.869	31.526	34.805	37.156	42.312
19	22.718	27.204	30.144	32.852	36.191	38.582	43.820
20	23.828	28.412	31.410	34.170	37.566	39.997	45.315
21	24.935	29.615	32.671	35.479	38.932	41.401	46.797
22	26.039	30.813	33.924	36.781	40.289	42.796	48.268
23	27.141	32.007	35.173	38.076	41.638	44.181	49.728
24	28.241	33.196	36.415	39.364	42.980	45.559	51.179
25	29.339	34.382	37.653	40.647	44.314	46.928	52.620
26	30.435	35.563	38.885	41.923	45.642	48.290	54.052
27	31.528	36.741	40.113	43.194	46.963	49.645	55.476
28	32.621	37.916	41.337	44.461	48.278	50.993	56.892
29	33.711	39.088	42.557	45.722	49.588	52.336	58.302
30	34.800	40.256	43.773	46.979	50.892	53.672	59.703
40	45.616	51.805	55.759	59.342	63.691	66.766	73.402
50	56.336	63.167	67.505	71.420	76.154	79.490	86.661
60	66.981	74.397	79.082	83.298	88.379	91.952	99.607
70	77.577	85.527	90.531	95.023	100.425	104.215	112.317
80	88.130	96.578	101.879	106.629	112.329	116.321	124.839
90	98.650	107.565	113.145	118.136	124.116	128.299	137.208
100	109.141	118.498	124.342	129.561	135.807	140.169	149.449

Table 10
VALUES OF THE F DISTRIBUTION FOR $\alpha = 0.05$

v_2 \ v_1	1	2	3	4	5	6	7	8	9	10	12	15	20	24	30	40	60	120	∞
1	161.4	199.5	215.7	224.6	230.2	234.0	236.8	289.9	240.5	241.9	243.9	245.9	248.0	249.1	250.1	251.1	252.2	253.3	254.3
2	18.51	19.00	19.16	19.25	19.30	19.33	19.35	19.37	19.38	19.40	19.41	19.43	19.45	19.45	19.46	19.47	19.48	19.49	19.50
3	10.13	9.55	9.28	9.12	9.01	8.94	8.89	8.85	8.81	8.79	8.74	8.70	8.66	8.64	8.62	8.59	8.57	8.55	8.53
4	7.71	6.94	6.59	6.39	6.26	6.16	6.09	6.04	6.00	5.96	5.91	5.86	5.80	5.77	5.75	5.72	5.69	5.66	5.63
5	6.61	5.79	5.41	5.19	5.05	4.95	4.88	4.82	4.77	4.74	4.68	4.62	4.56	4.53	4.50	4.46	4.43	4.44	4.36
6	5.99	5.14	4.76	4.53	4.39	4.28	4.21	4.15	4.10	4.06	4.00	3.94	3.87	3.84	3.81	3.77	3.74	3.70	3.67
7	5.59	4.74	4.35	4.12	3.97	3.87	3.79	3.73	3.68	3.64	3.57	3.51	3.44	3.41	3.38	3.34	3.30	3.27	3.23
8	5.32	4.46	4.07	3.84	3.69	3.58	3.50	3.44	3.39	3.35	3.28	3.22	3.15	3.12	3.08	3.04	3.01	2.97	2.93
9	5.12	4.26	3.86	3.63	3.48	3.37	3.29	3.23	3.18	3.14	3.07	3.01	2.94	2.90	2.86	2.83	2.79	2.75	2.71
10	4.96	4.10	3.71	3.48	3.33	3.22	3.14	3.07	3.02	2.98	2.91	2.85	2.77	2.74	2.70	2.66	2.62	2.58	2.54
11	4.84	3.98	3.59	3.36	3.20	3.09	3.01	2.95	2.90	2.85	2.79	2.72	2.65	2.61	2.57	2.53	2.49	2.45	2.40
12	4.75	3.89	3.49	3.26	3.11	3.00	2.91	2.85	2.80	2.75	2.69	2.62	2.54	2.51	2.47	2.43	2.38	2.34	2.30
13	4.67	3.81	3.41	3.18	3.03	2.92	2.83	2.77	2.71	2.67	2.60	2.53	2.46	2.42	2.38	2.34	2.30	2.25	2.21
14	4.60	3.74	3.34	3.11	2.96	2.85	2.76	2.70	2.65	2.60	2.53	2.46	2.39	2.35	2.31	2.27	2.22	2.18	2.13
15	4.54	3.68	3.29	3.06	2.90	2.79	2.71	2.64	2.59	2.54	2.48	2.40	2.33	2.29	2.25	2.20	2.16	2.11	2.07
16	4.49	3.63	3.24	3.01	2.85	2.74	2.66	2.59	2.54	2.49	2.42	2.35	2.28	2.24	2.19	2.15	2.11	2.06	2.01
17	4.45	3.59	3.20	2.96	2.81	2.70	2.61	2.55	2.49	2.45	2.38	2.31	2.23	2.19	2.15	2.10	2.06	2.01	1.96
18	4.41	3.55	3.16	2.93	2.77	2.66	2.58	2.51	2.46	2.41	2.34	2.27	2.19	2.15	2.11	2.06	2.02	1.97	1.92
19	4.38	3.52	3.13	2.90	2.74	2.63	2.54	2.48	2.42	2.38	2.31	2.23	2.16	2.11	2.07	2.03	1.98	1.93	1.88
20	4.35	3.49	3.10	2.87	2.71	2.60	2.51	2.45	2.39	2.35	2.28	2.20	2.12	2.08	2.04	1.99	1.95	1.90	1.84
21	4.32	3.47	3.07	2.84	2.68	2.57	2.49	2.42	2.37	2.32	2.25	2.18	2.10	2.05	2.01	1.96	1.92	1.87	1.81
22	4.30	3.44	3.05	2.82	2.66	2.55	2.46	2.40	2.34	2.30	2.23	2.15	2.07	2.03	1.98	1.94	1.89	1.84	1.78
23	4.28	3.42	3.03	2.80	2.64	2.53	2.44	2.37	2.32	2.27	2.20	2.13	2.05	2.01	1.96	1.91	1.86	1.81	1.76
24	4.26	3.40	3.01	2.78	2.62	2.51	2.42	2.36	2.30	2.25	2.18	2.11	2.03	1.98	1.94	1.89	1.84	1.79	1.73
25	4.24	3.39	2.99	2.76	2.60	2.49	2.40	2.34	2.28	2.24	2.16	2.09	2.01	1.96	1.92	1.87	1.82	1.77	1.71
26	4.23	3.37	2.98	2.74	2.59	2.47	2.39	2.32	2.27	2.22	2.15	2.07	1.99	1.95	1.90	1.85	1.80	1.75	1.69
27	4.21	3.35	2.96	2.73	2.57	2.46	2.37	2.31	2.25	2.20	2.13	2.06	1.97	1.93	1.88	1.84	1.79	1.73	1.67
28	4.20	3.34	2.95	2.71	2.56	2.45	2.36	2.29	2.24	2.19	2.12	2.04	1.96	1.91	1.87	1.82	1.77	1.71	1.65
29	4.18	3.33	2.93	2.70	2.55	2.43	2.35	2.28	2.22	2.18	2.10	2.03	1.94	1.90	1.85	1.81	1.75	1.70	1.64
30	4.17	3.32	2.92	2.69	2.53	2.42	2.33	2.27	2.21	2.16	2.09	2.01	1.93	0.89	1.84	1.79	1.74	1.68	1.62
40	4.08	3.23	2.84	2.61	2.45	2.34	2.25	2.18	2.12	2.08	2.00	1.92	1.84	1.79	1.74	1.69	1.64	1.58	1.51
60	4.00	3.15	2.76	2.53	2.37	2.25	2.17	2.10	2.04	1.99	1.92	1.84	1.75	1.70	1.65	1.59	1.53	1.47	1.39
120	3.92	3.07	2.68	2.45	2.29	2.17	2.09	2.02	1.96	1.91	1.83	1.75	1.66	1.61	1.55	1.50	1.43	1.35	1.25
∞	3.84	3.00	2.60	2.37	2.21	2.10	2.01	1.94	1.88	1.83	1.75	1.67	1.57	1.52	1.46	1.39	1.32	1.22	1.00

v_1 = degrees of freedom in the numerator
v_2 = degrees of freedom in the denominator

Table 11
CRITICAL VALUES FOR THE LINEAR CORRELATION COEFFICIENT

Degrees of Freedom	Probability	Degrees of Freedom	Probability
1	.997	21	.413
2	.950	22	.404
3	.878	23	.396
4	.811	24	.388
5	.754	25	.381
6	.707	26	.374
7	.666	27	.367
8	.632	28	.361
9	.602	29	.355
10	.576	30	.349
11	.553	35	.325
12	.532	40	.304
13	.514	45	.288
14	.497	50	.273
15	.482	60	.250
16	.468	70	.232
17	.456	80	.217
18	.444	90	.205
19	.433	100	.195
20	.423		

The probabilities listed are for $\alpha = 0.05$. If a calculated value of R_{yx} is larger or equal to the table value, there is a statistically significant correlation between x and y.

Table 12
CONTROL LIMITS FOR SMALL-SAMPLE CASE: np CHARTS

$\bar{p}$ \ n	5 LCL	5 UCL	6 LCL	6 UCL	7 LCL	7 UCL	8 LCL	8 UCL	9 LCL	9 UCL	10 LCL	10 UCL
0.001	-	2	-	2	-	2	-	2	-	2	-	2
0.002	-	2	-	2	-	2	-	2	-	2	-	2
0.003	-	2	-	2	-	2	-	2	-	2	-	2
0.004	-	2	-	2	-	2	-	2	-	2	-	2
0.005	-	2	-	2	-	2	-	2	-	2	-	2
0.010	-	2	-	2	-	2	-	2	-	3	-	3
0.020	-	3	-	3	-	3	-	3	-	3	-	3
0.030	-	3	-	3	-	3	-	3	-	3	-	4
0.040	-	3	-	3	-	3	-	4	-	4	-	4
0.050	-	3	-	3	-	4	-	4	-	4	-	4
0.060	-	3	-	4	-	4	-	4	-	4	-	5
0.070	-	4	-	4	-	4	-	4	-	5	-	5
0.080	-	4	-	4	-	4	-	5	-	5	-	5
0.090	-	4	-	4	-	4	-	5	-	5	-	5
0.100	-	4	-	4	-	4	-	5	-	5	-	5
0.110	-	4	-	4	-	5	-	5	-	5	-	5
0.120	-	4	-	4	-	5	-	5	-	5	-	6
0.130	-	4	-	5	-	5	-	5	-	5	-	6
0.140	-	4	-	5	-	5	-	5	-	6	-	6
0.150	-	4	-	5	-	5	-	5	-	6	-	6
0.160	-	4	-	5	-	5	-	6	-	6	-	6
0.170	-	5	-	5	-	5	-	6	-	6	-	7
0.180	-	5	-	5	-	5	-	6	-	6	-	7
0.190	-	5	-	5	-	6	-	6	-	6	-	7
0.200	-	5	-	5	-	6	-	6	-	7	-	7
0.210	-	5	-	5	-	6	-	6	-	7	-	7
0.220	-	5	-	5	-	6	-	6	-	7	-	7
0.230	-	5	-	6	-	6	-	6	-	7	-	7
0.240	-	5	-	6	-	6	-	6	-	7	-	8
0.250	-	5	-	6	-	6	-	7	-	7	-	8
0.260	-	5	-	6	-	6	-	7	-	7	-	8
0.270	-	5	-	6	-	6	-	7	-	7	-	8
0.280	-	5	-	6	-	6	-	7	-	8	-	8
0.290	-	5	-	6	-	7	-	7	-	8	-	8
0.300	-	5	-	6	-	7	-	7	-	8	-	8
0.310	-	5	-	6	-	7	-	7	-	8	-	8
0.320	-	6	-	6	-	7	-	7	-	8	-	9
0.330	-	6	-	6	-	7	-	7	-	8	-	9
0.340	-	6	-	6	-	7	-	7	-	8	-	9
0.350	-	6	-	6	-	7	-	8	-	8	-	9
0.360	-	6	-	6	-	7	-	8	-	8	-	9
0.370	-	6	-	6	-	7	-	8	-	8	-	9
0.380	-	6	-	7	-	7	-	8	-	9	-	9
0.390	-	6	-	7	-	7	-	8	-	9	-	9
0.400	-	6	-	7	-	7	-	8	-	9	-	9
0.410	-	6	-	7	-	7	-	8	-	9	-	9
0.420	-	6	-	7	-	7	-	8	-	9	-	9
0.430	-	6	-	7	-	7	-	8	-	9	-	10
0.440	-	6	-	7	-	8	-	8	-	9	-	10
0.450	-	6	-	7	-	8	-	8	-	9	-	10
0.460	-	6	-	7	-	8	-	8	-	9	0	10
0.470	-	6	-	7	-	8	-	8	-	9	0	10
0.480	-	6	-	7	-	8	-	8	-	9	0	10
0.490	-	6	-	7	-	8	-	9	-	9	0	10
0.500	-	6	-	7	-	8	-	9	-	9	0	10

Table 12
CONTROL LIMITS FOR SMALL-SAMPLE CASE: np CHARTS (continued)

n	11		12		13		14		15		16	
$\bar{p}$	LCL	UCL	LCL	UCL	LCL	UCL	LCL	UCL	LCL	UCL	LCL	UCL
0.001	-	2	-	2	-	2	-	2	-	2	-	2
0.002	-	2	-	2	-	2	-	2	-	2	-	2
0.003	-	2	-	2	-	2	-	2	-	2	-	2
0.004	-	2	-	2	-	2	-	2	-	2	-	2
0.005	-	2	-	2	-	2	-	2	-	2	-	2
0.010	-	3	-	3	-	3	-	3	-	3	-	3
0.020	-	3	-	3	-	3	-	3	-	4	-	4
0.030	-	4	-	4	-	4	-	4	-	4	-	4
0.040	-	4	-	4	-	4	-	4	-	4	-	5
0.050	-	4	-	4	-	5	-	5	-	5	-	5
0.060	-	5	-	5	-	5	-	5	-	5	-	5
0.070	-	5	-	5	-	5	-	5	-	5	-	6
0.080	-	5	-	5	-	5	-	6	-	6	-	6
0.090	-	5	-	5	-	6	-	6	-	6	-	6
0.100	-	5	-	6	-	6	-	6	-	6	-	7
0.110	-	6	-	6	-	6	-	6	-	7	-	7
0.120	-	6	-	6	-	6	-	7	-	7	-	7
0.130	-	6	-	6	-	7	-	7	-	7	-	7
0.140	-	6	-	7	-	7	-	7	-	7	-	8
0.150	-	6	-	7	-	7	-	7	-	8	-	8
0.160	-	7	-	7	-	7	-	8	-	8	-	8
0.170	-	7	-	7	-	7	-	8	-	8	-	8
0.180	-	7	-	7	-	8	-	8	-	8	-	9
0.190	-	7	-	7	-	8	-	8	-	9	-	9
0.200	-	7	-	8	-	8	-	8	-	9	-	9
0.210	-	7	-	8	-	8	-	9	-	9	-	9
0.220	-	8	-	8	-	8	-	9	-	9	-	9
0.230	-	8	-	8	-	9	-	9	-	9	-	10
0.240	-	8	-	8	-	9	-	9	-	10	-	10
0.250	-	8	-	8	-	9	-	9	-	10	-	10
0.260	-	8	-	9	-	9	-	9	-	10	-	10
0.270	-	8	-	9	-	9	-	10	-	10	-	11
0.280	-	8	-	9	-	9	-	10	-	10	-	11
0.290	-	9	-	9	-	10	-	10	-	10	-	11
0.300	-	9	-	9	-	10	-	10	-	11	-	11
0.310	-	9	-	9	-	10	-	10	-	11	-	11
0.320	-	9	-	9	-	10	-	10	-	11	0	12
0.330	-	9	-	10	-	10	-	11	-	11	0	12
0.340	-	9	-	10	-	10	-	11	-	11	0	12
0.350	-	9	-	10	-	10	-	11	0	12	0	12
0.360	-	9	-	10	-	11	-	11	0	12	0	12
0.370	-	10	-	10	-	11	0	12	0	12	0	12
0.380	-	10	-	10	0	11	0	12	0	12	0	13
0.390	-	10	-	10	0	11	0	12	0	12	0	13
0.400	-	10	0	11	0	11	0	12	0	12	0	13
0.410	-	10	0	11	0	11	0	12	0	12	0	13
0.420	-	10	0	11	0	11	0	12	0	13	0	13
0.430	-	10	0	11	0	12	0	12	0	13	1	14
0.440	-	10	0	11	0	12	0	12	1	13	1	14
0.450	-	11	0	11	0	12	0	12	1	13	1	14
0.460	0	11	0	11	0	12	0	13	1	13	1	14
0.470	0	11	0	11	0	12	0	13	1	13	1	14
0.480	0	11	0	11	0	12	0	13	1	14	1	14
0.490	0	11	0	11	0	12	0	13	1	14	1	14
0.500	0	11	0	12	0	13	0	13	1	14	2	15

Table 12
CONTROL LIMITS FOR SMALL-SAMPLE CASE: np CHARTS (continued)

n	17		18		19		20		21		22	
p̄	LCL	UCL	LCL	UCL	LCL	UCL	LCL	UCL	LCL	UCL	LCL	UCL
0.001	-	2	-	2	-	2	-	2	-	2	-	2
0.002	-	2	-	2	-	2	-	2	-	2	-	2
0.003	-	2	-	2	-	2	-	2	-	3	-	3
0.004	-	2	-	2	-	2	-	2	-	3	-	3
0.005	-	3	-	3	-	3	-	3	-	3	-	3
0.010	-	3	-	3	-	3	-	3	-	3	-	3
0.020	-	4	-	4	-	4	-	4	-	4	-	4
0.030	-	4	-	4	-	4	-	4	-	5	-	5
0.040	-	5	-	5	-	5	-	5	-	5	-	5
0.050	-	5	-	5	-	5	-	5	-	6	-	6
0.060	-	5	-	6	-	6	-	6	-	6	-	6
0.070	-	6	-	6	-	6	-	6	-	6	-	7
0.080	-	6	-	6	-	6	-	7	-	7	-	7
0.090	-	6	-	7	-	7	-	7	-	7	-	7
0.100	-	7	-	7	-	7	-	7	-	8	-	8
0.110	-	7	-	7	-	7	-	8	-	8	-	8
0.120	-	7	-	8	-	8	-	8	-	8	-	8
0.130	-	8	-	8	-	8	-	8	-	9	-	9
0.140	-	8	-	8	-	8	-	9	-	9	-	9
0.150	-	8	-	8	-	9	-	9	-	9	-	9
0.160	-	8	-	9	-	9	-	9	-	10	-	10
0.170	-	9	-	9	-	9	-	10	-	10	-	10
0.180	-	9	-	9	-	10	-	10	-	10	-	10
0.190	-	9	-	9	-	10	-	10	-	10	-	11
0.200	-	9	-	10	-	10	-	10	-	11	-	11
0.210	-	10	-	10	-	10	-	11	-	11	-	11
0.220	-	10	-	10	-	11	-	11	-	11	-	12
0.230	-	10	-	10	-	11	-	11	-	12	-	12
0.240	-	10	-	11	-	11	-	11	-	12	-	12
0.250	-	11	-	11	-	11	-	12	-	12	0	13
0.260	-	11	-	11	-	12	-	12	0	13	0	13
0.270	-	11	-	11	-	12	-	12	0	13	0	13
0.280	-	11	-	12	-	12	0	13	0	13	0	13
0.290	-	11	-	12	0	13	0	13	0	13	0	14
0.300	0	12	0	13	0	13	0	13	0	13	0	14
0.310	0	12	0	13	0	13	0	13	0	14	0	14
0.320	0	12	0	13	0	13	0	13	0	14	0	14
0.330	0	12	0	13	0	13	0	14	0	14	1	15
0.340	0	13	0	13	0	13	0	14	0	14	1	15
0.350	0	13	0	14	0	14	0	14	1	15	1	15
0.360	0	13	0	14	0	14	1	15	1	15	1	15
0.370	0	13	0	14	0	14	1	15	1	15	1	16
0.380	0	13	1	15	1	15	1	15	1	15	1	16
0.390	0	13	1	15	1	15	1	15	1	16	1	16
0.400	1	14	1	15	1	15	1	15	1	16	2	17
0.410	1	14	1	15	1	15	1	15	2	16	2	17
0.420	1	14	1	15	1	15	2	16	2	16	2	17
0.430	1	14	1	15	1	15	2	16	2	17	2	17
0.440	1	14	1	15	1	15	2	16	2	17	2	17
0.450	1	14	1	16	2	16	2	16	2	17	3	18
0.460	1	14	1	16	2	16	2	16	2	17	3	18
0.470	1	15	2	16	2	16	2	17	2	17	3	18
0.480	2	15	2	16	2	16	3	17	3	18	3	18
0.490	2	15	2	16	2	16	3	17	3	18	3	18
0.500	2	15	2	17	3	17	3	17	3	18	3	18

Table 12
CONTROL LIMITS FOR SMALL-SAMPLE CASE: np CHARTS (continued)

n	23		24		25		26		27		28	
p̄	LCL	UCL	LCL	UCL	LCL	UCL	LCL	UCL	LCL	UCL	LCL	UCL
0.001	-	2	-	2	-	2	-	2	-	2	-	2
0.002	-	2	-	2	-	2	-	2	-	2	-	2
0.003	-	2	-	2	-	2	-	2	-	2	-	2
0.004	-	3	-	3	-	3	-	3	-	3	-	3
0.005	-	3	-	3	-	3	-	3	-	3	-	3
0.010	-	3	-	3	-	3	-	3	-	3	-	3
0.020	-	4	-	4	-	4	-	4	-	4	-	4
0.030	-	5	-	5	-	5	-	5	-	5	-	4
0.040	-	5	-	5	-	5	-	5	-	5	-	5
0.050	-	6	-	6	-	6	-	6	-	6	-	6
0.060	-	6	-	6	-	7	-	7	-	6	-	6
0.070	-	7	-	7	-	7	-	7	-	7	-	7
0.080	-	7	-	7	-	7	-	8	-	7	-	7
0.090	-	8	-	8	-	8	-	8	-	8	-	8
0.100	-	8	-	8	-	8	-	8	-	9	-	8
0.110	-	8	-	9	-	9	-	9	-	9	-	9
0.120	-	9	-	9	-	9	-	9	-	9	-	9
0.130	-	9	-	9	-	10	-	10	-	10	-	10
0.140	-	9	-	10	-	10	-	10	-	10	-	10
0.150	-	10	-	10	-	10	-	11	-	10	-	11
0.160	-	10	-	10	-	11	-	11	-	11	-	11
0.170	-	10	-	11	-	11	-	11	-	12	-	12
0.180	-	11	-	11	-	11	-	12	-	12	-	12
0.190	-	11	-	11	-	12	-	12	-	12	-	12
0.200	-	11	-	12	-	12	-	12	-	12	-	13
0.210	-	12	-	12	-	12	-	13	-	13	0	13
0.220	-	12	-	12	0	13	0	14	-	13	0	14
0.230	-	12	0	13	0	13	0	14	0	14	0	14
0.240	0	13	0	13	0	14	0	14	0	14	0	14
0.250	0	13	0	14	0	14	0	14	0	14	0	15
0.260	0	13	0	14	0	14	0	14	0	15	1	15
0.270	0	14	0	14	0	14	0	15	0	15	1	16
0.280	0	14	0	14	0	15	0	15	1	16	1	16
0.290	0	14	0	14	0	15	1	16	1	16	1	16
0.300	0	14	0	15	0	15	1	16	1	16	1	17
0.310	1	15	0	15	1	16	1	16	1	16	1	17
0.320	1	15	1	16	1	16	1	16	1	17	1	17
0.330	1	15	1	16	1	16	1	17	1	17	2	18
0.340	1	16	1	16	1	16	1	17	1	17	2	18
0.350	1	16	1	16	2	17	2	18	2	18	2	18
0.360	1	16	1	16	2	17	2	18	2	18	2	18
0.370	1	16	2	17	2	17	2	18	2	18	3	19
0.380	2	17	2	17	2	18	2	18	3	19	3	19
0.390	2	17	2	17	2	18	2	18	3	19	3	19
0.400	2	17	2	17	2	18	3	19	3	19	3	20
0.410	2	17	2	18	2	18	3	19	3	19	3	20
0.420	2	17	2	18	3	19	3	19	3	20	3	20
0.430	3	18	3	18	3	19	3	19	4	20	4	21
0.440	3	18	3	19	3	19	4	20	4	20	4	21
0.450	3	18	3	19	3	19	4	20	4	21	4	21
0.460	3	18	3	19	4	20	4	20	4	21	4	21
0.470	3	18	3	19	4	20	4	20	4	21	5	22
0.480	3	19	4	20	4	20	5	21	5	22	5	22
0.490	4	19	4	20	4	20	5	21	5	22	5	22
0.500	4	19	4	20	5	21	5	21	5	22	5	22

Table 12
CONTROL LIMITS FOR SMALL-SAMPLE CASE: np CHARTS (continued)

n	29		30		31		32		33		34	
p̄	LCL	UCL	LCL	UCL	LCL	UCL	LCL	UCL	LCL	UCL	LCL	UCL
0.001	-	2	-	2	-	2	-	2	-	2	-	2
0.002	-	2	-	2	-	2	-	2	-	2	-	2
0.003	-	3	-	3	-	3	-	3	-	3	-	3
0.004	-	3	-	3	-	3	-	3	-	3	-	3
0.005	-	3	-	3	-	3	-	3	-	3	-	3
0.010	-	4	-	4	-	4	-	4	-	4	-	4
0.020	-	4	-	4	-	5	-	5	-	5	-	5
0.030	-	5	-	5	-	5	-	5	-	5	-	6
0.040	-	6	-	6	-	6	-	6	-	6	-	6
0.050	-	6	-	7	-	7	-	7	-	7	-	7
0.060	-	7	-	7	-	7	-	7	-	7	-	8
0.070	-	8	-	8	-	8	-	8	-	8	-	8
0.080	-	8	-	8	-	8	-	9	-	9	-	9
0.090	-	9	-	9	-	9	-	9	-	9	-	9
0.100	-	9	-	9	-	9	-	10	-	10	-	10
0.110	-	10	-	10	-	10	-	10	-	10	-	10
0.120	-	10	-	10	-	10	-	11	-	11	-	11
0.130	-	10	-	11	-	11	-	11	-	11	-	11
0.140	-	11	-	11	-	11	-	12	-	12	-	12
0.150	-	11	-	11	-	12	-	12	-	12	-	12
0.160	-	12	-	12	-	12	-	12	-	13	-	13
0.170	-	12	-	12	-	13	-	13	-	13	0	14
0.180	-	12	-	13	-	13	-	13	0	14	0	14
0.190	-	13	-	13	0	14	0	14	0	14	0	15
0.200	0	14	0	14	0	14	0	14	0	15	0	15
0.210	0	14	0	14	0	14	0	15	0	15	0	15
0.220	0	14	0	14	0	15	0	15	0	15	0	16
0.230	0	15	0	15	0	15	0	15	0	16	1	16
0.240	0	15	0	15	0	15	1	16	1	17	1	17
0.250	0	15	1	16	1	16	1	16	1	17	1	17
0.260	1	16	1	16	1	17	1	17	1	17	1	17
0.270	1	16	1	16	1	17	1	17	1	17	1	18
0.280	1	16	1	17	1	17	1	17	1	18	2	19
0.290	1	17	1	17	1	17	2	18	2	19	2	19
0.300	1	17	1	17	2	18	2	18	2	19	2	19
0.310	1	17	2	18	2	18	2	19	2	19	2	19
0.320	2	18	2	18	2	19	2	19	2	19	3	20
0.330	2	18	2	18	2	19	2	19	3	20	3	20
0.340	2	18	2	19	2	19	3	20	3	20	3	21
0.350	2	19	2	19	3	20	3	20	3	21	3	21
0.360	2	19	3	20	3	20	3	20	3	21	4	22
0.370	3	20	3	20	3	20	3	21	4	22	4	22
0.380	3	20	3	20	4	21	4	21	4	22	4	22
0.390	3	20	3	20	4	21	4	22	4	22	4	22
0.400	3	20	4	21	4	21	4	22	4	22	5	23
0.410	4	21	4	21	4	22	4	22	5	23	5	23
0.420	4	21	4	21	4	22	5	23	5	23	5	23
0.430	4	21	5	22	5	22	5	23	5	23	6	24
0.440	4	21	5	22	5	23	5	23	6	24	6	24
0.450	5	22	5	22	5	23	6	24	6	24	6	25
0.460	5	22	5	22	5	23	6	24	6	24	7	25
0.470	5	22	6	23	6	24	6	24	7	25	7	25
0.480	5	22	6	23	6	24	6	24	7	25	7	26
0.490	6	23	6	23	6	24	7	25	7	25	7	26
0.500	6	23	6	24	7	25	7	25	7	25	8	26

501

Table 12
CONTROL LIMITS FOR SMALL-SAMPLE CASE: np CHARTS (continued)

n → p̄	35 LCL	35 UCL	36 LCL	36 UCL	37 LCL	37 UCL	38 LCL	38 UCL	39 LCL	39 UCL	40 LCL	40 UCL
0.001	-	2	-	2	-	2	-	2	-	2	-	2
0.002	-	2	-	2	-	2	-	2	-	2	-	2
0.003	-	3	-	3	-	3	-	3	-	3	-	3
0.004	-	3	-	3	-	3	-	3	-	3	-	3
0.005	-	3	-	3	-	3	-	3	-	3	-	3
0.010	-	4	-	4	-	4	-	4	-	4	-	4
0.020	-	5	-	5	-	5	-	5	-	5	-	5
0.030	-	6	-	6	-	6	-	6	-	6	-	6
0.040	-	6	-	6	-	7	-	7	-	7	-	7
0.050	-	7	-	7	-	7	-	7	-	7	-	8
0.060	-	8	-	8	-	8	-	8	-	8	-	8
0.070	-	8	-	8	-	9	-	9	-	9	-	9
0.080	-	9	-	9	-	9	-	9	-	10	-	10
0.090	-	10	-	10	-	10	-	10	-	10	-	10
0.100	-	10	-	10	-	10	-	11	-	11	-	11
0.110	-	11	-	11	-	11	-	11	-	11	-	12
0.120	-	11	-	11	-	12	-	12	-	12	-	12
0.130	-	12	-	12	-	12	-	12	-	13	-	13
0.140	-	12	-	12	-	13	-	13	-	13	-	13
0.150	-	13	-	13	-	13	-	13	0	14	0	14
0.160	-	13	0	14	0	14	0	14	0	15	0	15
0.170	0	14	0	14	0	15	0	15	0	15	0	15
0.180	0	14	0	15	0	15	0	15	0	15	0	15
0.190	0	15	0	15	0	15	0	16	0	16	0	16
0.200	0	15	0	15	0	16	0	16	0	16	1	17
0.210	0	16	0	16	0	16	1	17	1	17	1	17
0.220	0	16	1	17	1	17	1	17	1	17	1	18
0.230	1	17	1	17	1	17	1	18	1	18	1	18
0.240	1	17	1	17	1	18	1	18	1	18	2	19
0.250	1	17	1	18	1	18	2	19	2	19	2	19
0.260	1	18	1	18	2	19	2	19	2	19	2	20
0.270	1	18	2	19	2	19	2	19	2	20	2	20
0.280	2	19	2	19	2	19	2	20	2	20	3	21
0.290	2	19	2	19	2	20	2	20	3	21	3	21
0.300	2	19	2	20	3	21	3	21	3	21	3	22
0.310	3	20	3	21	3	21	3	21	3	22	3	22
0.320	3	21	3	21	3	21	4	22	4	22	4	23
0.330	3	21	3	21	4	22	4	22	4	23	4	23
0.340	3	21	4	22	4	22	4	23	4	23	4	23
0.350	4	22	4	22	4	22	4	23	4	23	5	24
0.360	4	22	4	22	4	23	4	23	5	24	5	24
0.370	4	22	4	23	4	23	5	24	5	24	6	25
0.380	5	23	5	23	5	24	5	24	6	25	6	25
0.390	5	23	5	24	5	24	6	25	6	25	6	26
0.400	5	23	5	24	6	25	6	25	6	25	6	26
0.410	5	24	5	24	6	25	6	25	6	26	7	27
0.420	5	24	6	25	6	25	7	26	7	26	7	27
0.430	6	25	6	25	7	26	7	26	7	27	7	27
0.440	6	25	6	25	7	26	7	26	7	27	8	28
0.450	6	25	7	26	7	26	8	27	8	28	8	28
0.460	7	26	7	26	8	27	8	27	8	28	8	28
0.470	7	26	7	26	8	27	8	28	8	28	9	29
0.480	7	26	8	27	8	27	8	28	9	29	9	29
0.490	8	27	8	27	9	28	9	28	9	29	10	30
0.500	8	27	9	28	9	28	9	29	9	29	10	30

Table 12
CONTROL LIMITS FOR SMALL-SAMPLE CASE: np CHARTS (continued)

n	41		42		43		44		45		46	
p̄	LCL	UCL	LCL	UCL	LCL	UCL	LCL	UCL	LCL	UCL	LCL	UCL
0.001	-	2	-	2	-	2	-	2	-	2	-	2
0.002	-	3	-	3	-	3	-	3	-	3	-	3
0.003	-	3	-	3	-	3	-	3	-	3	-	3
0.004	-	3	-	3	-	3	-	3	-	3	-	3
0.005	-	3	-	3	-	3	-	3	-	3	-	3
0.010	-	4	-	4	-	4	-	4	-	4	-	4
0.020	-	5	-	5	-	5	-	5	-	5	-	5
0.030	-	6	-	6	-	6	-	6	-	6	-	6
0.040	-	7	-	7	-	7	-	7	-	7	-	7
0.050	-	8	-	8	-	8	-	8	-	8	-	8
0.060	-	8	-	9	-	9	-	9	-	9	-	9
0.070	-	9	-	9	-	9	-	10	-	10	-	10
0.080	-	10	-	10	-	10	-	10	-	10	-	10
0.090	-	10	-	11	-	11	-	11	-	11	-	11
0.100	-	11	-	11	-	11	-	12	-	12	-	12
0.110	-	12	-	12	-	12	-	12	-	12	-	13
0.120	-	12	-	13	-	13	-	13	-	13	-	13
0.130	-	13	-	13	-	13	-	14	-	14	-	14
0.140	-	13	0	14	0	15	0	15	0	15	0	15
0.150	0	15	0	15	0	15	0	15	0	15	0	15
0.160	0	15	0	15	0	15	0	15	0	16	0	16
0.170	0	15	0	16	0	16	0	16	0	16	1	17
0.180	0	16	0	16	0	16	1	17	1	17	1	18
0.190	0	16	1	17	1	17	1	17	1	18	1	18
0.200	1	17	1	17	1	18	1	18	1	18	1	18
0.210	1	18	1	18	1	19	1	18	1	19	1	19
0.220	1	18	1	18	2	19	2	19	2	20	2	20
0.230	1	18	2	19	2	20	2	20	2	20	2	20
0.240	2	19	2	20	2	20	2	20	2	20	2	21
0.250	2	20	2	20	2	21	3	21	3	21	3	22
0.260	2	20	2	20	3	21	3	21	3	22	3	22
0.270	3	21	3	21	3	22	3	22	3	22	3	22
0.280	3	21	3	21	3	22	3	22	4	23	4	23
0.290	3	21	3	22	3	23	4	23	4	23	4	23
0.300	3	22	4	23	4	23	4	23	4	24	4	24
0.310	4	23	4	23	4	24	4	24	4	24	5	25
0.320	4	23	4	23	5	24	5	25	5	25	5	25
0.330	4	23	5	24	5	25	5	25	5	25	6	26
0.340	5	24	5	24	5	25	5	25	6	26	6	26
0.350	5	24	5	25	5	26	6	26	6	26	6	27
0.360	5	25	5	25	6	26	6	26	6	27	6	27
0.370	6	26	6	26	6	27	7	27	7	27	7	28
0.380	6	26	6	26	7	27	7	27	7	28	7	28
0.390	6	26	7	27	7	27	7	28	7	28	8	29
0.400	7	27	7	27	7	28	7	28	8	29	8	29
0.410	7	27	7	27	8	28	8	29	8	29	9	30
0.420	7	27	8	28	8	29	8	29	9	30	9	30
0.430	8	28	8	28	8	30	9	30	9	30	9	30
0.440	8	28	9	29	9	30	9	30	9	30	10	31
0.450	9	29	9	29	9	30	9	30	10	31	10	31
0.460	9	29	9	30	9	31	10	31	10	31	11	32
0.470	9	29	10	30	10	31	10	31	11	32	11	32
0.480	10	30	10	30	10	32	11	32	11	32	12	33
0.490	10	30	10	31	11	32	11	32	12	33	12	33
0.500	11	31	11	31	11	32	11	32	12	33	12	34

Table 12
CONTROL LIMITS FOR SMALL-SAMPLE CASE: np CHARTS (continued)

p̄	n=47 LCL	n=47 UCL	n=48 LCL	n=48 UCL	n=49 LCL	n=49 UCL	n=50 LCL	n=50 UCL
0.001	-	2	-	2	-	2	-	2
0.002	-	3	-	3	-	3	-	3
0.003	-	3	-	3	-	3	-	3
0.004	-	3	-	3	-	3	-	3
0.005	-	3	-	3	-	3	-	3
0.010	-	4	-	4	-	4	-	4
0.020	-	5	-	5	-	5	-	6
0.030	-	6	-	7	-	7	-	7
0.040	-	7	-	7	-	8	-	8
0.050	-	8	-	8	-	8	-	9
0.060	-	9	-	9	-	9	-	9
0.070	-	10	-	10	-	10	-	10
0.080	-	11	-	11	-	11	-	11
0.090	-	11	-	12	-	12	-	12
0.100	-	12	-	12	-	12	-	13
0.110	-	13	-	13	-	13	-	13
0.120	-	13	-	14	-	14	-	14
0.130	-	14	0	15	0	15	0	15
0.140	0	15	0	15	0	15	0	16
0.150	0	16	0	16	0	16	0	16
0.160	0	16	0	16	0	17	0	17
0.170	1	17	1	18	1	18	1	18
0.180	1	18	1	18	1	18	1	18
0.190	1	18	1	18	1	19	1	19
0.200	1	19	1	19	1	19	2	20
0.210	2	20	2	20	2	20	2	20
0.220	2	20	2	20	2	20	2	21
0.230	2	20	2	21	2	21	3	22
0.240	2	21	3	22	3	22	3	22
0.250	3	22	3	22	3	22	3	23
0.260	3	22	3	22	4	23	4	24
0.270	4	23	4	23	4	24	4	24
0.280	4	23	4	24	4	24	4	24
0.290	4	24	4	24	5	25	5	25
0.300	5	25	5	25	5	25	5	26
0.310	5	25	5	25	5	26	5	26
0.320	5	25	6	26	6	27	6	27
0.330	6	26	6	27	6	27	6	27
0.340	6	27	6	27	7	28	7	28
0.350	6	27	7	28	7	28	7	28
0.360	7	28	7	28	7	28	8	29
0.370	7	28	8	29	8	29	8	30
0.380	8	29	8	29	8	29	8	30
0.390	8	29	8	29	9	30	9	31
0.400	9	30	9	30	9	31	9	31
0.410	9	30	9	31	9	31	10	32
0.420	9	30	9	31	10	32	10	32
0.430	10	31	10	32	10	32	11	33
0.440	10	31	10	32	11	33	11	33
0.450	10	32	11	33	11	33	12	34
0.460	11	33	11	33	12	34	12	34
0.470	11	33	12	34	12	34	13	35
0.480	12	34	12	34	13	35	13	35
0.490	12	34	13	35	13	35	14	36
0.500	13	35	13	35	13	35	14	36

University of Tennessee, 1986.

Table 13
MOLINA'S TABLES: c CHARTS[a]

					Number of Defects							
c̄	0	1	2	3	4	5	6	7	8	9	10	11
0.02	980	1000										
0.04	961	999	1000									
0.06	942	998	1000									
0.08	923	997	1000									
0.10	905	995	1000									
0.15	861	990	999	1000								
0.20	819	982	999	1000								
0.25	779	974	998	1000								
0.30	741	963	996	1000								
0.35	705	951	994	1000								
0.40	670	938	992	999	1000							
0.45	638	925	989	999	1000							
0.50	607	910	986	998	1000							
0.55	577	894	982	998	1000							
0.60	549	878	977	997	1000							
0.65	522	861	972	996	999	1000						
0.70	497	844	966	994	999	1000						
0.75	472	827	959	993	999	1000						
0.80	449	809	953	991	999	1000						
0.85	427	791	945	989	998	1000						
0.90	407	772	937	987	998	1000						
0.95	387	754	929	984	997	1000						
1.00	368	736	920	981	996	999	1000					
1.10	333	699	900	974	995	999	1000					
1.20	301	663	879	966	992	998	1000					
1.30	273	627	857	957	989	998	1000					
1.40	247	592	833	946	986	997	999	1000				
1.50	223	558	809	934	981	996	999	1000				
1.60	202	525	783	921	976	994	999	1000				
1.70	183	493	757	907	970	992	998	1000				
1.80	165	463	731	891	964	990	997	999	1000			
1.90	150	434	704	875	956	987	997	999	1000			
2.00	135	406	677	857	947	983	995	999	1000			
2.20	111	355	623	819	928	975	993	998	1000			
2.40	91	308	570	779	904	964	988	997	999	1000		
2.60	74	267	518	736	877	951	983	995	999	1000		
2.80	61	231	469	692	848	935	976	992	998	999	1000	
3.00	50	199	423	647	815	916	966	988	996	999	1000	
3.20	41	171	380	603	781	895	955	983	994	998	1000	
3.40	33	147	340	558	744	871	942	977	992	997	999	1000
3.60	27	126	303	515	706	844	927	969	988	996	999	1000
3.80	22	107	269	473	668	816	909	960	984	994	998	999
4.00	18	92	238	433	629	785	889	949	979	992	997	999
4.20	15	78	210	395	590	753	867	936	972	989	996	999
4.40	12	66	185	359	551	720	844	921	964	985	994	998
4.60	10	56	163	326	513	686	818	905	955	980	992	997
4.80	8	48	143	294	476	651	791	887	944	975	990	996
5.00	7	40	125	265	440	616	762	867	932	968	986	995
5.20	6	34	109	238	406	581	732	845	918	960	982	993
5.40	5	29	95	213	373	546	702	822	903	951	977	990
5.60	4	24	82	191	342	512	670	797	886	941	972	988
5.80	3	21	72	170	313	478	638	771	867	929	965	984

Appendix

Table 13
MOLINA'S TABLES: c CHARTS[a] (continued)

					Number of Defects							
c̄	0	1	2	3	4	5	6	7	8	9	10	11
6.00	2	17	62	151	285	446	606	744	847	916	957	980
6.20	2	15	54	134	259	414	574	716	826	902	949	975
6.40	2	12	46	119	235	384	542	687	803	886	939	969
6.60	1	10	40	105	213	355	511	658	780	869	927	963
6.80	1	9	34	93	192	327	480	628	755	850	915	955
7.00	1	7	30	82	173	301	450	599	729	830	901	947
7.20	1	6	25	72	156	276	420	569	703	810	887	937
7.40	1	5	22	63	140	253	392	539	676	788	871	926
7.60	1	4	19	55	125	231	365	510	648	765	854	915
7.80	0	4	16	48	112	210	338	481	620	741	835	902
8.00	0	3	14	42	100	191	313	453	593	717	816	888
8.50	0	2	9	30	74	150	256	386	523	653	763	849
9.00	0	1	6	21	55	116	207	324	456	587	706	803
9.50	0	1	4	15	40	89	165	269	392	522	645	752
10.00	0	0	3	10	29	67	130	220	333	458	583	697
10.50	0	0	2	7	21	50	102	179	279	397	521	639
11.00	0	0	1	5	15	38	79	143	232	341	460	579
11.50	0	0	1	3	11	28	60	114	191	289	402	520
12.00	0	0	1	2	8	20	46	90	155	242	347	462
12.50	0	0	0	2	5	15	35	70	125	201	297	406
13.00	0	0	0	1	4	11	26	54	100	166	252	353
13.50	0	0	0	1	3	8	19	41	79	135	211	304
14.00	0	0	0	0	2	6	14	32	62	109	176	260
14.50	0	0	0	0	1	4	10	24	48	88	145	220
15.00	0	0	0	0	1	3	8	18	37	70	118	185
16.00	0	0	0	0	0	1	4	10	22	43	77	127
17.00	0	0	0	0	0	1	2	5	13	26	49	85
18.00	0	0	0	0	0	0	1	3	7	15	30	55
19.00	0	0	0	0	0	0	1	2	4	9	18	35
20.00	0	0	0	0	0	0	0	1	2	5	11	21
21.00	0	0	0	0	0	0	0	0	1	3	6	13
22.00	0	0	0	0	0	0	0	0	1	2	4	8
23.00	0	0	0	0	0	0	0	0	0	1	2	4
24.00	0	0	0	0	0	0	0	0	0	0	1	3
25.00	0	0	0	0	0	0	0	0	0	0	1	1

Table 13
MOLINA'S TABLES: c CHARTS[a] (continued)

$\bar{c}$	Number of Defects											
	12	13	14	15	16	17	18	19	20	21	22	23
3.80	1000											
4.00	1000											
4.20	1000											
4.40	999	1000										
4.60	999	1000										
4.80	999	1000										
5.00	998	999	1000									
5.20	997	999	1000									
5.40	996	999	1000									
5.60	995	998	999	1000								
5.80	993	997	999	1000								
6.00	991	996	999	999	1000							
6.20	989	995	998	999	1000							
6.40	986	994	997	999	1000							
6.60	982	992	997	999	999	1000						
6.80	978	990	996	998	999	1000						
7.00	973	987	994	998	999	1000						
7.20	967	984	993	997	999	999	1000					
7.40	961	980	991	996	998	999	1000					
7.60	954	976	989	995	998	999	1000					
7.80	945	971	986	993	997	999	1000					
8.00	936	966	983	992	996	998	999	1000				
8.50	909	949	973	986	993	997	999	999	1000			
9.00	876	926	959	978	989	995	998	999	1000			
9.50	836	898	940	967	982	991	996	998	999	1000		
10.00	792	864	917	951	973	986	993	997	999	999	1000	
10.50	742	825	888	932	960	978	988	994	997	999	999	1000
11.00	689	781	854	907	944	968	982	991	995	998	999	1000
11.50	633	733	815	878	924	954	974	986	992	996	998	999
12.00	576	682	772	844	899	937	963	979	988	994	997	999
12.50	519	628	725	806	869	916	948	969	983	991	995	998
13.00	463	573	675	764	835	890	930	957	975	986	992	996
13.50	409	518	623	718	798	861	908	942	965	980	989	994
14.00	358	464	570	669	756	827	883	923	952	971	983	991
14.50	311	413	518	619	711	790	853	901	936	960	976	986
15.00	268	363	466	568	664	749	819	875	917	947	967	981
16.00	193	275	368	467	566	659	742	812	868	911	942	963
17.00	135	201	281	371	468	564	655	736	805	861	905	937
18.00	92	143	208	287	375	469	562	651	731	799	855	899
19.00	61	98	150	215	292	378	469	561	647	725	793	849
20.00	39	66	105	157	221	297	381	470	559	644	721	787
21.00	25	43	72	111	163	227	302	384	471	558	640	716
22.00	15	28	48	77	117	169	232	306	387	472	556	637
23.00	9	17	31	52	82	123	175	238	310	389	472	555
24.00	5	11	20	34	56	87	128	180	243	314	392	473
25.00	3	6	12	22	38	60	92	134	185	247	318	394

Table 13
MOLINA'S TABLES: c CHARTS[a] (continued)

	Number of Defects											
$\bar{c}$	24	25	26	27	28	29	30	31	32	33	34	35
11.50	1000											
12.00	999	1000										
12.50	999	999	1000									
13.00	998	999	1000									
13.50	997	998	999	1000								
14.00	995	997	999	999	1000							
14.50	992	996	998	999	999	1000						
15.00	989	994	997	998	999	1000						
16.00	978	987	993	996	998	999	999	1000				
17.00	959	975	985	991	995	997	999	999	1000			
18.00	932	955	972	983	990	994	997	998	999	1000		
19.00	893	927	951	969	980	988	993	996	998	999	999	1000
20.00	843	888	922	948	966	978	987	992	995	997	999	999
21.00	782	838	883	917	944	963	976	985	991	994	997	998
22.00	712	777	832	877	913	940	959	973	983	989	994	996
23.00	635	708	772	827	873	908	936	956	971	981	988	993
24.00	554	632	704	768	823	868	904	932	953	969	979	987
25.00	473	553	629	700	763	818	863	900	929	950	966	978

	Number of Defects							
$\bar{c}$	36	37	38	39	40	41	42	43
20.00	1000							
21.00	999	999	1000					
22.00	998	999	999	1000				
23.00	996	997	999	999	1000			
24.00	992	995	997	998	999	999	1000	
25.00	985	991	994	997	998	999	999	1000

[a]Table gives 1,000 times the probability of obtaining c or fewer defects from a population with an average number of defects equal to $\bar{c}$.

Source: Adapted from E. L. Grant and R. S. Leavenworth, *Statistical Quality Control*, Fifth Edition, McGraw-Hill, New York, New York, 1980.

BIBLIOGRAPHY

American Society for Quality Control, Chemical and Process Industries Division, Chemical Interest Committee, *Quality Assurance for the Chemical and Process Industries -- A Manual of Good Preactice*, ASQC, Milwaukee, Wisconsin, 1987.

American Supplier Institute, *Statistical Thinking for Manufacturing Process Control*, ASI, Romulus, Michigan, 1984.

AT&T, *Statistical Quality Control Handbook*, Delmar Printing, Charlotte, North Carolina, 1956.

Box, E. P. G., Hunter, W. G., and Hunter, J. S., *Statistics for Experimenters*, John Wiley & Sons, Inc., New York, New York, 1978.

Daniels, A. C. and Rosen, T. A., *Performance Management*, Performance Management Publications, Inc., Tucker, Georgia, 1986.

Deming, W. E., *Out of Crisis*, Massachusetts nstitute of Technology, Center for Advanced Engineering Study, Cambridge, Massachusetts, 1982.

DuPont, *Cumulative Sum Statistical Control*, Applied Statistics Group, E. I. du Pont de Nemours & Co., 1987.

DuPont, *CUSUM, A Series of Articles by James M. Lucas*, Applied Statistics Group, E. I. du Pont de Nemours & Co., 1987.

Ford Motor Company, *Continuing Process Control*, Statistical Methods Office, ations Support Staff, 1984.

Gilbert, T. G., *Human Competence*, McGraw-Hill, New York, New York, 1978.

Grant, E. L. and Leavenworth, R. S., *Statistical Quality Control*, McGraw-Hill, New York, New York, 1980.

Hagan, J. T., Editor, *Principles of Quality Cost*, American Society for Quality Control, Milwaukee, Wisconsin, 1986.

Himmelblau, D. M., *Process Analysis by Statistical Methods*, John Wiley & Sons, Inc., New York, New York, 1970.

Hunter, J. S., "The Exponentially Weighted Moving Average Chart," *Journal of Quality Technology*, Vol. 18, No. 4, October, 1986.

Hunter, J. S., *The Technology of Quality*, Ellis R. Ott Foundation, Edison, New Jersey, 1988.

Ishikawa, I., *Guide to Quality Control*, Asian Productivity Organization, 1982.

Bibliography

Juran, J. M., Gryna, F. M., and Bingham, R. S., *Quality Control Handbook*, McGraw-Hill, New York, New York, 1979.

Miller, L. M., *Leading Management Change*, L. M. Miller & Company, Atlanta, Georgia, 1987.

University of Tennessee, *The Institute for Productivity through Quality*, Management Development Program, Knoxville, Tennessee, 1986.

University of Tennessee, *The Design of Experiments Institute for Productivity through Quality*, Management Development Program, Knoxville, Tennessee, 1987.

Wheeler, D. J. and Chambers, D. S., *Understanding Statistical Process Control, Statistical Process Controls, Inc.*, Knoxville, Tennessee, 1986.

510

INDEX

511